Constructing Race

Constructing Race helps unravel the complicated and intertwined history of race and science in the United States. Tracy Teslow explores how physical anthropologists in the twentieth century struggled to understand the complexity of human physical and cultural variation, and how their theories were disseminated to the public through art, museum exhibitions, books, and pamphlets. In their attempts to explain the history and nature of human peoples, anthropologists persistently saw both race and culture as critical components. This is at odds with a broadly accepted account that suggests racial science was fully rejected by scientists and the public following World War II. This book offers a corrective, showing that both race and culture informed how anthropologists and the public understood human variation from 1900 through the decades following the war. The book offers new insights into the work of Franz Boas, Ruth Benedict, and Ashley Montagu, as well as less well-known figures, including Harry Shapiro, Gene Weltfish, and Henry Field.

Tracy Teslow is Associate Professor of History at the University of Cincinnati. She has received prestigious fellowships from the Center for the Study of Diversity in Science, Technology and Medicine at MIT and the Andrew Mellon Foundation.

Constructing Race

The Science of Bodies and Cultures in American Anthropology

TRACY TESLOW

University of Cincinnati

CAMBRIDGE
UNIVERSITY PRESS

CAMBRIDGE
UNIVERSITY PRESS

32 Avenue of the Americas, New York, NY 10013-2473, USA

Cambridge University Press is part of the University of Cambridge.

It furthers the University's mission by disseminating knowledge in the pursuit of education, learning, and research at the highest international levels of excellence.

www.cambridge.org
Information on this title: www.cambridge.org/9781107011731

First published 2014

Printed in the United States of America

A catalog record for this publication is available from the British Library.

Library of Congress Cataloging in Publication Data
Teslow, Tracy, 1964–
Constructing race : the science of bodies and cultures in American anthropology / Tracy Teslow.
pages   cm
Includes bibliographical references and index.
ISBN 978-1-107-01173-1 (hardback)
1. Physical anthropology – United States – History – 20th century.   2. Race – Social aspects – United States – History – 20th century.   3. Somatotypes – United States – History – 20th century.   4. Race awareness – United States – History – 20th century.   5. Racism in anthropology – United States – History – 20th century.   6. Century of Progress International Exposition (1933–1934 : Chicago, Ill.) – Exhibitions.   I. Title.
GN50.45.U6T47   2014
305.80097309′04–dc23   2013039525

ISBN 978-1-107-01173-1 Hardback

Contents

v

Figures

Acknowledgments

This book has been a long time in the making. Over the many years of its gestation, first as a dissertation, and now finally as a monograph, I have incurred many debts to teachers and mentors, colleagues and archivists, family and friends. All along the way, I also have been the grateful recipient of generous institutional support from my universities and from institutes and foundations without whose support the lonely historian would have a hard time researching and writing narratives of any meaning or scope. I have no doubt I will fail to remember all who deserve my sincere thanks. I hope they will know who they are, and will know that I am grateful for the many, many enlightening conversations, presentations, seminars, and chance encounters that have molded the form and content of my ideas.

The list of institutions that have supported this project is a long one, and it begins with the University of Chicago, where I wrote the dissertation that became the foundation of this book. The vigorous intellectual climate, the cross-disciplinary culture, and the brilliant faculty and students I encountered there formed an ideal environment in which I was encouraged to pursue questions down whatever avenue they led, and challenged to think subtly and deeply. At Chicago my project was supported by the Morris Fishbein Center for the History of Science, as well as the Center for the Study of Race, Politics and Culture. Chief among the faculty who deserve acknowledgment are my dissertation advisor, Robert Richards, who provided wise counsel, intellectual guidance, and unfailing good humor, and George Stocking, a member of my dissertation committee, but more importantly an intellectual model and mentor. Thanks also to Neil Harris, another Chicago historian who served on my dissertation committee and whose work on museums inspired my own. Other faculty who were critical to my intellectual growth at Chicago included George Chauncey, Lorraine Daston, William Wimsatt, Leora Auslander, Barbara Stafford, and W. J. T. Mitchell. Thanks also to the members of the Human Sciences Workshop, whose lively intellectual debates and thoughtful critiques of my work were enormously valuable. Elizabeth Bitoy, the long-serving Fishbein Center secretary, also deserves special mention for creating, along with Robert Richards, an inviting academic home for history of science students. While a student at Chicago, I also received generous support from the Dibner Institute for the History of Science and Technology.

Two fellowships provided the precious gift of time. A year-long postdoctoral fellowship at the Center for the Study of Diversity in Science, Technology and Medicine at the Massachusetts Institute of Technology under the directorship of Evelynn Hammonds was an invaluable opportunity to continue my work at one of the most vibrant centers of science studies in the world. My conversations with Evelynn Hammonds, Senior Fellow Patricia Seed, and the many colleagues in the study of science, technology, and medicine, not only at MIT, but also from Harvard, Wellesley, and Brandeis universities, helped me think beyond the original boundaries of my dissertation toward the broader ramifications of my topic. Similarly, I am grateful to the College of Liberal Arts and the Department of History at the University of Illinois, Urbana-Champaign, for the post-doctoral fellowship from the Andrew W. Mellon Foundation that allowed me to spend two years among another vibrant community of scholars. The time afforded me to rethink my project and move it forward was a gift for which I am truly thankful.

My current institutional home, the University of Cincinnati, has provided critical support as I turned my dissertation into this monograph. The Charles Phelps Taft Research Center and the University Research Council provided generous grants to support research trips that enabled me to extend the reach of my project. The Taft Center and my department also supported conference and seminar travel that allowed me to confer with colleagues outside Cincinnati. To my colleagues in the Department of History, I owe a variety of debts, not least for their generosity and patience in mentoring me from a green, newly minted PhD into a full member of the professoriate, with all the rewards and challenges that implies. This book has profited in particular from the insights, cajoling, advice, and friendship of Hilda Smith, Maura O'Connor, Sigrun Haude,

Elizabeth Frierson, John K. Alexander, Chris Phillips, David Stradling, Wendy Kline, Jason Krupar, Mark Lause, David Ciarlo, and Barbara Ramusack. Willard Sunderland, who served as head of the department as I was finishing this book, deserves special thanks for his inimitable combination of enthusiasm, patience, advice, and badgering. His unflagging belief in me and this project bolstered me through the final ups and downs, and I will be ever grateful. Our department administrator, Hope Earls, deserves special note; she has been indispensable.

Colleagues outside my various institutional homes also have been crucial in shaping how I have thought about the many facets of this book. Chief among them is Linda Kerber, in whose seminar this project first took shape. Her support for my work has been unstinting, as both a mentor and friend. Philip Kitcher, whose provocative queries and intellectual respect prodded me down this path many years ago, also deserves my sincere gratitude. Gregg Mitman took my project seriously from its earliest stages and encouraged my interest in the links between science and visual culture. I'm grateful for stimulating conversations with and professional support from Alice Conklin and Michelle Brattain, who share my interests in race and science. Other generous colleagues who have read drafts, commented at conferences, shared their work, written letters of support, invited presentations and publication, or simply talked about shared interests include Steven Allison, Lee Baker, David Bindman, Kevin Boyle, Juliet Burba, Antoinette Burton, Steven Conn, Kenneth Gonzales-Day, Matthew Pratt Guterl, Bradley Hume, John P. Jackson, Sally Gregory Kohlstedt, Henrika Kuklick, Sharon Macdonald, Kris Morrissey, Yolanda Moses, Philip J. Pauly, Jenny Reardon, Samuel Redman, David Roediger, Fatimah Tobing Rony, Helaine Silverman, Alaka Wali, and Nadine Weidman. Audiences and colleagues at meetings of the History of Science Society, Organization of American Historians, National Council for Public History, the Getty Center for the History of Art and the Humanities, and the International Society for the History, Philosophy and Social Studies of Biology asked questions and offered comments that have influenced my thinking. In addition, I'm indebted to various "anonymous" readers whose comments have helped me improve this monograph.

Every historian owes an enormous debt to the knowledgeable librarians and archivists who manage the documents upon which we erect our narratives. At the Field Museum, Anthropology Department registrar Janice Klein, librarian Ben Williams, museum archivist Armand Esai, and photo archivist Nina Cummings have been unfailingly helpful. Similarly, I benefited from the able and generous work of the staff in

Special Collections at the American Museum of Natural History, especially librarian Tom Baione and archivist Barbara Mathe, research librarian Gregory Raml, and Julie Kasper, as well as the assistance of Anthropology Department registrars Belinda Kaye and Kristen Mable. At the Smithsonian Institution Archives, historian Pamela Henson was especially helpful, while the staff at the National Anthropological Archives helped make my time there very productive. I'm similarly grateful to the librarian at the Archives and Special Collections of Vassar College, where Ruth Benedict's papers are housed. At the Cranbrook Institute Archives, archivists Leslie Edwards and Cheri Gay, as well as Anthropology Department coordinator and museum educator Cameron Wood, helped me find crucial documentation and photographs of the 1944 *Races of Mankind* exhibition. My sincere thanks to Beth Guynn, Kristin Hammer, Kathlin Ralston Knutsen, Don Anderle, and Wim de Wit for their assistance, and for the grants that made my trips possible, at the Getty Research Institute of the Getty Center for the History of Art and the Humanities, where Malvina Hoffman's papers reside. I also want to thank Juliana Field, Derek Ostergard, Harriet Shapiro, and James Shapiro for granting me the right to reproduce photographs and other material in this book.

I am indebted as well to my editors and their assistants at Cambridge University Press. Eric Crahan embraced this book and shepherded my manuscript through the contracting process, for which I will be ever grateful. After he departed for other publishing pastures, Deborah Gershenowitz has seen it through to publication. Thanks also to their assistants, Jason Przybylski, Abby Zorbaugh, and Dana Bricken, who answered many and sundry questions from the author.

Finally, to my friends and family, I offer thanks for sticking with me all these years, for laughter and solace, for a bed and a meal when I needed it, and for asking – or not asking – how it was going. I doubt I will ever truly repay A. Holly Shissler for all the advice, commiseration, meals, endless conversation about this project, and a multitude of other things (including dealing with that old Toyota). I am grateful for old friends who have known me long before I ever embarked on an academic's life, who want the best for me and encourage me, but when it comes right down to it, don't really care about this book. To old friends who help me keep it all in perspective, thank you, especially Rachael Bergan, Anne Tschida, Mary Upham, and Catherine Popowits. Pamela Baker is a good friend, a good colleague, and a darn good cook. Thanks also to Jim Murray, for his encouragement and support over many years. Ann Kelley gave me excellent advice when I wasn't sure returning to school for a doctorate was sensible or doable. I met Rajeev Samantrai at the orientation for incoming history graduate students at the University of Chicago, where we had the first of many, many conversations about graduate school, history, and life. I miss his incisive mind, humor, and cheerful cynicism. To my friends in graduate school at Chicago, many who remain in my life, my genuine thanks for everything – the ideas, the fun, the support. I want to thank especially Cheryce (Kramer) von Xylander, Paul White, John Ceccatti, Greg Mikkelson, John Huss, Jeff Smith, David Ciepley, Elizabeth Liebman, Sean Anderson, David Walton, Kathy Cooke, David Valone, Vincent and Andrea (Williams) Wan, Evalyn Tennant, Brian Ogilvie, Eric Caplan, Sheri Lyons, Andre Wakefield, Karl Galle, and Nicolette Warisse Sosulski. Thanks also to Eric Kupferberg, a fellow history of science graduate student, at MIT, who became a good friend in graduate school and beyond.

Among my family, I thank my mother, Mary Lee Short, my sister Elizabeth Teslow, and my father, Ray Waechter, for their unquestioning love and support. So too, the love and support of my extended family – aunts and uncles, cousins and step-sisters – has also meant a great deal to me. Love and thanks to Jim Waechter and Paul Olson, Suki Kwak, HaeNa Waechter, Vera and Parker Waechter, Parker B. (Barry) and M'Lissa Waechter, Sandy Retzlaff, Charlie Retzlaff, Renee Ruff, Terri and Michael Dilley, Carol Waechter and Susan and Brandt Merrild. In memory, thanks to my grandparents, June and Lee Short, Anne and Raymond (Bud) Waechter, who meant so much to me, and who would have loved to see this book.

Race, Anthropology, and the American Public

An Introductory Essay

In 1933, the Field Museum of Natural History in Chicago opened an exhibition to wide publicity and enormous attendance, timed to coincide with the World's Fair, right across the street. The new exhibit, a blockbuster in today's terms, was not devoted to Native Americans, American anthropology's signature object, or to dinosaurs, a Field Museum specialty. Instead, the marquee exhibit was the *Races of Mankind*, an exposition on race arrayed in a series of 101 life-size bronze sculptures by prominent sculptor Malvina Hoffman, each man, woman, and child depicting a distinct racial type, each the scientific doppelgänger to the diversity of Americans outside the museum doors and the spectacle of living "exotics" next door at the World's Fair. At the center of the exhibit stood a massive sculpture depicting the unity of humanity, symbolized by three idealized, but racialized, male figures, topped by a globe.

Created in a nexus of competing anthropological ideas about the nature and significance of race, the *Races of Mankind* promoted a vision of humanity that was both humanist and racialist, a panoply of human diversity refracted through the prism of racial typology. Presented to the public as insightful art and rigorous science, Hoffman's sculptures and the *Races of Mankind* exhibit boldly presented the discipline of physical anthropology to the public and asserted its relevance to American life. Coming on the heels of eugenic triumphs in immigration restriction, racial unrest in Chicago and around the nation, and the onset of the Depression, the exhibition offered a stable and familiar world order, frozen in bronze, at once unified and diverse in its preservation of endangered "primitives" and its reinforcement of the natural place of Europeans and Americans at the top of the racial heap.

In its time, this exhibit represented a stirring synthesis of art and anthropology, yet today it is all but forgotten. How did it slide into obscurity? Why has the history of anthropology and racial science in the interwar years been largely overlooked, and when addressed, frequently misconstrued, its complexities flattened? What impact did all this have on popular understanding about the nature of race? By 1968, the *Races of Mankind* exhibit was roundly denounced as racist and dismantled. Yet in the 1990s, Malvina Hoffman's sculptures, now scattered decoratively around the museum, stripped of their scientific status, were hailed by another generation as sensitive depictions of human multicultural diversity. How can we understand these shifting intersections of race and science in American life? What role have anthropology and its popularization played in American racial formation?

These are the fundamental questions at the heart of *Constructing Race*. Throughout the first decades of the twentieth century, until roughly the 1950s, racial science remained a critical part of professional and popular anthropology. It was a racial science characterized not only by forms of racial essentialism and biological determinism, but also by attempts to grasp human diversity in cultural, historicized terms. This study explores how physical anthropologists struggled to understand variation in bodies and cultures in a crucial yet understudied period, how they represented race to the public, and how their efforts contributed to an American formulation of race that has remained rooted in both bodies and cultures, heredity, and society.

At the core of this book is the argument that mid-century anthropology, along with other racial sciences, has been misconstrued in ways that have kept us from fully appreciating the causes and effects of American racial formation. The failure to see fully the range of racial conceptualizations, as well as the related failure to understand fully the complex interplay between cultural and biological theory, has hampered our ability to make sense of the persistent contradictions and complexity of the American racial landscape.

Much scholarship perpetuates a teleological narrative in which progressive scientists, chiefly Franz Boas and his students, championed a cultural understanding of human variation against pernicious essentialist, racist conceptions of difference. By contrast, this study reveals a much more complex and contested picture of racialist theorizing among anthropologists, not least by Boas himself, who was a leading racial scientist, in addition to being one of the foremost cultural anthropologists in America and a crusader against racism. In the interwar era, race was regarded among physical anthropologists (and most Americans) as an essential, biological component of human identity. Within that consensus, however, there were serious debates over the nature of race, heredity, identity, classifications, and scientific methods. Moreover, the somatic emphasis of racial anthropology did not preclude anthropologists from also understanding race in cultural and historical terms. This study offers answers to questions that remain unanswered: How did Franz Boas, the father of American cultural anthropology and a champion against racism, regard race and racial science? What *was* the relationship between culture and race in anthropology before 1950? How and why do racial essentialism and biological determinism persist despite strenuous efforts to reject them? The chapters that follow demonstrate three central arguments about race and racial science between 1900 and 1960: pre–World War II anthropologists, including Franz Boas, struggled with a racial science that comprehended human variation in both biologically essentialist and culturally nuanced ways; these ideas were communicated to a vast American public through a variety of media over a period of decades; and the result is an American formulation of race that has remained fundamentally rooted in both bodies and cultures. Through an examination of select anthropologists, critical exhibitions, and exemplary texts, this book illuminates how physical anthropologists in the twentieth century promoted a vision of race rooted in bodies *and* cultures, a vision that shaped popular perceptions of race in America for generations.

A Distorted Past

The historiographic treatment of Franz Boas is particularly emblematic of the lens through which racial science has too often been viewed. The picture of Boas as the father of cultural relativism and a champion against scientific racism that developed in the late twentieth century was a consequence of developments within anthropology in the postwar period, as well a reflection of the way anthropologists and others selectively amplified the first serious efforts by a scholar outside anthropology to offer an historical account of American anthropology and Franz Boas.

It may come as a surprise to readers accustomed to accounts of Boas as the author of the modern, relativistic culture concept that in the years after his death in 1942 he was dismissed by many anthropologists, most notably his own students Alfred Kroeber and Clyde Kluckhohn, as a diligent data collector but no theoretician, someone who failed to systematize his work, and who, in his relentless, a-theoretical empiricism, actually stunted development of the culture concept and the field of anthropology.[1] A broader view of the history of anthropology helps explain why Boas was first dismissed as insufficiently theoretical in the 1950s, and then later embraced as the critical figure in the development and promulgation of the culture concept. In the 1940s and 1950s, American anthropology experienced a resurgence of theories about human cultural and social formations that emphasized commonalities and universal qualities, functions, or processes, and was especially concerned to put anthropology on a fully "scientific" footing. Functionalism prospered at the University of Chicago; cultural ecology thrived at the Smithsonian Institution's Institute of Social Anthropology under Julian Steward; Leslie White promoted an evolutionary culture theory that emphasized the discovery of universal cultural laws.[2] At odds with the evolutionists and others promoting a quest for universal laws of cultural development, Boasian anthropologists like Kroeber and Kluckhohn defended historical particularism by rooting the culture concept, not in Franz Boas's articulation of it across his decades of work, but further in the past, in English anthropologist Edward Tylor's 1871 definition of culture and its subsequent elaboration by a variety of anthropologists.[3]

Historian George Stocking's work on Boas and his development of the culture concept profoundly changed the earlier, dismissive view. Even though Stocking was careful to characterize Boas as a "transitional figure" who "retained strong residual elements" of the nineteenth-century evolutionary "commitment to 'progress in civilization,'" he nonetheless made the point that it was Boas, and not Tylor, who in the face of the enormous diversity of human traditions and social practices had articulated a systematic critique of cultural evolution. According to Stocking, it was left to his students to fully elaborate his discussion of "cultures"

[1] George W. Stocking, Jr., "Franz Boas and the Culture Concept in Historical Perspective," *American Anthropologist*, New Series, vol. 68, No. 4 (Aug. 1966), pp. 867–882; Alfred Kroeber and Clyde Kluckhohn, "Culture: A Critical Review of Concepts and Definitions," *Papers of the Peabody Museum of American Archeology and Ethnology*, Harvard University, vol. 47, no. 1, 1952. See also Stocking, "Matthew Arnold, E. B. Tylor, and the Uses of Invention," *American Anthropologist*, vol. 65 (1963), pp. 783–799; *Race, Culture and Evolution* (Chicago: University of Chicago Press, [1968] 1982).

[2] Regna Darnell, *And Along Came Boas: Continuity and Revolution in Americanist Anthropology* (Amsterdam and Philadelphia: J. Benjamins, 1998); William J. Peace, *Leslie A. White: Evolution and Revolution in Anthropology* (Lincoln: University of Nebraska Press, 2004), pp. 164–169.

[3] Stocking, "Franz Boas and the Culture Concept"; E.[dward] B.[urnett] Tylor, *Primitive Culture: Researches into the Development of Mythology, Philosophy, Religion, Art, and Custom* (London: J. Murray, 1871).

into the modern culture concept.[4] The irony was that, in response to the postwar resurgence of evolutionism and scientism in anthropology, some influential Boasian anthropologists had promoted cultural relativism and historical particularism in part by dismissing Boas's critical contribution. Following Stocking's intervention in anthropologists' disciplinary history, and the eclipse of evolutionary and other universalizing theoretical orientations, Boas took on a founder's mantle in anthropology similar to that afforded Charles Darwin among modern evolutionary biologists.[5]

The embrace of Boas as the father of the culture concept was accompanied by a pronounced lack of interest in his racial science. Following World War II and the humanist turn that prompted disciplinary organizations, as well as international organizations such as the United Nations Educational, Scientific and Cultural Organization (UNESCO), to speak out against racism, Boas's antiracism was much more palatable, and more congruent with the politics of cultural relativism, than his racial science. By the 1970s, the treatment of Boas was enmeshed in the epistemological crisis that led humanists and social scientists, particularly anthropologists,

---

[4] Stocking, "Franz Boas and the Culture Concept," pp. 878–879.

[5] Anthropologists have been notably interested in their own disciplinary history, and in particular in examining the Boasian legacy. Numerous accounts of Boas's anthropology and of the history of anthropology have been authored by anthropologists themselves since Boas died in 1942. These include A. Irving Hallowell, "The History of Anthropology as an Anthropological Problem," *Journal of the History of the Behavioral Sciences*, vol. 1 (1965), pp. 24–38; Robert Lowie, "Boas Once More," *American Anthropologist*, vol. 58 (1956), pp. 159–164 and "Reminiscences of Anthropological Currents in America Half a Century Ago," *American Anthropologist*, vol. 58 (1956), pp. 995–1016; Alfred Kroeber, "The Place of Boas in Anthropology," *American Anthropologist*, vol. 58 (1956), pp. 151–159 and "A History of the Personality of Anthropology," *American Anthropologist*, vol. 61 (1959), pp. 398–404; Melville Herskovits, *Franz Boas: The Science of Man in the Making* (New York: Charles Scribner's Sons, 1953); Margaret Mead and Ruth Bunzel, eds., *The Golden Age of American Anthropology* (New York: George Bazilier, 1960), as well as Mead's unconventional biography of Ruth Benedict, *An Anthropologist at Work: Writings of Ruth Benedict* (Boston: Houghton Mifflin Company, 1959) and the later work, *Ruth Benedict* (New York: Columbia University Press, 1974), and her autobiography, *Blackberry Winter* (New York: Morrow, 1972); Marvin Harris, *The Rise of Anthropological Theory* (New York: Thomas Crowell, 1968); Leslie White, "The Ethnography and Ethnology of Franz Boas," *Texas Memorial Museum Bulletin*, vol. 6 (1963); Stephen O. Murray, "The Non-Eclipse of Americanist Anthropology during the 1930s and 1940s," Lisa Valentine and Regna Darnell, eds., *Theorizing the Americanist Tradition* (Toronto: University of Toronto Press, 1999). Regna Darnell has produced more than thirty books, articles, and edited volumes on the history of American anthropology, including *Along Came Boas* and *Theorizing the Americanist Tradition*, cited earlier, and *Invisible Genealogies: A History of Americanist Anthropology*, Critical Studies in the History of Anthropology (Lincoln and London: University of Nebraska Press, 2001).

to question positivism and claims to universal knowledge. Historian of anthropology Regna Darnell perhaps put it best when she noted: "The real Boas tends to disappear amidst the apotheosis of angst-ridden anthropological reflexivity."[6] For anthropologists and those who study the history of anthropology, the last quarter of the twentieth century saw a profound turn inward toward self-reflexiveness and an equally profound self-consciousness and discomfort with the traditional project of anthropology, its methods, and assumptions.[7] For anthropologists, history has functioned in part as identity formation, reconstructing a lineage that justifies or supports current commitments, particularly for a discipline acutely uncomfortable with its participation, direct and indirect, in racial injustice and its legacy around the world. All history is written from a situated vantage point, but a discipline's practitioners are especially burdened with the weight of their intellectual and methodological genealogy when attempting to reconstruct a fraught past. At the same time, cultural relativism and the multiplication of culture into cultures, from singular to plural – a part of the anthropological legacy – is so deeply and widely embraced that even the most self-reflexive scholars sometimes fail to see it as a historically evolving worldview. The corollary is an often adamant, ironically nearly reflexive, rejection of biology or heredity as useful or interesting explanatory frameworks for understanding humanity. Those few who espouse such notions often have been simply dismissed as racists, part of a genuinely repugnant tradition of slaveholders, racist eugenicists, and Nazis.

But I would argue that it is vital to understand the past of racial science as the range of practice that it was, from work that was regarded at the time as wholly sound and legitimate to racist propaganda, and the relation of all of it to the society in which it was produced, in this case a society that in the 1920s passed strict immigration restrictions, saw a resurgence of the Ku Klux Klan, and witnessed race riots, entrenched Jim Crow laws, and eugenic denunciation of hereditary "defectives." Historians should not ignore scientific and intellectual contexts that existed in the past because they no longer seem to hang together as a legitimate whole, and retrieve

6 Regna Darnell, "Review: Reenvisioning Boas and Boasian Anthropology," *American Anthropologist*, New Series, vol. 102, no. 4 (Dec. 2000), pp. 896–899.
7 Regna Darnell, *Invisible Genealogies*; Julia Liss, "Diasporic Identities: The Science and Politics of Race in the Work of Franz Boas and W. E. B. Du Bois, 1894–1919," *Cultural Anthropology*, vol. 13, no. 2 (May 1998), pp. 127–166; James Clifford and George E. Marcus, eds., *Writing Culture: The Poetics and Politics of Ethnography* (Berkeley: University of California Press, 1986).

only the science, the ideas that make the story coherent from a presentist point of view. Indeed, recapturing the fullest possible picture of how science and scientists functioned within society serves an epistemological and historical purpose in counteracting to some degree the powerful ideological and rhetorical force of science itself, which continually recasts itself, through reconstructed histories of its great researchers and fundamental ideas, as a process outside of history and society. Anthropology, among the most self-conscious of the sciences, is surely less guilty of this than some. But even anthropology has often constructed a history that served current purposes and theoretical commitments more than the historical record. The failure to face unpalatable pasts squarely has too often led otherwise thoughtful, perspicacious scholars to abandon rigorous analysis of the ideas, motives, and choices of intelligent, politically savvy historical characters in favor of limp excuses like capitulation to the zeitgeist.

Despite extensive historical and anthropological attention paid to Boas by scholars and anthropologists, the scholarship (with some important exceptions) has not comprehensively – even adequately – addressed Boas as a physical anthropologist, but rather, and revealingly, the contrary. Boas is a key figure in the story of physical anthropology prior to World War II not only for the ways he, and others, notably Harry Shapiro, defined the problem of race and its solutions, but also because in delineating his views and practices we confront prevailing historiographies, both visions of the past created by historians writing the history of anthropology, as well as in the history anthropology tells itself, that have been significantly at odds with what seems to have been, for lack of a better term, true. Much of the historiography has elided or distorted the actual character of his physical anthropology, and specifically his interest and belief in race. Much of the historical work that treats Boas fails to confront the full nature of his views, perhaps because it does not seem to comport with the kind of figure many historians, anthropologists, and others want him to be, one crucial to a historiographical narrative about the ascendance of culture and the decline of scientific racism.[8] One rarely

---

[8] In introductory remarks to selections of Boas's early writings on race, George Stocking made a similar point, noting that Boas has been recognized principally as a critic of racism, but that early arguments that "were conditioned by the racist milieu in which he wrote," which brought together various aspects of anthropology and which revealed early conception of culture embedded in his critique of racial determinism had been overlooked. George W. Stocking, Jr., "Racial Capacity and Cultural Determinism," *The Shaping of American Anthropology, 1883–1911: A Franz Boas Reader*, Midway Reprint,

reads about Boas's racial science,[9] except in cursory, often apologetic or dismissive terms, whereas his credentials as an innovator in the practice of cultural anthropology are widely recounted. Our understanding of physical anthropology and the history of racial science is distorted because the view of race and the biological study of human difference that Boas, and later Harry Shapiro, espoused has been neglected or flattened, and the tensions within even the most apparently typological racial science forgotten.

In gaining a deeper understanding of mid-century racial science, we can begin to see the outlines of a broader story about how race disappeared from anthropology in very specific ways, replaced by or transformed into other concepts and categories, and how the transformation of physical anthropology into biological anthropology both contributed to and reflected similar major shifts in modern social and political discourse in America.[10] By looking more carefully at the actual landscape of theory

---

(Chicago and London: The University of Chicago Press, [1974] 1989), pp. 219–221. For a similar critique, see John S. Allen, "Franz Boas's Physical Anthropology: The Critique of Racial Formalism Revisited," *Current Anthropology*, vol. 30, no. 1 (Feb. 1989), pp. 79–84.

9   There are a number of important exceptions to this. George Stocking's work is a prominent and crucial exception, along with more recent work by Liss, "Diasporic Identities," and "The Cosmopolitan Imagination: Franz Boas and the Development of American Anthropology" (PhD diss., University of California, Berkeley, 1990). Also notable are Darnell, *Invisible Genealogies*; Lee D. Baker, *From Savage to Negro: Anthropology and the Construction of Race, 1896–1954* (Berkeley and Los Angeles: University of California Press, 1998); Vernon J. Williams, Jr., *Rethinking Race: Franz Boas and His Contemporaries* (Lexington: The University of Kentucky Press, 1996); and Nancy Leys Stepan, "Race, Gender, Science and Citizenship," *Gender & History*, vol. 10, no. 1 (Apr. 1998), pp. 26–52. Elazar Barkan discusses Boasian racial science and interwar racial science more generally at length in *The Retreat of Scientific Racism: Changing Concepts of Race in Britain and the United States Between the World Wars* (Cambridge: Cambridge University Press, 1992), although this work relies heavily on an unfortunate dichotomization of scientists into "racists" and "egalitarians," a retrospective oversimplification that obscures more than it illuminates.

10 Other scholars who have begun to reexamine twentieth-century racial science include: Mitchell B. Hart, "Jews and Race: An Introductory Essay," Mitchel B. Hart, ed., *Jews and Race: Writings on Identity and Difference, 1880–1940* (Waltham, MA: Brandeis University Press, 2011), pp. xiii–xxxix; Joanne Meyerowitz, "'How Common Culture Shapes the Separate Lives': Sexuality, Race, and Mid-Twentieth Century Social Constructionist Thought," *Journal of American History*, vol. 96, no. 4 (Mar. 2010), pp. 1057–1084; Gavin Schaffer, *Racial Science and British Society, 1930–1962* (New York: Palgrave Macmillan, 2008); Anthony Q. Hazard, Jr., "Postwar Anti-racism: The United States, UNESCO and 'Race', 1945–1968" (PhD diss., Temple University, 2008); Michael Yudell, "Making Race: Biology and the Evolution of the Race Concept in 20th Century American Thought" (PhD diss., Columbia University, 2008); Michelle

and practice surrounding the study of race in the decades before and after World War II, we can begin to see where the real disjunctions and actual continuities lie between various pasts and presents. With a fuller understanding of Boas's views, and of those who came after him and followed his lead, Harry Shapiro, Ruth Benedict, and Ashley Montagu prominently among them – as well as those who did not, such as Henry Field, Arthur Keith, and Earnest Hooton – we can begin to see continuity with views about race that are more typical of the post–World War II period in the United States, the very views that seem to have made it difficult to lucidly comprehend the racial science Boas and others practiced. This book corrects the historiographical tilt toward culture in the history of anthropology by arguing that anthropology was foundational to American racial formation precisely because it promoted both racial essentialism and cultural relativism. Taking the science of race seriously, understanding it as its practitioners did, in all its complexities, contradictions, and shifting emphases, illuminates not only how Americans used to think about race and culture, but why we still think about it the way we do. A study of early and mid-century physical anthropology can help explain how and why race and science seem to have reemerged with such force in the late twentieth and early twenty-first century.

Race in America, Race in Science

In the 1990s, biological determinism linked to race seemed to have reemerged. Physicians promised better health via race-targeted drugs like BiDil. Companies offering DNA tests prompted genealogical odysseys as celebrities like Oprah Winfrey searched for their origins. Long-festering debates over race and IQ exploded back into public view with *The*

Brattain, "Race, Racism, and Antiracism: UNESCO and the Politics of Presenting Science to the Postwar Public," *The American Historical Review*, vol. 112, no. 5 (Dec. 2007), pp. 1386–1413; Keith Wailoo and Stephen Pemberton, *The Troubled Dream of Genetic Medicine: Ethnicity and Innovation in Tay-Sachs, Cystic Fibrosis and Sickle Cell Disease* (Baltimore: The Johns Hopkins University Press, 2006); Richard H. King, *Race, Culture, and the Intellectuals, 1940–1970* (Baltimore and London: The Johns Hopkins University Press, 2004); William H. Tucker, *The Funding of Scientific Racism: Wickliffe Draper and the Pioneer Fund* (Urbana: University of Illinois Press, 2002); John P. Jackson, *Social Scientists for Social Justice: Making the Case against Segregation* (New York: New York University Press, 2001); Kenan Malik, *The Meaning of Race: Race History and Culture in Western Society* (New York: Macmillan Press, 1996).

Bell Curve.[11] Many who thought racial essentialism and biological determinism had been safely discredited decades earlier viewed these developments with alarm, and puzzlement. From a longer historical perspective, however, we can see recent developments as yet another stage in an ongoing dynamic in the United States between predominantly biological and predominantly social solutions to pressing problems. The discourses of nature and nurture, biology and society, have been consistently part of American culture for more than a century.

The complex mix of biologically essentialist explanations and historically or culturally grounded theorizing that one finds in the work of Franz Boas, Ruth Benedict, Harry Shapiro, or at the Field Museum in the interwar period is distinctive of its era but also remarkable in the ways it reaches across the supposed gulf of World War II (and the Evolutionary Synthesis) to illuminate tensions and connections between the biological and the cultural in the latter half of the twentieth century. Indeed, *inter*war racial science was much like *post*war racial science in its complex brew of biology and society, the promiscuous intermingling of bodies and cultures. Rather than seeing the postwar reaction against race and racism as either decisive or unique, we should view it as one era in an ongoing construction of human variation that has always consisted of a volatile, unstable mix of cultures and bodies.

The construction of race in science has never been an either/or proposition. It has never been a question of race *or* culture, heredity *or* society, bodies *or* minds. The discourses of human variation since the eighteenth century have always incorporated visions of bodies, capabilities, and cultures as a means to explain diversity and justify hierarchy. That is not to say that there has been easy consensus. For an entity that supposedly encompasses a set of patently evident natural kinds, race has been a profoundly unstable scientific object, subject to constant contestation and in need of continual reconstruction. The science of race has been marked

---

[11] On Bidil, see Jonathan Kahn, *Race in a Bottle: The Story of BiDil and Racialized Medicine in a Post-Genomic Age* (New York: Columbia University Press, 2012); Wailoo and Pemberton, *The Troubled Dream of Genetic Medicine*; Harriet A. Washington, *Medical Apartheid: The Dark History of Medical Experimentation on Black Americans from Colonial Times to the Present* (New York: Harlem Moon, 2006). Also see www.bidil. com. On Oprah Winfrey's search for her ancestors through DNA analysis, and the making of the PBS documentary, see Henry Louis Gates, Jr., *Finding Oprah's Roots, Finding Your Own* (New York: Random House, Crown Publishing Group, 2007). Richard J. Herrnstein and Charles Murray, *The Bell Curve: Intelligence and Class Structure in American Life* (New York: Free Press, 1994).

by persistent debates about methods, types, and implications. This was true in the interwar period no less than the immediate postwar period, with the eras before and after World War II each marked by only limited consensus. Although the nature of consensus clearly shifted away from racial essentialism and hereditarian racial typology between 1900 and 1970, this must be seen within a broader historical perspective of the waxing and waning of hereditarian, biologically deterministic, and essentialist views. The conviction that science will ultimately solve the riddle of human diversity is one that has persisted since at least the application of comparative anatomy to the "problem" of African Americans and Native Americans in antebellum America. As new scientific theories and techniques emerged, they have invariably been applied to the conundrum of race, from Johann Friedrich Blumenbach's zoological classifications in the eighteenth century, to Samuel Morton's craniometry and Francis Galton's composite photographs in the nineteenth century, to genetics in the twentieth century. And yet, despite the growing authority of the sciences within American society, there has been no period of U.S. history when essentialism, hereditarianism, and white supremacy have not been challenged. In periods when a hereditarian vision has been most widely embraced, we still find critiques and persistent questioning. Conversely, humanist, cultural accounts of human diversity never fully drive out biological essentialism or hereditarian claims. Indeed, from the perspective of a historian of racial science, what is striking is the unrelenting persistence of "racial" questions.[12] Concerted efforts following World War II to reject "scientific racism" and promote cultural relativism (and later to promote an account of race as a social construction without biological meaning) can be seen as a moderately effective attempt to combat divisive biological, hereditarian essentialism by erecting a sturdy boundary between "race" and "culture" that had not existed earlier in American history and that has been eroding again with the resurgence of genetics. The history of racial and cultural studies within anthropology is not one of successive Kuhnian paradigms, with Boasian cultural relativism displacing hereditarian, deterministic "scientific racism," so much as tandem discourses with distinct institutional and methodological histories

---

[12] Faye Harrison has noted the persistence of race, but her concern was not so much its persistent scientific reconstruction as it was the culture and politics of racism. See "The Persistent Power of 'Race' in the Cultural and Political Economy of Racism," *Annual Review of Anthropology*, vol. 24 (1995), p. 48.

that have been frequently allied as the practice of anthropology developed in the United States.[13] Much attention has been paid by scholars to the nuances of the culture concept and cultural theorizing in anthropology; the racial science of anthropologists deserves a similarly careful examination. As historian Bronwen Douglas has forcefully argued, we need more critical studies "grounded in rigorous vernacular reading of the original works of Euro-American thinkers whose broad, labile gamut of positions on human differences is often collapsed under the homogenizing rubric of racism."[14] Lumping together all anthropological racial science under the umbrella of "scientific racism" obscures rather than illuminates precisely the vexed concepts and practices we seek to understand.

In the first four decades of the twentieth century, between "race" and "culture," race was the more firmly established concept within and outside anthropology. During the interwar period, for the American public, common sense held that bodies and societies were deeply, causally intertwined. "Civilization," or the lack of it, was commonly viewed as a manifestation of race, and a given racial group's inherent capacities. In this view, "race" and "ethnology" were merely different components of the "civilization" process. Franz Boas articulated a competing theory that stressed the fluid, historically contingent, environmentally conditioned

---

[13] A variety of scholars have questioned the extent to which anthropology fits a Kuhnian paradigm of scientific development, and some have questioned the idea that "culture" replaced "race" in anthropology in any straightforward Kuhnian sense. In part this is attributable to the "four field" nature of American anthropology, in which the overall discipline of anthropology coalescing out of a variety of prior fields, including anatomy and medicine, philology, and natural history and philosophy, developed via four fairly distinct subfields: linguistics, archaeology, ethnology (later cultural anthropology), and physical (later biological) anthropology. Some also have argued that Kuhn's theory of paradigms has been most fruitfully applied to the natural sciences, especially the "hard" experimental sciences (physics, chemistry, etc.), and less successfully to the human, social, and historical sciences, which some labeled "pre-paradigmatic." These distinctions are still a matter of debate. Kuhn himself made no claim that his rubric applied to the social sciences. Thomas S. Kuhn, *The Structure of Scientific Revolutions*, second enlarged edition (Chicago: University of Chicago Press, [1962] 1970). For a similar critique of anthropology's development, see Regna Darnell, *And Along Came Boas*, especially pp. xvi, 1–7, 271–297; Lee D. Baker, *From Savage to Negro*. On the question of the utility or applicability of Kuhnian paradigms to anthropology, see, for example, Lawrence A. Kuznar, *Reclaiming a Scientific Anthropology* (Lanham, MD: AltaMira Press, 2008); Steve Fuller, *Thomas Kuhn: A Philosophical History for Our Times* (Chicago: University of Chicago Press, 2000).

[14] Bronwen Douglas, "Foreign Bodies in Oceania," in Bronwen Douglas and Chris Ballard, eds., *Foreign Bodies: Oceania and the Science of Race 1750–1940* (ANU E Press, The Australian National University, 2008), retrieved from http://epress.anu.edu.au/foreign_bodies/html/frames.php.

process of human cultural formation, and rejected the idea of universally fixed stages of social, technological, and political development espoused by the evolutionists. The pluralistic culture concept propounded by Boas had been articulated by him before 1900, but was not fully assimilated into the practice of ethnology until the 1920s, and not into the rest of the social sciences and beyond until toward the end of the interwar period. The general public did not broadly embrace it until a good deal later. A broad public acceptance of the culture concept is tricky to pin down, but it seems to have taken a long while before the idea that diverse cultural forms were equally valid took hold. Lois Banner cites Margaret Mead's assertion in 1928 that she felt she still had to explain it for her reading public. Ruth Benedict also went to great lengths in both *Patterns of Culture* in 1934 and *Race: Science and Politics* in 1940 to explain what she meant by "culture." Arguably, it remained a matter of contention in 1955 when Edward Steichen mounted "The Family of Man" at the Metropolitan Museum of Art, a humanist vision of brotherhood illustrated by cultural variations on universal themes.[15]

In the nineteenth century and the first decades of the twentieth century, most of the American public and most scientists who studied human variation believed in racial typology. They thought race was something essentially biological and static, and at their most rigid adhered to the idea that not only groups of people but individuals could be accurately categorized. But even within this general consensus there were varied approaches. Whereas Henry Field at the Field Museum and his British mentor, Arthur Keith, promoted a traditional typological view of fixed racial types, Franz Boas and Harry Shapiro at the American Museum of New York approached the problem of human variation with a more holistic view. Boas and Shapiro, like Field and Keith, believed that race

---

[15] On acceptance of the Boasian culture concept within anthropology and the social sciences, see John S. Gilkeson, Jr., "The Domestication of 'Culture' in Interwar America, 1919–1941," in JoAnne Brown and David K. van Keuren, eds., *The Estate of Social Knowledge* (Baltimore and London: The Johns Hopkins University Press, 1991), pp. 153–174; George W. Stocking, Jr., *Delimiting Anthropology: Occasional Essays and Reflections* (Madison: University of Wisconsin Press, 2001), pp. 23, 32, 272; George W. Stocking, Jr., "Ideas and Institutions in American Anthropology," *The Ethnographer's Magic and Other Essays in the History of Anthropology* (Madison: University of Wisconsin Press, 1992), especially p. 164; Regna Darnell, *Along Came Boas*, p. xii; Lois W. Banner, *Intertwined Lives: Margaret Mead, Ruth Benedict, and Their Circle* (New York: Alfred A. Knopf, 2003), 195; the text Banner cites for Margaret Mead's assertion is the preface to the 1973 edition of *Coming of Age in Samoa* (New York: William Morrow, 1973), n.p.

was a feature of human biology. But Boas and Shapiro regarded it as malleable as well, associated with groups of people who moved through space and time and who existed everywhere in cultures that created environments that shaped the form of human bodies. By the late 1930s, many scientists, especially geneticists and physical anthropologists, objected to shoddy scientific and popular racial ideologies that trafficked in overgeneralization and what they viewed as folk knowledge or personal bias. Boas and Shapiro objected that race and racial characteristics were not biologically well understood, had been inadequately studied (in part because of serious methodological problems), and were improperly theorized (e.g., no one had any idea which characteristics were evolutionarily or hereditarily significant, nor just what the role of environment was, although Boas's study of changes in the form of the head had demonstrated that there was one). The problem, from their point of view, was poor science, not a fundamental conceptual error in the object of study. They opposed racism, not race.

Late-twentieth-century visions of human diversity rejected biological aspects altogether and came to see race as exclusively a cultural/social phenomenon that has no biological existence. But even this social constructionist definition often acknowledges that race is "real" in the sense that, as a social category, race remains a powerful factor in human relations and American society. By the 1960s, much work in genetics, anthropology, and human biology tried to redefine race in evolutionary terms, looking at population groups over time, not seeing race as static but rather as a function (as Shapiro termed it in 1939) of migration and environment. Although a generally biological vision of race predominated in the pre–WWII period and a culturally constructed one in the latter twentieth century, a multiplicity of views have existed throughout the century, shifting in complex relation to changing concepts and methods in both the sciences and wider society. All these investigators, using significantly different notions of "race," have sought an answer to the same persistent question: why is there so much variety among human beings?

Among cultural anthropologists, the slow diffusion of the culture concept into American society influenced the strategy they adopted in their attempts to combat prejudice and discrimination. By the 1930s, many anthropologists increasingly shared a sense that race had become distorted in the popular (and frequently in the scientific) mind. Misapprehensions about race had already led to all kinds of objectionable action, repercussions that only grew more dire as Nazis consolidated their power in Germany. Anthropologists felt an urgent need to address what they saw as gross distortions, misunderstandings, and misapplications. Most adhered to a liberal humanist, positivist belief that "truth" and knowledge would undo prejudice. Following profound theoretical interests in individual psychology and the relationship of individuals to their cultures, many cultural anthropologists believed that the root problem was individual racial prejudice. Thus, most of their solutions were focused on disabusing individuals of their biases and misinformation. They did not consistently offer an analysis of structures of racism, power, or wealth, and even when they did, such as Ruth Benedict's work in *Race: Science and Politics*, they did not see those structures of racism as related to the "fact" of race, except through the political exploitation of bias. They did not view the race concept itself as inherently problematic, or likely to engender the disparities and discrimination they attacked. By the end of the interwar period, anthropologists were stressing the separate nature of race and culture, describing the "facts" of human physical variation on the one hand and the limitless diversity of culture, which varied independently of race, on the other. For Boasian ethnologists, interwar efforts to rethink race occurred simultaneously with efforts to change Americans' perception of society and culture, shifting it away from social evolutionary hierarchies of primitive and civilized and toward the plural, relativistic culture concept. The effect, which would have far-reaching, unintended consequences, was to reify race while relativizing culture. The new postwar "common sense" held that race was merely a matter of inherited physical characteristics, while culture was an entirely separate process by which a group of people constructed a social life.

George Stocking has argued that the firm boundary erected by anthropologists between culture and race has been "defended" since it was first embraced in the 1920s for theoretical and methodological reasons, but also for ideological and political ones.[16] The challenge for the historian of race and racial science is not unlike that of postcolonial scholars attempting to provide subtle, nuanced accounts of an imperial world too easily cast into victims and villains, colony and metropole, and more perilously, us and them. Like the landscape of race and racial science in the United States, the landscape of race and power in colonial contexts was complex, and often cut against the tidy conceptual and historical categories that help scholars parse the past but which too often flatten or obscure the thing we most desire to understand. Race has always been a protean concept, slippery and flexible, used in contradictory and inconsistent ways,

---

[16] Stocking, *Delimiting Anthropology*, p. 315.

interwoven with politics, class, and economics, gender and sex, social relations, and material cultures, in myriad permutations. The historian is perpetually in danger of oversimplifying a complex story, imagining that the past was less complex than the present. Ann Laura Stoler captured the challenges and promise of historical work in these complex pasts when she argued that for the scholar of colonialism, "Replacing colonialism's stick figures with actors who combined goodwill and sympathy for the dispossessed with racist beliefs underscores that colonial regimes were not less complex racially inflected social and political configurations than are ours today."[17] The historian of racial science in the twentieth century faces a similar challenge.

The history I recount in this book may be obscure to many today, but that is not because the anthropologists, institutions, and practices I discuss were equally obscure in the past. The failure to remember *The Races of Mankind* at the Field Museum of Natural History, Ruth Benedict's extensive exposition of race, Franz Boas's decades-long pursuit of racial science, or Harry Shapiro's career is not because these scientists and institutions were obscure, their works unknown. I did not have to dig in dark corners of the library or plumb the depths of the archive to unearth the science I discuss here. It was there, had been there all along, for anyone who cared to look. *The Races of Mankind* sculptures still grace the halls of the Field Museum. *Patterns of Culture* and *Race: Science and Politics* are still in print. Franz Boas published dozens of his papers on race in an anthology of his work shortly before he died in 1942; indeed, his last words were, "I have a new theory about race."[18] It seems that historians and other scholars have, until very recently, had a blind spot for mid-century racial science.

Perhaps that is because the anthropology of race in the interwar and immediate postwar period contradicts a satisfying narrative of progressive antiracism in which physical anthropology was a practice vanquished and best forgotten. Like a narrative of WWII as the "good war," the transplantation of cultural relativism where "scientific racism" once lurked provided a welcome resolution to a tumultuous, often violent and painful,

---

[17] "Matters of Intimacy as Matters of State: A Response," Ann Laura Stoler, *The Journal of American History*, vol. 88, no. 3 (Dec., 2001), pp. 893–897, quote p.896; see also Stoler, "Racial Histories and Their Regimes of Truth," *Political Power and Social Theory*, vol. 11 (1997), pp. 183–206.
[18] Margaret Mead, *An Anthropologist at Work: Writings of Ruth Benedict* (Boston: Houghton Mifflin Company, 1959), p. 355.

past. The desire to forget the racial science of physical anthropologists is wedded to a liberal humanist ideology that once believed, and I think still hopes, that science is self-correcting, that "good" science will prevail over "bad," just as freedom and equality will ultimately prevail over racism and other oppressive ideologies and practices. Indeed, cultural relativism's triumph over "scientific racism" ranks with oxygen replacing phlogiston and the heliocentric universe as canonical case studies used to demonstrate that whatever might be the subjective limits to objective knowledge and however illusory the idea of progress, science can correct at least some profound misapprehensions. In these narratives, it is science itself, sometimes fueled by a progressive philosophy and often practiced by members of the oppressed classes, that forces repressive, reactionary ideologues to confront the inconsistencies, errors, and hypocrisy of racist, essentialist accounts of human variation that constructed the "other" as naturally, immutably less capable and deserving. Race is revealed as a social construct, not a natural discovery. The problem with this narrative is not that it is utterly wrong – it is not – but what it privileges and what it obscures. As Ann Stoler has argued, racial discourses are characterized by their "polyvalent mobility," their remarkable flexibility and adaptability, their ability to "draw on the past as they harness themselves to new visions." Ideologically malleable, the concept of "race" has been employed by historical actors on all sides, not merely by those in power or advocating a "racist" perspective. Scholars' preoccupation with "debunking" race and the science of race, she argued, left us with a body of literature that relied on an oversimplified account.[19] Too often in such narratives, racial scientists become "stick figures," and the complexities of nonlinear conceptual, social, political, and professional relations get flattened or ignored altogether. According to Stoler's critique: "Histories of racisms that narrate a shift from the fixed and biological to the cultural and fluid impose a progression that poorly characterizes what racisms looked like... and therefore have little to say about what distinguishes racisms today."[20] Since Stoler published her incisive analysis, much of the work on race and public health, eugenics, sociology, and psychology has adopted the more nuanced view she advocated, and has explored the "polyvalent mobility" of race that she

---

[19] Stoler, "Racial Histories and Their Regimes of Truth," especially pp. 190–192, 195–196, 198.
[20] Ibid., p. 198.

identified.[21] But the history of anthropology, especially the core narrative about culture and "scientific racism," has largely resisted a similar analysis until very recently. It is this unwitting blindness, an "epistemology of ignorance," to use Charles Mills's evocative and apt phrase, born of a desire to be free of the odious burden of race, that leads directly to

[21] See Warwick Anderson, *Colonial Pathologies: American Tropical Medicine, Race, and Hygiene in the Philippines* (Durham, NC: Duke University Press, 2006); Nayan Shah, *Contagious Divides: Epidemics and Race in San Francisco's Chinatown* (Berkeley: University of California Press, 2001); Keith Wailoo, *Dying in the City of the Blues: Sickle Cell Anemia and the Politics of Race and Health* (Durham, NC: University of North Carolina Press, 2001); Evelynn Hammonds, *The Nature of Difference: Sciences of Race in the United States from Jefferson to Genomics* (Cambridge: MIT Press, 2008), Susan Reverby, *Tuskegee's Truths: Re-Thinking the Tuskegee Syphilis Study* (Chapel Hill: University of North Carolina Press, 2000); Jenny Reardon, *Race to the Finish: Identity and Governance in an Age of Genomics* (Princeton: Princeton University Press, 2004); Wendy Kline, *Building a Better Race: Gender, Sexuality, and Eugenics from the Turn of the Century to the Baby Boom* (Berkeley: University of California Press, 2001); Alexandra Stern, *Eugenic Nation: Faults and Frontiers of Better Breeding in the United States* (Berkeley: University of California Press, 2005); Jackson P. Jackson, *Social Scientists for Social Justice: Making the Case against Segregation* (New York: New York University Press, 2001); Nadine Weidman, *Constructing Scientific Psychology: Karl Lashley's Mind-Brain Debates* (Cambridge: Cambridge University Press: 2004); William H. Tucker, *The Funding of Scientific Racism*.

Recent work on the history of race in anthropology sensitive to its complexities includes Donna Haraway, "Universal Donors in a Vampire Culture: It's all in the Family – Biological Kinship Categories in the Twentieth-Century United States," *Modest-Witness@Second-Millennium.FemaleMan©-Meets-OncoMouse*[TM]: *Feminism and Technoscience* (New York: Routledge, 1997), pp. 213–266; John P. Jackson, Jr., "'In Ways Unacademical': The Reception of Carleton S. Coon's *The Origin of Races*," *Journal of the History of Biology*, vol. 34 (2001), pp. 247–285; Michelle Brattain, "Race, Racism and Antiracism"; Andrew Zimmerman, *Anthropology and Anti-Humanism in Imperial Germany* (Chicago: University of Chicago Press, 2001); C. C. Mukhopadhyay and Yolanda T. Moses, "Reestablishing 'Race' in Anthropological Discourse," *American Anthropologist*, 99(3), 1997, pp. 517–533; Faye Harrison, "Introduction: Expanding the Discourse on 'Race,'" *American Anthropologist*, 100(3), 1999, pp. 609–631, and "Persistent Power"; Alan Goodman, Deborah Heath, and M. Susan Lindee, eds., *Genetic Nature/Culture: Anthropology and Science beyond the Two-Culture Divide* (Berkeley: University of California Press, 2003); Michaela di Leonardo, *Exotics at Home: Anthropologies, Others, American Modernity* (Chicago: University of Chicago Press, 1998), and "Human Cultural Diversity," delivered at the Race and Human Variation: Setting an Agenda for Future Research and Education conference, 2004; Henrika Kuklick, ed., *A New History of Anthropology* (Blackwell Publishing, 2008); Andrew Evans, *Anthropology at War: World War I and the Science of Race in Germany* (University of Chicago, 2010); and Alice Conklin, *In the Museum of Man: Race, Anthropology and Empire in France, 1850–1950* (Ithaca: Cornell University Press, 2013). Another work that shares my historicist concerns about twentieth-century anthropology is Andrew P. Lyons and Harriet D. Lyons, *Irregular Connections: A History of Anthropology and Sexuality* (Lincoln: University of Nebraska Press, 2004).

the surprise and disgust with which *The Bell Curve* was greeted by so many.[22]

The importance of capturing the complexities of a racial science that, whatever its currency in the mid-twentieth century, has been effectively marginalized for the last generation lies not only in correcting the historical record, but also in improving our understanding of the present. Even a glancing acquaintance with everyday life in the United States should disabuse us about the efficacy with which culture replaced race. Race continues to intermingle with culture in ways that are virtually inexplicable had the ideas promoted by racial scientists been vanquished as thoroughly as the standard narrative suggests. An illustration of this conundrum has been discussed by Australian anthropologist Gillian Cowlishaw, who has written about anthropologists' efforts to cope with race and racism in the modern world. Cowlishaw was trained in the early 1970s in Australia, a period when cultural relativism had become the consensus position in the field and had been widely embraced in Western societies. Although the analytical concept of "race" had been replaced by terminology such as "traditional" by the time she entered graduate school, Cowlishaw argued that categories such as "'traditional' filled precisely the same semantic space previously occupied by the Aboriginal race."[23] This semantic shift, she argued, did not lead to any diminishment of racism toward the Aboriginal people of Australia, but merely obscured the continuing significance of their racialized identity and hobbled anthropologists' ability to respond effectively to the highly racialized and racist social, political, and economic world of contemporary Australia. In her view, the utter displacement of race with culture in anthropology "had the effect of confusing the interpretation of a world where race and culture, however conceptualized, had become so intertwined that attempts to separate their functioning were futile." She argued that anthropologists simply accepted a biological definition of race and rejected it, leaving it uncontested.[24] Cowlishaw warned that "while speaking of race may appear to reproduce racial categories" – a fear that motivated anthropologist Ashley Montagu and geneticist Julian Huxley to urge abandoning the term in

---

[22] Charles W. Mills, *The Racial Contract* (Ithaca, NY: Cornell University Press, 1997); Shannon Sullivan and Nancy Tuana, eds., *Race and Epistemologies of Ignorance* (Albany, NY: State University of New York Press, 2007).

[23] Gillian Cowlishaw, "Censoring Race in 'Post-colonial' Anthropology," *Critique of Anthropology*, vol. 20, no. 2 (2000), pp. 101–123 (p. 108).

[24] Ibid., p. 111.

the 1940s – "not speaking about race allows racial differentiation to flourish unchallenged."[25]

It is not only modern anthropologists and other scholars who grapple with the meaning and implications of race and culture. Earlier anthropologists struggled to make sense of human diversity in the context of their times, too. Close reading of anthropologists' texts, lectures, and imagery reveals scientists' struggles to consciously create concepts and methods in accord with their discipline, the larger world, and their own instincts and proclivities, what Tom Holt has aptly termed a "sediment" of prior theoretical and personal commitments that constrain conceptual possibilities.[26] Changes in the conception of race were sometimes a response by anthropologists to a disjunction between what they "knew" to be true and what their process gave them as "facts" about the world. In the late nineteenth century, Paul Topinard began to question the existence of race when measurements failed to yield categories and standards by which individuals and groups could be reliably classed, but retreated to a search for hypothetical pure races to explain the origins of such heterogeneous people, in part because his "common sense" experience of the world told him that races existed.[27] Similarly, Arthur Keith advised Henry Field that there was no reason to burden his Field Museum exhibition project with an enormous program of measurement and statistical analysis, as Harvard anthropologist Earnest Hooton urged, because it was plain to anyone who walked down the street that races existed.[28] For scientists who studied race, it was a problem of how to make their work congruent with their experience of and vision for the world, as well as a problem of how to orient their work with that of other sciences – and because the subject was race – also a social, cultural, and political problem, one whose complexities cannot be encompassed in a set of dichotomous epithets such as "racist" or "egalitarian." *Constructing*

[25] Ibid., p. 101.

[26] On the idea of "sediment," see Tom Holt, *The Problem of Race in the Twenty-First Century* (Cambridge, MA: Harvard University Press, 2000), p. 23, cited in Brattain, "Race, Racism and Anti-Racism." On personal and theoretical constraints on knowledge production, see, for example, William B. Provine, "Geneticists and the Biology of Race Crossing," *Science*, New Series, vol. 182, no. 4114 (Nov. 23, 1973), pp. 790–796; and William B. Provine and Elizabeth S. Russell, "Geneticists and Race," *American Zoologist*, vol. 26, no. 3 (1986), pp. 857–887.

[27] Stocking, *Race, Culture and Evolution*, p. 59.

[28] Reported to Malvina Hoffman by Stanley Field, September 9, 1931, Box 3, Malvina Hoffman Collection, 850042-1, Special Collections, The Getty Center for the History of Art and the Humanities, Los Angeles, California (hereafter MHC).

Race illuminates how physical anthropologists in the middle of the twentieth century promoted a vision of race rooted in bodies and cultures, a vision that shaped popular perceptions of race in America for generations.

Constructing Race: Argument and Organization

This book originates in a desire to understand the tenacity and vicissitudes of biological, racial ways of thinking that have persisted despite the late-twentieth-century "cultural turn." How do we explain the growth of sociobiology, the appeal of evolutionary psychology, the return of IQ, the fascination with DNA analyses of ancestry for celebrities, commoner, and dog alike? These trends suggest that ideas about race that had supposedly been put to rest with the ascendance of the anthropological culture concept endured, and that the broad embrace of cultural explanations for social phenomena has been incomplete. As I researched anthropology, race, and science, I began to wonder if the intellectual puzzle was not only racial theorizing itself, but also the intellectual framework with which Americans had narrated the undoubted shift toward culture over the course of the twentieth century. Perhaps it was not a matter of rejection or replacement so much as transformation or ascendance. The desire to explain diverse bodies and cultures as a result of natural or biological processes looked like a persistent theme in American history, not an anomaly. I saw a narrative of waxing and waning, not replacement and triumph.

This book reflects my own interest in both intellectual and cultural history. The result is an intellectual history that grapples with how key ideas have been translated into a variety of cultural forms for broad popular audiences. As an intellectual historian, I am deeply interested in understanding how ideas develop. As a cultural historian, I am interested in how these ideas have been expressed not only through texts or speech, and not only for elite audiences, but also through the variety of cultural forms aimed at broad audiences in the United States. My goal in this book is to uncover and analyze ways of thinking about and publicly presenting race and culture that have been overlooked or obscured. To accomplish this, I focus on key figures and events that played critical roles in the public face of anthropology, both professional and popular. I am interested in how anthropologists disseminated and defended their understandings of race and culture among themselves, and how they popularized those views. This book is not an attempt to exhaustively recount the history of the discipline of physical anthropology in America, nor an attempt to comprehensively document the intellectual trajectories of its protagonists, nor an exhaustive accounting of race in museums or other popular venues. Rather, my goal is to capture the essential role anthropologists, especially physical anthropologists, have had in the public formation of race in the United States in the twentieth century, especially through the venues most closely allied to anthropology – natural history museums.

Anthropologists grappling with human diversity before and after WWII employed notions of race that included both naturalized, quantified bodies and concepts of culture, both racialized peoples and efforts to understand diverse cultural and environmental contexts. The fact that this racial theorizing was not decisively replaced by cultural relativism but instead persisted in tandem with it for most of the twentieth century had profound consequences for popular understanding of human diversity. Despite concerted efforts by anthropologists to confine the notion of race to bodies, and to stress culture as the source of variation among societies and their denizens, race and culture remained intermingled, entwined in both professional and popular discourse. The failure to reject race itself along with racism left confusion about the nature and source of human variation, and hardened a biological definition of differences.

This book offers a perspective on the trajectory of race and culture in twentieth-century America that spans the WWII. The racial dimensions of that conflict, particularly the horrors of Nazi racial policies, played a crucial role in the formation of race and the sciences of human variation, but it was only one catalyst in a postwar shift away from racial essentialism. By the 1940s, a number of factors combined to create a climate in which it seemed politically, scientifically, and socially necessary to assert a humanist framework over and against a racialist, and frequently racist, one. The history of racial formation, and the salient place of anthropology and other sciences in that formation, long predate WWII, and persist after it in ways that intersect with global political, economic, and military events, but were not governed by them. The profound political, social, and economic shifts associated with WWII, civil rights activism, and postcolonial movements coincided with equally profound shifts in biological and anthropological theory and practice. Although racial essentialism was increasingly subordinated to a humanist cultural framework in the postwar era, it persisted in competing formulations of human diversity despite concerted efforts to eliminate it. I argue that the history of nineteenth- and twentieth-century racial essentialism and biological determinism is not one of long ascent followed by steep decline, but rather a story of waxing and waning in relation to complex societal conditions, not least the development of new scientific theories and tools with which to attack the persistent puzzle of human variation.

This transformation of civilization into "cultures" has been regarded as a progressive story of the evolution of a scientific discipline. The story of racial anthropology is a messier, less appealing tale in which anthropologists (along with many others) persisted in attending to physical differences long after many in society had become uncomfortable with it. As a result, despite its important role in constructing race in the United States, racial anthropology has been forgotten, dismissed, or mischaracterized. This neglect reflects the conflicted way Americans have thought about race in recent decades, asserting that it does not exist or does not matter while but still clinging to the supposed reality and significance of physical differences. There has been a profound split consciousness about race in the United States, evident throughout the twentieth century. It is simply not true that racial science and race thinking were vanquished by the success of Boasian cultural anthropology (nor by the rejection of eugenics following its nadir under National Socialism, nor as a result of African Americans' persistent efforts to attain equality and civil rights, nor the success of postcolonial movements, nor a host of other things that were supposed to have contributed to a thorough rejection of "scientific racism").

My reconsideration of American physical anthropology and the place of racial science in professional and popular discourse begins chronologically and conceptually with one of the central figures in the development of American anthropology and American racial science, Franz Boas. The second chapter redresses serious omissions and misreadings of Boas's anthropology. Focused on the construction of a public racial discourse, my analysis draws on dozens of Boas's professional and popular publications and lectures, spanning his earliest physical anthropological work in the 1890s to his last pronouncements against racism before his death in 1942. I focus, in particular, on two sets of studies Boas conducted on problems central to American and European racial science: race mixing and heredity of racial characteristics. These studies not only demonstrate convincingly the extent to which Boas was committed to the scientific exploration of race, they illuminate the varied and complex ways that Boas attempted to understand human diversity in terms of biology, culture, and environment.

Boas has rightly been seen as a key figure in the development of a modern way of thinking about race and culture because he challenged assumptions about fixity. He challenged the commonly accepted nineteenth-century view that racial types as well as cultural stages were immutable kinds, the product of the inevitable, teleological expression of human evolution. But contrary to the usual vision of Boas, his challenge did not entail the rejection of race or a science of race. Rather, as this book seeks to demonstrate, Boas, and physical anthropologists following his lead, viewed culture *and* race as adaptations to novel sets of environmental constraints. *Both* race and culture were contingent, varied products of history, the natural environment, and the social environment. Boas transformed the cultural evolutionist's singular "culture" into the relativist, comparative, plural "cultures." And instead of racial typologists' fixed classifications, Boas emphasized racial variation, its malleability over time and space, and the complexity of factors that contributed to distinctive formations. His innovation was not in divorcing race from culture, or in rejecting race altogether, but in rejecting simplistic causal explanations and the idea that either race or culture could be easily delineated or catalogued like beetles pinned in a collector's cabinet. Indeed, one of Boas's signal differences from his racial scientist peers was his distinct lack of interest in the elaborate classificatory projects that had been the raison d'être of racial science since the eighteenth century. Although Boas never rejected the possibility of distinguishing among people along racial lines, he was more interested in comparative projects that might illuminate the complexities of underlying processes and their results. As Boas, his students, other social scientists, and a wider world became increasingly concerned about racism and its consequences over the course of the mid-twentieth century, the nuances of Boasian racial science and the idea that both race and culture reflected adaptations to changing conditions became obscured, lost in a battle to excise racism from race.

While Boas and others pursued environmental, comparative studies of race and culture, typological and evolutionary traditions continued in racial science. The field of physical anthropology was unsettled by a lack of consensus over methods, classifications, and even the nature of race itself. Practitioners debated whether and how to apply statistical analysis to their measurements of racial characteristics, how to best measure key traits, which tools to employ, which traits were the key ones and how fixed they were, the extent of variability in a racial group, how to interpret their results (e.g., was extensive variability a marker of race mixing, or not?), how many races and subraces there were, and what factors contributed to racial differences. Despite the methodological and conceptual ferment within physical anthropology, and within racial science more broadly, throughout the first half of the twentieth century most practitioners and much of the American public clung to at least some elements of typological racial essentialism. The idea that "race" described relatively stable naturally occurring kinds was a deeply entrenched professional and popular view. The racial types so commonly employed by scientists in the late nineteenth and early twentieth centuries were defined and classified on the basis of anatomical or other readily observed or measured characteristics, although evidence of behavioral, cognitive, and moral qualities was often employed, as well. Heredity was regularly invoked, although genetic analysis was often missing or rudimentary, often a matter of simple genealogical analysis or elementary statistical analysis of recurring traits. Implied or overt hierarchies, assertions of superiority, evaluations of progress, and association of various aptitudes with somatic types remained common well into the twentieth century among physical anthropologists and in popular accounts of race and racial science. Many racial scientists continued to see human racial and cultural variation as divisible into primitive and civilized classes, but not all were prepared to dictate social policy on that basis. Earnest Hooton rejected the idea that individuals could be reliably identified as a particular racial type, but believed that racial types existed across a population, and that hierarchical classifications could be constructed on that basis. Conversely, although Berthold Laufer also accepted the idea of racial types as natural kinds, he rejected hierarchies of superior and inferior types.

In Chapters 3 and 4, I turn to the *Races of Mankind* museum exhibition, mounted in 1933, for a detailed examination of contending theory and practice in interwar physical anthropology. The exhibition also provides a striking example of anthropologists' efforts to disseminate their conclusions to a wide popular audience. Chapter 3 explores the anthropological racial science that lay behind this landmark exhibition, delving into the racial theories of Field Museum anthropologists Berthold Laufer and Henry Field, and key outside advisors Arthur Keith, Earnest Hooton, and L. H. Dudley Buxton. The chapter explains how conflicted anthropologists offered Field Museum visitors a vision of race in the *Races of Mankind* that was rooted simultaneously in the biological essentialism of racial typology, the cultural currency of ethnic stereotypes, and the idealism of human unity. Chapter 4 analyzes the contested process through which the *Races of Mankind* was conceived and mounted at the

Field Museum in the 1920s and 1930s. Exhibit documentation, extensive correspondence, and contemporary publications reveal the tangled mix of racial and cultural theorizing, philosophical and methodological disagreement, compromise and convenience that lay behind the supposedly straightforward science of race presented to the American public by the Field Museum. The contested, contingent, socially embedded process revealed in the creation of the *Races of Mankind* demonstrates quite clearly how deeply entangled race and culture was in interwar anthropology, even in an exhibition explicitly devoted solely to the form of human bodies. The exhibition combined the aesthetic appeal and detailed, ethnographic naturalism of Malvina Hoffman's bronze sculptures with the purported empirical rigor of physical anthropology and the scientific authority of the natural history museum to create a powerful vehicle for delivering to the American public a vision of race that promoted both division and unity.

Chapter 5 demonstrates how Harry Shapiro, an influential but now largely forgotten physical anthropologist, put Boasian principles of racial science into practice in the 1930s to refute shoddy, deterministic studies used to promote discrimination. Like Boas before him, Shapiro did not reject biological notions of race, nor anthropometric methods, in favor of a cultural understanding of race. Rather, both Boas and Shapiro insisted that only by combining biological, cultural, environmental, and linguistic understandings of people would the answers to persistent questions about human differences be answered. Shapiro, a specialist on Polynesia, saw the tools of the physical anthropologist – their notorious calipers, charts of eye and nose shape, skin color, and hair form – as a critical piece in an overall strategy that combined consideration of culture and physical environment with the study of bodies. In this Boasian racial science, we can see the roots of the cultural construct embraced by social scientists after WWII, as well as the rudiments of a science of bodies abandoned with the repudiation of racial science as scientific racism.

Chapter 5 focuses on Shapiro's two most extensive research projects, completed in the 1930s at the height of his active research career as curator of physical anthropology at the American Museum of Natural History in New York. Like Franz Boas a generation earlier, Shapiro tackled two perennial problems in racial science. He self-consciously modeled his projects after Franz Boas's groundbreaking researches on hybridity and the impact of environments on bodily forms. In Hawaii, Shapiro recreated Boas's famous immigrant study, this time with Japanese immigrants, their Hawaiian offspring, and relatives back home. Like Boas, Shapiro found significant environmental effects on the descendants' physiques, and hoped his work would change the practice and assumptions of racial science. In his most famous project, and the one with the most popular impact, Shapiro studied the descendants of HMS *Bounty* mutineers and their Tahitian wives on Pitcairn Island. Shapiro was convinced these Pitcairn Islanders were a natural experiment in racial mixing whose bodies and lives would refute ugly, ill-founded assertions of theorists such as Charles Davenport, whose study of Jamaica imputed hereditary inferiority to the children of mixed parentage. Shapiro's popular account of his study, *Heritage of the Bounty*, constructed for the public a vision of Polynesian intermarriage that combined elements of racial science and cultural sensitivity characteristic of Boas.

In 1942, in his popular book *Man's Most Dangerous Myth*, anthropologist M. F. Ashley Montagu famously argued that race was nothing more than a historically contingent, socially constructed fallacy. I conclude my study with a consideration of the professional and popular discourse about race and culture during WWII and the decades following it. Chapters 6 and 7 examine the degree to which Montagu and others were successful in repudiating "scientific racism" and the concept of race during and after World War II. I argue that contrary to a common narrative that highlights work such as Montagu's and the 1950 UNESCO "Statement on Race," early efforts to reject racism frequently retained essentialist and biologically determinist notions of race and racial types. Thus, rather than a complete rejection of race and racial science during and shortly following the war, I find that the mix of cultural and biological perspectives continues, sometimes in concert and sometimes in tension. To demonstrate this, I examine a range of works produced in the 1940s and 1950s, work deliberately written and designed to communicate with the American public about race and racism.

Chapter 6 looks at the shifting wartime and postwar scientific, political, and social climate, and examines the increasing popularity of concepts such as intercultural understanding and the study of diverse cultures as an antidote to the threat of racism. This chapter focuses on influential works by Boas student and colleague Ruth Benedict. In addition to her widely read books *Patterns of Culture* and *Race: Science and Politics*, I also analyze "The Races of Mankind," a collaboration between Benedict and anthropologist Gene Weltfish, produced initially as a pamphlet and subsequently re-created as an exhibition at the Cranbrook Institute of Science in Michigan. Meant as a humanist intervention against wartime prejudice, particularly racial hierarchy, the work tackled racism as the

product of ignorance and individual bias. Although Benedict and Weltfish argued for fundamental human equality, attributing differences to
culture and physical environment, this "Races of Mankind," like the
Field Museum's, retained the notion of basic human racial types and
deployed ethnic stereotypes to communicate ideas about cultural diversity. In Benedict's work, the anthropological culture concept was disseminated alongside a delimited, increasingly biologized, concept of race, not
as a replacement for it.

Chapter 7 continues my analysis of ideas about race and culture within
the public space, among anthropologists and other scientists and for the
American people. By the 1940s and 1950s, despite nearly universal opposition to Nazism and Nazi racial doctrine, American and international
scientists found it difficult to reach perfect unity on the subject of race.
Provoked into crafting more explicit, sometimes collective, statements on
the nature of race, culture, and society, they debated the conclusions of
racial science, the nature of race, its relation to culture and environment,
the nature of racism, and how to combat it. Anthropologists were not
simply attempting to intervene in society to promote social justice and
combat discrimination. They were also attempting to reorient their discipline, grappling with how to discuss and study race (and heredity) in
changing political, international, and scientific circumstances. And despite
anthropologists' growing anxiety about racism, racial misconceptions,
and what they regarded as misguided or malevolent misuse of their work,
most were not interested in abandoning a biological, hereditary definition
of race. To demonstrate this, I examine a range of works, including Ashley
Montagu's antiracist classic *Man's Most Dangerous Myth: The Fallacy
of Race*, as well as his lesser-known postwar guidebook, *An Introduction
to Physical Anthropology*, the UNESCO "Statements on Race," whose
pronouncements on the concept of race and racism reflected divisions
between biological and social scientists, and Harry Shapiro's work in the
postwar period, including his popular text *Race Mixture* and his 1961
exhibition for the American Museum of Natural History, *The Biology of
Man*.

By the 1960s and 1970s, a cursory examination of biology, anthropology, and other sciences would seem to reveal the virtual disappearance
of race. But it had not disappeared. Although the *Races of Mankind* at
the Field Museum was dismantled, rejected by anthropologists and civil
rights activists as outmoded, scientific interest in human variation persisted. Concerns about human diversity that had once been expressed
in the language of racial essentialism and biological determinism were

translated into anxiety about world population, interest in human evolution and migration, and arguments about persistent gaps in achievement. In human genetics and medical research, the racialized body flourished. In society more broadly, liberal humanist efforts to repudiate race as a concept and racism as an individual failing met with mixed success. Attempts to reorient racial stereotyping into a celebration of ethnic cultural diversity often only masked a persistent biological determinism. Liberal humanist pronouncements, such as those by UNESCO or Benedict and Weltfish, were insufficient to eradicate a deeply rooted belief in the reality of racial differences, in part because those popular pronouncements often continued to promote a basic form of racial essentialism themselves, but also because they failed to address the complex contradiction at the heart of their assertion about race in the United States – biological determinism is a fallacy, but socially constructed race is a very real, very powerful force with both invidious and empowering effects. In the late twentieth century, the very real repercussions of the concept of race in the social, economic, and political lives of the American people continued to coexist uneasily with a "common sense" belief in the biological reality of race that had been promoted, alongside culture, for decades. I argue that racial and cultural theories explored in *Racial Science* formed the foundation for the shift toward cultural explanations of human diversity, as well as for the continuing tension with persistent biological essentialism and determinism.

A Few Words about Terminology

"Race" is a profoundly inadequate term. Our failure as English speakers to devise a richer language to distinguish among the very many entities to which the term "race" has been applied suggests either a shocking lack of imagination or, perhaps more likely, a rather sinister but unwitting ingenuity that casts the umbrella of one small word and its supposed conceptual clarity – an immutable natural kind – over a remarkably messy heterogeneity of objects. In the course of more than 400 years, race has referred to a group of horses; a family lineage; an animal or plant species or variety; animal or plant groups above the level of species; a people or nation; large groups of people transcending tribal or national boundaries; the entirety of humanity; and even the class of all men or all women; in addition to the more common recent usage to denote inherited physical characteristics shared with a (often ill-defined) group and to denote a social class, often associated with particular physical features but not

limited by that.[29] A paltry few variations on "race" – "racial," "racist,"
"racialized," "racialist" – and "ethnic," a twentieth-century innovation,
represent the universe of terms. The advent of "ethnic," which was supposed to clarify matters, has not. It has become as confused in its usage as
"race." Often, ethnic simply replaces race, without any change in intent.
It is used as a polite way to refer to what is still also conventionally
referred to as "race" when the speaker or writer means essential, usually
apparent, bodily differences, and their attendant individual and social
ramifications. As Walter Benn Michaels noted in *The Trouble with Diversity*, faced with the fallacy of biological race, his students "just stopped
talking about black and white and Asian races and started talking about
black and European and Asian cultures instead."[30] Artist Kara Walker,
whose provocative phantasmagorical cutouts evoking nineteenth-century
slavery raise complex questions about race, sexuality, gender, and American society, has argued that Americans do not really want to abandon
the race concept. Echoing Ben Michaels, she has argued that Americans'
obsession with race is a form of identity. "I think really the whole problem with racism and its continuing legacy in this country is that we simply
love it. Who would we be without the 'struggle?'"[31] Matthew Jacobsen
has argued that "ethnicity" was used as a tool to de-racinate whites by
creating non-racialized groups among a broad "Caucasian" racial type,
a trend that he argues was coincident with the mid-twentieth-century
process of separating race and culture while reifying race along color
lines.[32] Historically, and in practice, the biological and cultural remain
intermingled, despite terminological and ideological attempts to sunder
them.

In most historical and popular uses, race has connoted both heredity,
as somatic difference, and cultural difference, although the formulation

[29] The *OED* lists 62 adjectival permutations on the word "race" ("race line," "race man,"
   "race question," etc). The term itself has six definitions, the earliest dating from 1547.
   Zoological definitions and application to identifiable groups of people (biologically or
   otherwise) date from the late sixteenth century. Use of the term to designate a supposedly
   physically distinct group of people dates from the early eighteenth century. *Oxford
   English Dictionary* (Oxford and New York: Oxford University Press, 1989).
[30] Walter Benn Michaels, *The Trouble with Diversity: How We Learned to Love Identity and Ignore Inequality* (New York: Metropolitan Books, part of Henry Holt and
   Company, 2006), pp. 5–7.
[31] Elizabeth Armstrong, "Interview with Kara Walker," in Richard Flood et al., eds.,
   no place (like home) (Minneapolis: Walker Art Center, 1997).
[32] Matthew Frye Jacobson, *Whiteness of a Different Color: European Immigrants and the
   Alchemy of Race* (Cambridge, MA: Harvard University Press, 1998).

"race and culture" suggests that culture only resides in the second term and that "race," by binary opposition, denotes only bodies and ancestry. But when meanings of "race" are taken apart, we see that "culture" is deeply embedded in nearly all its permutations. Cultural or social understanding of race often includes bodies and heredity, if only because the social construction of race was originally erected upon morphological – or purportedly morphological – differences. Since Ashley Montagu introduced "ethnic" as way to denote groups without a biological component, there have been efforts to use race as a purely sociological notion. But this has been very hard to do without importing essentialist or hereditarian connotations. Race is a resilient, protean concept that is constantly redefined and renegotiated. It is one of the sturdiest, most promiscuous products of modernity, a persistent, frequently invidious tool deployed in endless permutations and elaborations, repeatedly remade in shifting configurations of politics, law, economics, culture, and science. I understand "race," not as a natural category or biological essence, but as a force in American life, an ideological system rooted in the pursuit of group and individual advantage that produces and is sustained by political, cultural, social, economic, and scientific practices and their material consequences.[33] This book is an effort to understand how anthropologists' public discourse about race, among themselves and with the American people, contributed to this persistent duality of race in the United States.

---

[33] Ian Haney Lopez, "Race and Colorblindness after *Hernandez* and *Brown*," Paper presented at *Race and Human Variation: Setting an Agenda for Future Research and Education*, American Anthropological Association, September 12–14, pp. 1–14; Barbara J. Fields, "Ideology and Race in American History," J. M. Kousser and J. M. McPherson, eds., *Region, Race, and Reconstruction: Essays in Honor of C. Vann Woodward* (New York: Oxford University Press, 1982), pp. 143–177.

2

Franz Boas and Race

History, Environment, Heredity

Revisiting Boas: Race and Culture

Franz Boas was one of the most prominent anthropologists of his day, in both cultural and physical anthropology, and arguably one of the most influential anthropologists in American history. Throughout his career, Boas explicitly critiqued the questions racial science pursued, the sorts of methods that were required to answer those questions, and the failings of much of the work that was being done within the discipline. Contrary to the common image of Boas, he spent the whole of his career both practicing and critiquing racial science. His own compilation of his key publications, intended to "prove the validity of my view," opened with an extensive section on "Race." Published in 1940, shortly before his death, the twenty articles and addresses on race, published from 1892 to 1939, total nearly one-third of the book.[1] I argue in this chapter that our understanding of physical anthropology, the history of racial science, and the work of Boas are incomplete and distorted because we have failed to study and come to terms with a view of race and the biological study of human difference that Boas espoused. His work exemplified many of the ongoing tensions between and intermingling of race and culture in anthropology and in twentieth-century America.

From the vantage point of the decades following the Second World War, the physical anthropology practiced by Franz Boas employed discomfiting methods and assumptions. Preoccupied with the elusive task of

---

[1] Franz Boas, *Race, Language, and Culture* (New York: Macmillan Company, [1940] 1948).

quantifying and elucidating human physical variety, the kind of work he and other respected anthropologists engaged in for decades is now seen as a benighted, outdated, often racist practice. But this view lumps together a variety of beliefs and approaches that were condemned for their common agreement that race was a biologically valid and interesting object of study, defined as a scientific problem by the effort to develop coherent racial categories or types. It is true that in the early and mid-twentieth century, virtually all physical anthropologists (and many others) believed that race was an essential human quality. And they all incorporated anthropometric measurement among the methodologies used to elucidate questions of race. But some constructed their study of races according to an essentially romantic typology (characteristic of nineteenth-century physical anthropology) in which groups of people shared essentially unchanging attributes that, once observed, measured, and catalogued, would enable anthropology to sort human diversity into a genealogical taxonomy of types, each of which could be represented by those individuals who most perfectly embodied the traits of their race. For Boas, race could only be deduced from a historically, culturally informed statistical study of the type and variability of characteristics found among groups of people living in particular locales. No individual could typify a racial type because the type existed as a distribution of variations on biologically, evolutionarily adaptive traits. In his view, race was malleable, associated with groups of people who move through space and time, and who existed everywhere in cultural and natural environments that shaped the form of their bodies.

Boas's work exemplifies the complexities of anthropology in the first decades of the twentieth century. He grappled with all the methodological and conceptual problems that beset the study of human diversity – hereditary, somatic, linguistic, cultural. His body of work was no simple "rejection of scientific racism," no anachronistic conflation of race and racism, but rather was a decades-long struggle to comprehend human variation while contesting and combating prejudice and discrimination. Boas's project combined various aspects of American anthropology to comprehend human difference of all kinds. Rather than divorcing biological from cultural studies of humanity, as so many have asserted, Boas worked tirelessly to see that these studies were joined. He thought that one could not be understood without the other, and it was his insistence on the importance of culture, broadly understood as environment, that constituted his primary critique and contribution to physical anthropology and the study of race. His demonstration that physical characteristics were changeable only reinforced his belief that rigorous studies of bodies, heredity, and the environments in which people lived might eventually yield genuine understanding of human variations and possibilities.

Franz Boas's 1911 study of changes in head form among descendants of immigrants has been recognized by historians for the groundbreaking study that it was, a study that rigorously[2] undermined both the time-honored and unquestioned use of head shape for distinguishing races and the equally unquestioned belief that such physical characters were stable and unchanging across space and time. Boas showed that environment mattered, that physical forms could change, and that no one really knew for sure what true "racial characters" were. His work suggested an evolutionary reading of human difference that made rigid, simplistic typological classifications untenable and showed why scientists had had so much trouble sorting their measurements into any kind of coherent categories. By all rights, it should have stood physical anthropology on its head. The study did create a great deal of discussion, but then, for most physical anthropologists, it was back to business as usual. In the decades between Boas's study and Harry Shapiro's work in the late 1920s and 1930s, critiques of racial science grew in number and pointedness across

---

[2] Boas's study of changing bodily form among immigrants was methodologically and statistically rigorous and sophisticated for its time. In recent years, there has been a renewed interest in reexamining his study and its results, sometimes applying modern statistical manipulations to his data to test his conclusions. This renewed interest in discrediting Boas's study, a study which anthropologists and historians have seen as a critical step in the repudiation of hereditarian science and philosophy, seems to coincide with the resurgence of genetic explanations for behavior, mind, and social problems. Perhaps not surprisingly, there continues to be disagreement over Boas's methods and his conclusions. Corey Shepard Sparks, "Reassessment of Cranial Plasticity in Man: A Modern Critique of Changes in Bodily Form of Descendants of Immigrants" (unpublished Masters thesis, Knoxville: University of Tennessee, Aug. 2001); R. L. Holloway, "Head to Head with Boas: Did He Err on the Plasticity of Head Form?", *Proceedings of the National Academy of Sciences*, vol. 99 (2002), pp. 14622–14623; Corey Sparks and Richard Janz, "A Reassessment of Human Cranial Plasticity: Boas Revisited," *Proceedings of the National Academy of Sciences*, vol. 99, no. 23 (2002), pp. 14636–14639; Clarence C. Gravelee, H. Russell Bernard, and William R. Leonard, "Heredity, Environment, and Cranial Form: A Re-Analysis of Boas's Immigrant Data," *American Anthropologist*, vol. 105, no. 1 (2003), pp. 123–136; Clarence C. Gravelee, H. Russell Bernard, and William R. Leonard, "Boas's *Changes in Bodily Form*: The Immigrant Study, Cranial Plasticity, and Boas's Physical Anthropology," *American Anthropologist*, vol. 105, no. 2 (2003), pp. 326–332; Corey Sparks and Richard Janz, "Changing Times, Changing Faces: Franz Boas's Immigrant Study in Modern Perspective," *American Anthropologist*, vol. 105, no. 2 (2003), pp. 333–337; John H. Relethford, "Boas and Beyond: Migration and Craniometric Variation," *American Journal of Human Biology*, vol. 16 (2004), pp. 379–386.

a range of disciplines, particularly as more was understood about the nature of genetics.[3] But nevertheless, in 1939, twenty-nine years after Boas published his first results, Harry Shapiro still lamented the lack of change in physical anthropological method and the kinds of conclusions that were still erroneously being drawn. In the historiography of racial science, Boas's work is now seen for the conceptual break that it was. Most physical anthropologists in the first decades of the twentieth century, lacking the benefit of hindsight and preoccupied with concerns different than those of later anthropologists and historians, persevered with their practices, theoretical commitments, and assumptions. In this sense, Boas's work and his persistent critiques of the methods and results in physical anthropology are more important for their influence on younger colleagues such as Shapiro and others – who were intellectually and professionally formed in the interwar period and later participated in the postwar transformation of physical anthropology and racial science – than for their effects on the racial studies of his peers.

Boas did not reject race altogether, nor did he reject biological studies of humanity. He was avidly interested in studies of human growth, evolution, and variation until the end of his life. But neither did he believe in biological essentialism that promiscuously linked apparent physical differences to moral or intellectual traits, nor did he regard physical differences as any kind of marker of inferiority.[4] He regarded broad racial groups as real biological entities and believed that there were boundaries to the plasticity of human form. He thought physical anthropologists had some hope of sorting out human physical variation using rigorous scientific methods, including statistics and genetic theory, particularly if they

---

[3] Elazar Barkan discusses biologists' critiques of racial science at length in *The Retreat of Scientific Racism: Changing Concepts of Race in Britain and the United States between the World Wars* (Cambridge: Cambridge University Press, 1992).

[4] In his early work, prior to his study of changing head form among descendants of immigrants, Boas did adhere to certain hierarchical notions of civilized and primitive societies and suggested that "Negroes" might be developmentally inferior to whites, views that changed over the years. By the period of my study, he had developed his more relativistic understanding of cultures, as well as sophisticated critiques of racial science. For selections of Boas's early writings and introductory essays concerning these shifts and Boas's early work, see George W. Stocking, Jr., *A Franz Boas Reader: The Shaping of American Anthropology, 1893–1911* (Chicago: University of Chicago Press, 1974). For a biography of Boas in his formative years and early career, see Douglas Cole, *Franz Boas: The Early Years, 1858–1906* (Vancouver, BC: Douglas & McIntyre; Seattle: University of Washington Press, 1999). Also, Regna Darnell, *And Along Came Boas: Continuity and Revolution in Americanist Anthropology* (Amsterdam and Philadelphia: J. Benjamins, 1998).

could refrain from unwarranted speculation about causes. He engaged in
a debate, not only with other physical anthropologists but also a variety
of biological and social scientists, with whom he shared a basic belief in
the heredity of human differences and the reality of race. In the end, only
part of the interwar Boasian project, as he, and later Shapiro, conceived
it prevailed. Biological essentialism and simplistic typological racial classification were ultimately marginalized, but so, too, was a historicized,
rigorously evolutionary examination of human physical difference. This
chapter seeks to restore to the history of anthropology the complexities of
a Boasian vision in which race was neither exclusively somatic nor entirely
cultural, but rather was an inextricably interwoven – if analytically sepa-
rable – set of critical factors: history, heredity, culture, and environment.

American Anthropology and Its Subject

In America, anthropology is a discipline that has been uneasy with the
bodies of its object. The study of bodies predated the disciplinary coales-
cence of anthropology in the mid- and late nineteenth century (as did the
study of "civilization," as well as language, and human prehistory). The
preoccupation with elaborating narratives about kinds, both evolutionary
and racial, allied the anthropological study of bodies to practices whose
domain had been delimited earlier: especially medicine, but also anatomy
and paleontology. Some anthropologists embraced growing and diversi-
fying disciplines of biology, particularly those concerned with evolution,
heredity, genetics, and statistics. Others ignored them or scorned them.
The shifting terminology for the practice of studying human bodies –
craniology, somatology, physical anthropology, biometrics, biological
anthropology, forensic anthropology – betrays its fluctuating locus in contemporary constellations of scientific disciplines, unsteady self-definition,
and uneasy relation to the study of a broader human context.

　　Edward Burnett Tylor famously defined anthropology's purview as
"Culture, or Civilization," which, "taken in its wide ethnographic sense,
is that complex whole which includes knowledge, belief, art, morals,
law, custom, and any other capabilities and habits acquired by man as a
member of society."[5] In the United States, although it defined itself as a

---

[5] Quoted in George W. Stocking, Jr., *Race Culture and Evolution, Essays in the History
of Anthropology* (Chicago and London: University of Chicago Press, 1982), p. 73, from
E. B. Tylor, *Primitive Culture: Researches into the Development of Mythology, Philosophy, Religion, Art and Custom*, 2 vols. (London: 1871, I), p. 1.

single discipline, in contrast to other fields that also took the human as their object – sociology, political economy, psychology – anthropology has functioned more as a loose coalition than a unified discipline.[6] Franz Boas himself remarked on the "multifarious" origins of anthropology.[7] Disparate fields of inquiry – philology, interest in the texts and objects of ancient civilizations, natural historical inquiries into humans and their environments – were drawn together by scholars in the nineteenth century as the American anthropological fields of archaeology, linguistics, ethnology, and physical anthropology began to coalesce into self-conscious scientific disciplines. In the United States, as in Europe, the nineteenth century was a period in which the empirical, rationalizing impulses of the Enlightenment pursuit of knowledge, the institutionalization of knowledge making in royal societies and publishing venues, and the university model of education, particularly in its German incarnation, were increasingly marshaled into bounded disciplines.

American physical anthropology, along with the other anthropological fields, developed in a very different context from its European counterparts. In a land that had once been colonized by European powers, and populated by indigenous peoples – as well as the descendants of Africans forcibly imported as laborers and immigrants from all parts of Europe, China, Mexico, and other parts of the world – anthropologists encountered their "subject" in their own country. This had a number of repercussions for anthropology in general and the study of race in particular. The most obvious way that the American context shaped the development of

---

[6] The distinction between anthropology and sociology was a particularly troublesome one, a subject of debate and contest since their inception. Questions of where the boundary lies between these disciplines have been rooted in thorny questions about scientific practice and theorizing, the nature of human culture and society (and the validity of making a distinction between culture and society), and the establishment of domains of knowledge making that are beyond the scope of this book. In practice, institutions and individuals did not always observe a neat distinction. For example, William Ripley was regarded by many physical anthropologists as an authority on racial measurement and classification. His work on *The Races of Europe* was widely cited in anthropology (including by Henry Field, who used it as a reference when planning the *Races of Mankind* exhibit at the Field Museum of Natural History). But the subtitle to his famous work was "A Sociological Study." Ripley himself bridged the fields, being both an assistant professor of sociology at the Massachusetts Institute of Technology and a lecturer in anthropology at Columbia University, when his influential text was published.

[7] Franz Boas, "The History of Anthropology," Address at the International Congress of Arts and Sciences, St. Louis, Sep. 1904, as published in *Congress of Arts and Science*, ed. H. J. Rogers, 8 vols. (Boston: Houghton Mifflin, 1906), vol. 5, p. 468–482; also published in *Science*, vol. 20 (1904), pp. 513–524., reprinted in Stocking, ed., *The Shaping of American Anthropology*, pp. 23–36.

anthropology was the presence of ostensibly "vanishing" natives. American ethnologists regarded Native Americans as their purview, a practice reinforced by government support and institutional structures such as the Bureau of American Ethnology, and a focus reflected in natural history museum collections and exhibits to this day. For those interested in human physical difference, America was a nation filled with people of dramatically different appearance and heritage, which "tended to make every citizen, if not an ethnologist, at least a speculator on matters of race."[8]

The preoccupation of American citizens and anthropologists with the racial and ethnic complexion of the nation was a product of more than idle empirical interest. Thomas Patterson has ably argued that American anthropology and race relations in the nineteenth century were shaped by a violent, contentious history: colonists who "waged genocide" against indigenous peoples, westward expansion into contested Indian and Mexican lands, a southern economy based in slave labor, and incorporation of European immigrants into emerging and industrialized cities who were frequently regarded as inferior and a threat.[9] Brad Hume has argued that American anthropology developed in conjunction with a society conflicted about the Enlightenment ideal of universal human nature and rights that flowed from it, and the diverse peoples who sought equal membership in a society founded on those principles. Hume argues that anthropology participated in a national project to "define, monitor and protect valid citizens" and to "distinguish them from non-citizens." Applying its science to the question of who was capable of citizenship, and the rights and privileges that accompanied it, anthropology "naturalized" women, blacks, Native Americans, and criminals in order to exclude them.[10]

---

[8] William Stanton, *The Leopard's Spots: Scientific Attitudes towards Race in America, 1815–59* (Chicago: University of Chicago Press, 1960), p. 10.
[9] Thomas C. Patterson and Frank Spencer, "Racial Hierarchies and Buffer Races," *Transforming Anthropology*, vol. 5, nos. 1–2, pp. 20–27.
[10] Brad Hume, "Quantifying Characters: Polygenist Anthropologists and the Hardening of Heredity," *Journal of the History of Biology*, vol. 41, 1 (Mar. 2008); Hume, "The Naturalization of Humanity in America, 1776–1865," (PhD diss., Indiana University, 2000). There is quite a bit of literature addressing ideas about race and "Anglo-Saxonism" in the early republic, including discussions of Jefferson and other "founding fathers'" ideas about race and the governing ideals of the new nation. See also inter alia, Thomas F. Gossett, *Race: The History of an Idea in America* (New York: Shocken Books, 1963); Reginald Horsman, *Race and Manifest Destiny: The Origins of American Racial Anglo-Saxonism* (Cambridge, MA: Harvard University Press, 1981); Audrey Smedley, *Race in North America: Origin and Evolution of a Worldview* (Boulder: Westview Press, 1993); Matthew Frye Jacobson, *Whiteness of a Different Color: European Immigrants and the*

Romantic Typologies and the Reality of Race

This typological orientation, rooted in the eighteenth–century tax-onomies of Carolus Linneaus and Johann Friedrich Blumenbach, proved tenacious. Donna Haraway has traced the "romantic typological approach," committed to "scientizing racial becoming through a teleol-ogy of racial/moral/spiritual/intellectual development" from major late nineteenth-century figures such as Paul Broca, Paul Topinard, Joseph Deniker, and William Ripley, all the way to Arthur Keith, Earnest Hooton, and, ultimately, Carleton Coon, who stubbornly clung to the old typological traditions well into the 1960s.[11] Even in the 1920s and early 1930s, anthropologists of prominence such as Hooton or Alfred Haddon considered Joseph Deniker and William Ripley, whose works were published in 1899 and 1900, as authoritative sources, if not correct in all respects. Ripley and Deniker themselves drew heavily on French sci-entists Paul Topinard and Paul Broca, retaining their basic assumptions as well as much of the terminology.

Paul Broca, widely regarded as the founder of physical anthropology, has been credited with introducing statistical notions and methods of measurement, including special instruments, to the study of race. He insti-tutionalized the practice of anthropometry, the "hallmark of anthropol-ogy in the nineteenth century,"[12] through the Société d'Anthropologie de Paris, the Laboratoire d'Anthropologie, and the École d'Anthropologie, known collectively as "Broca's school."[13] Broca, and later his student Paul Topinard, themselves following Johann Friedrich Blumenbach and Georges-Louis Leclerc Comte de Buffon, elaborated a scheme of racial types based on a small number of basic families or stocks. Broca, Topinard, and other European physical anthropologists assumed that

---

Alchemy of Race (Cambridge, MA: Harvard University Press, 1998); Matthew Guterl, *The Color of Race in America, 1900–1940* (Cambridge, MA: Harvard University Press, 2001).

[11] Donna Haraway, "Remodeling the Human Way of Life: Sherwood Washburn and the New Physical Anthropology, 1950–1980," in Stocking, *Bones, Bodies, Behavior*, pp. 217–218. For an excellent discussion of nineteenth- and twentieth-century theories of race, including Broca, Topinard, Ripley, Deniker, and especially Franz Boas, see Stocking, *Race, Culture and Evolution*, as well as *Bones, Bodies, Behavior* and "Ideas and Institutions in American Anthropology." See also Frank Spencer, ed., *History of American Physical Anthropology, 1930–1980* (New York: Academic Press, 1982) and Barkan, *The Retreat of Scientific Racism*.

[12] John S. Haller, *Outcasts from Evolution: Scientific Attitudes of Racial Inferiority, 1859–1900* (Carbondale: Southern Illinois University, 1995), p. 7.

[13] Spencer, *History of Physical Anthropology*, pp. 5–6.

differences in human appearance and physical structure, especially in the skull and brain, were "virtually primordial." They believed cultural differences were a product of these differences, and through careful and extensive measurement of modern mixed populations, contemporary "types" that reflected purported original races could be distinguished. Like the notion of types in biological studies, a racial type was understood as a set of defining characteristics that individuals embodied to a greater or lesser degree, within narrow limits, and which marked members of one race off from all others. Borrowing from earlier American polygenists, Topinard relied on the form of the head, measured by the cephalic index, as the primary, fixed diagnostic racial characteristic.

Yet despite these rigid criteria and determinate hierarchies, even in the nineteenth century doubts about the material reality of race emerged. Broad classifications such as Caucasian, Negroid, and Mongoloid were based largely on skin color, and only gross and obvious distinctions were required. But as anthropologists attempted to make increasingly fine distinctions, the eye was no longer a sufficient instrument for detection of race. Instead, various measurements and instruments were created to quantify increasingly arcane features of human bodies. They measured the circumference of skulls, standing and sitting height, various aspects of facial geometry, such as the distance between the eyes and the length of the nose – eventually, virtually any aspect of the body that could be measured or quantified was. And yet, as more and more detailed information piled up, neatly bounded races retreated further and further. By the late nineteenth century, those who attempted the scientific study of race despaired at ever sorting out the heterogeneity they found among people. Historian George W. Stocking, Jr., related William Ripley's account of Otto Ammon's frustration at not being able to locate, among his thousands of photographs and measured subjects, a suitably typical "Alpine." "All his round-headed men were either blond, or tall, or narrow-nosed, or something else that they should not be," Ammon lamented.[14] Thousands and thousands of measurements had not yielded a single, widely accepted classification of human races, but rather a multiplicity of competing ones. Part of the problem was a lack of consistent methodology among practitioners – different investigators measured different traits in different ways, so that it became extremely difficult to compare and reconcile their findings. But eventually, investigators began to suspect

---

[14] Stocking, "Polygenist Thought in Post-Darwinian Anthropology," *Race, Culture and Evolution*, p. 58.

that the problem was a more fundamental one. The paradoxical result of increased quantification was increased skepticism about the reality of race at the individual level, the very heart of the "romantic typology."

As George Stocking has analyzed, Topinard came to believe that modern populations were too mixed to yield the sort of well-defined races he and other physical anthropologists had been seeking to delineate. Topinard effectively gave up the search for modern pure races, asserting that race was "an abstract conception, a notion of continuity in discontinuity." He did not abandon the idea that pure races once existed – indeed, he asserted that the three primary races were so distinct and foundational to subsequent human diversity that they were more akin to separate species than to branches from a common ancestor.[15] But although Topinard clung to the idea of pure races, he was gradually forced to admit that it was next to impossible to find them among living people. Confronted with heterogeneous "mixed races," Topinard redefined the location of the racial essence he and others regarded as the root of racial types. Bowing to recalcitrant data that refused to coalesce into tidy racial categories, he retreated to a search for pure races in the past, out of whom modern people were derived and whose existence might be abstracted from evidence among the living. The racial traits that had once characterized pure races were now hopelessly scattered among the world's mixed peoples. The best that physical anthropology could hope for was a laborious tabulation of the distributions of traits and contemporary "types" based on them. According to Topinard, the originary pure races and their continuity with present populations would remain "impossible to demonstrate." But if Topinard gave up, finally, on race as embodied in contemporary individuals, he did not abjure the reality of race: "[W]e cannot deny them, our intelligence comprehends them, our mind sees them, our labor separates them out; if in thought we suppress the intermixtures of peoples, their interbreedings, in a flash we see them stand forth – simple, inevitable, a necessary consequence of collective heredity."[16] Broca and Topinard's physical anthropology, along with other European work, was synthesized for a large American audience by William Ripley and Joseph Denicker, in accounts that evaded Topinard's methodological concerns and instead reflected his and their profound sense that races were real. By the end of the nineteenth century, in the United States and Britain, Francis

[15] Ibid., p. 57.
[16] Ibid., p. 59. See also C. Loring Brace, "Race in American Physical Anthropology" in Spencer, ed., *The History of American Physical Anthropology*, pp. 11–30.

Galton, a eugenicist, and Karl Pearson, a statistician – along with others frustrated by the failure of physical anthropology to successfully classify races – were developing new, more statistically powerful, quantitative techniques to find the races and types that had eluded Topinard and the rest of nineteenth-century anthropologists.

One Species or More? Monogenism and Polygenism

In the United States, in the first half of the nineteenth century, debates over the nature and legitimacy of slavery, native nations engulfed by an expanding America, and the natural history of human variation were multiple sides of a single discourse struggling to define the significance and direction of a new nation self-consciously imagining itself as a "beacon on the hill" for all of humanity. These debates were simultaneously enjoined in a contentious discourse that pitted a primarily theological narrative of natural history against an emerging evolutionary perspective. Looked at through the nexus of race, these debates came together in the thorny and provocative question of whether human beings, despite their differences, were but one species, unified through descent from a single origin, either providential or evolutionary – monogenism – or whether human differences were so profound that they could be understood only as evidence of the fundamental separateness of human beings, either specially created by God or separately evolved from prehuman ancestors – polygenism.

The monogenist/polygenist debate was a deeply divisive one with high stakes, whose effects were felt long after changes in science, religion, and society had made the original positions and rationales obsolete. The monogenist position was originally rooted in the biblical creation story, in which Ham, Shem, and Japheth, direct descendants of Adam and Eve, became the founders of the three principal races. (Blumenbach, for example, understood his five races as a providential creation.[17]) As theories of evolution developed in the late eighteenth and early nineteenth century, the monogenist position was also adopted by those who argued for a single, nonhuman ancestor for all of humanity. This position took on a great deal more force following Charles Darwin's 1859 publication of *The Origin of Species*, which finally marshaled an enormous, virtually irrefutable variety of evidence to demonstrate the reality of descent with modification, offered natural selection as a plausible mechanism for change, something that had been crucially lacking in earlier theories, and,

---

[17] Haller, *Outcasts from Evolution*, pp. 70–74.

also – importantly for the study of human difference – posited a gradual process that elapsed over astonishingly long periods of time.[18] In contrast, the polygenist position was taken up predominantly by many natural scientists and physicians who saw it as the more empirically grounded, less theologically driven interpretation, and by slave owners and others interested in defining non–Anglo-Saxons as irremediably separate and inferior.

Although the monogenist position was not fundamentally incompatible with a hierarchical vision of races (e.g., God might have created a Great Chain of Being in which some human beings were superior to others; the races might have evolved at different rates from a common ancestor or might include degenerate branches; "unnatural" mixing might have "polluted" the original races, producing "mongrels"[19]), the polygenist position more easily made difference seem inevitable, primordial, and fixed. Historians have regarded Samuel George Morton, his student, Josiah Clarke Nott, and Nott's coauthor, George R. Gliddon, as the key figures in the monogenist/polygenist debate in the history of American physical anthropology. Morton, Nott, and Gliddon formed the "American School" of anthropology, a school defined by its association with polygeny and their attempts to prove it, principally through a project of quantification and measurement, primarily of skulls and brain capacity in Morton's case, and through the study of racial mixing and its supposed deleterious results by Nott and Gliddon. Morton produced two huge volumes – *Crania Americana* and *Crania Aegyptiaca* – purporting to demonstrate, through deployment of prodigious amounts of data, that whites exceeded both blacks and Native Americans (the latter filling an intermediate position) in brain capacity, and therefore also in intelligence and the capacity for achievement and understanding. Morton's work was regarded at the time as providing excellent evidence of inherent differences between races that revealed nature's hierarchy. His work also further legitimated the use of the skull and brain as a locus for defining racial difference, and of quantitative methods.[20]

---

[18] The hundreds of thousands and millions of years posited for evolution to proceed is often referred to as "deep time." The shock of Darwin's timeline (following the lead of Charles Lyell in his geological researches) arose from its contrast to much shorter historical timelines, such as Bishop Ussher's estimate that man was created in 4004 BC. Evolution has never fully been accepted in all quarters of the United States, evidenced by the Scopes "Monkey" trial in the 1920s, modern-day creationists, and most recently the resurrection of the argument from design.

[19] Haller, *Outcasts from Evolution*, p. 72.

[20] Stephen Jay Gould, *The Mismeasure of Man* (New York: Norton, 1981), p. 53. Morton had gathered together a collection of human skulls of unprecedented size (it

In addition to naturalizing bodily differences into a polygenetic hierarchy, the American School participated in a debate about the significance of human hybridity. No one could question that human beings of very different appearance found no barriers to procreation. Following studies of animal breeding, and the wide familiarity, in a predominantly agricultural society, with the results of barnyard experiments in crossing species, human hybridity became a contested domain between monogenists and polygenists. Monogenists, regarding human beings as a single species, theorized that if two varieties mixed, the offspring, although a "mongrel," would be just as fertile as its parents. Polygenists, believing that humans existed as separate species, theorized that crosses between disparate human species would predominantly produce offspring who sported various "disharmonies" and were infertile, analogous to crosses among different but closely related animal species (such as the mule produced by mating the horse and donkey). Nott and Gliddon's popular volumes *Types of Mankind* and *Indigenous Races of the Earth* compiled studies to support the polygenist point of view. Although the strident anti-biblical rhetoric of some polygenist literature alienated more devout participants in the "unity" controversy, the theoretical and methodological orientations that Morton, Nott, and Gliddon advanced were themselves remarkably fertile. Although growing acceptance of *The Origin of Species* put polygenists at a disadvantage in the decades following 1859, Josiah Nott continued to promote his theories of primordial racial difference.[21] Indeed, the polygenism advocated by the American School in the early and mid-nineteenth century had lasting effects well into the late nineteenth and early twentieth centuries in both the United States and Europe, long after the original participants had died and the theological, scientific, and American racial context of the original debates had been transformed.[22]

---

was commonly referred to as the American Golgotha), and spent many hours developing techniques for accurately gauging brain capacity by filling the skulls, first with white mustard seed then, when that proved unreliable, with lead shot.

[21] As did zoologist Louis Agassiz, who embraced polygenism for both racial and antievolutionary reasons. Edward Lurie, *Louis Agassiz: A Life in Science* (Chicago: University of Chicago Press, 1960), pp. 238–239; Haller, *Outcasts from Evolution*, pp. 69–86. For Nott's views, see Josiah Nott and George Gliddon, *Types of Mankind* (Philadelphia: Lipponcott, Grambo & Co., 1854).

[22] See George W. Stocking, Jr., "The Persistence of Polygenist Thought in Post-Darwinian Anthropology," in *Race, Culture and Evolution*, pp. 4–68, for what is still one of the best discussions of the nature and ramifications of the American School in anthropology and American life.

The Trouble with Typology

In approaching human variation as a set of racial types, anthropologists and other racial scientists faced the same taxonomic problem biologists confronted in sorting plant and animal kingdoms into species with corresponding type specimens. In biology, in order to classify species and differentiate one from another, an individual is selected to typify a species. This representative is the model against which all other individuals are compared to evaluate whether or not they merit inclusion in the species. The type specimen generally represents a form median in most of its characters; in evolutionary biology a certain amount of variation is required for natural selection, and thus individuals classed as members of a species exhibit a defined, but limited, range of variation in characters. Of course, in practice, classification of species is not so straightforward. Identification of a type specimen presumes an investigator already has a firm notion about which traits are critical to define the species at hand, as well as the likely range of variation in those traits, and the boundaries that separate species, all of which is frequently subject to debate and dispute. In practice, some traits are considered more crucial than others for defining a species and may admit of less variation for that purpose. Moreover, taxonomies change as ideas about natural selection and evolution change, and as new scientific tools are developed to study and compare organisms (such as molecular and genetic analyses in the latter half of the twentieth century). Added to all this is the crucial criterion of "fit": the sense that biologists, or anthropologists, have about what a systematic classification should look like, about what the "real" relationships are among the entities being studied, be they humans or mollusks. This gut sense of what is true about the world not infrequently conflicts with the taxonomic systems that their criteria and techniques generate. When their scientific method results in a classification that does not "look right," biologists will go back to their specimens and their tools to tinker with and reexamine them, looking to make their process fit expectations born of their expertise. Racial scientists, believing deeply in the reality of race, did much the same thing. Although biologists – and anthropologists – strove for rigor in definition of characters and variation, there has always been an enormous amount of room for subjectivity, hunches, and aesthetic preferences in the construction of zoological classifications.[23]

---

[23] For an excellent discussion of the complexities and nuances of typology and classification, and what she calls the "applied metaphysics" of the type specimen, see Lorraine Daston,

Scientific efforts to categorize human races have inevitably run into difficulty in demarcating types because the physical characteristics used to define race are not distributed in ways that map neatly onto the groups scientists have wished to label, and because human beings by and large do not exist in isolated breeding populations the way most organisms classified as species or subspecies do. The evident differences among humans in a wide range of characters represent a graded variation that does not collapse into discreet categories. Indeed, Field Museum anthropologist Berthold Laufer conceded the difficulty in defining racial characteristics and the imprecision of racial categories, admitting that "in speaking of white, yellow, black, and red men we follow merely a popular terminology and take surface impressions for granted, while as a matter of fact the color variability of the complexion in individuals is almost infinite, and no one is strictly white or yellow or black or red."[24] Nonetheless, most physical anthropologists prior to the 1950s believed science would eventually solve these racial puzzles and humans could be meaningfully sorted into discrete categories.

A New Methodology: Biometrics

Biometrics grew out of nineteenth-century work by Samuel Morton, Francis Galton, Karl Pearson, and others who tried to move the study of human race away from anecdotal and impressionistic evidence, such as travelers' or missionaries' accounts; arguments that stressed linguistic, geographical, or biblical theories; and, most importantly, the reliance on purely descriptive accounts of human morphology. Instead, biometrics approached the human body as a set of quantifiable characteristics, or traits, that could be measured, and the measurements analyzed statistically to yield averages, distributions, indices, and frequencies. The hope was that patterns would emerge that would finally enable physical anthropology to successfully sort and classify the human diversity that was at once patent and elusive.[25] It was a tool to reveal Topinard's continuity

"Type Specimens and Scientific Memory," *Critical Inquiry* 31 (Autumn 2004), pp. 153–182.

[24] Henry Field, *The Races of Mankind*, Popular Series, Anthropology Leaflet 30, Fourth Edition (Chicago: Field Museum of Natural History, 1942), p. 6.

[25] This biometric reliance upon statistical analysis to reveal hidden patterns and qualities has more recently been combined with the study of genetic markers in the continuing search for the origins of human variation, patterns of human migrations, and delineation of contemporary variety. For an example of this work, see a widely read popularization

in discontinuity, to reveal racial type where only unruly diversity seemed apparent. Galton, Pearson, and later practitioners such as Franz Boas and Earnest Hooton, assumed the existence of discreet racial types and racial traits that characterized them, and believed that those types could only be given scientific status as genuine human classes after being delineated and described statistically. Their assumption was that race was a heritable human feature that expressed itself across populations rather than independently in each individual. Contrary to an earlier typological approach, advocates of the metrical, statistical approach to defining race disavowed the idea that an anthropologist could find an individual "typical" of their race. What was typical of a racial type was a statistical entity – indices, averages, frequencies, ranges, and distributions of variation, rates of change – that, by definition, could not be embodied in a single individual.[26] Under this rubric, a race was defined as a limited range of variation among a distinctive set of characteristics within a given population. Individuals who fell within this distribution could be said to belong to a particular race, or type, but could not be said to be typical of it.

Within the range of biometrical practice, there were two broad approaches. In one approach, biometrical description could be used to quantify characteristics that were already defined among groups that had been preselected for study. This was the most common, and tautological, use of biometrics. The races that statistical analysis of large numbers of measurements were supposed to find had already been defined – the only people measured had already been selected for their membership in the race being studied – and thus in the most basic sense, the outcome of the

---

in Michael J. Bamshad and Steve E. Olson, "Does Race Exist?" *Scientific American* (Dec. 2003), pp. 78–85, and its professional counterpart, Michael J. Bamshad, et al., "Human Population Genetic Structure and Inference of Group Membership," *American Journal of Human Genetics*, vol. 72, no. 3 (Mar. 2003), pp. 578–589.

[26] Francis Galton attempted to extract the typical from the heterogeneity of individuals through composite photography, a technique used to produce visual evidence of the assumed reality of underlying racial types. See Francis Galton, "Composite Portraits, Made by Combining Those of Many Different Persons into a Single Resultant Figure," *The Journal of the Anthropological Institute of Great Britain and Ireland*, vol. 8 (1879), pp. 132–144. For scholars' assessments of Galton's practice, see Allan Sekula, "The Body and the Archive," *October*, vol. 39 (Winter 1986), pp. 3–64; Daniel Akiva Novak, *Realism, Photography, and Nineteenth Century Fiction* (Cambridge and New York: Cambridge University Press, 2008); M. G. Bulmer, *Francis Galton: Pioneer of Heredity and Biometry* (Baltimore: The Johns Hopkins University Press, 2003); Sharrona Pearl, *About Faces: Physiognomy in Nineteenth-Century Britain* (Cambridge, MA: Harvard University Press, 2010).

analysis was predetermined. Because the belief that races existed as natural categories was so strong, for many anthropologists the application
of biometrics to populations already defined was not obviously problematic, as the key conceptual problem was not the existence of discrete
races, but rather which measured traits, and in what combinations, were
characteristic of the various races. The problem was to precisely and
thoroughly quantify races and their particular suites of characteristics
that were already understood to exist.

Another form of biometrics shared the understanding of race as a
feature distributed across populations, rather than residing in each individual, but approached the problem of characterizing races in a less tautological manner. In this method, statistically significant numbers of people
would be measured and the results analyzed statistically to see what could
be determined, how the measurements would sort out, and if anything
seemed to coalesce into groupings that could be termed races, or given
skepticism about the accessibility of genuine racial types to anthropological method, into various well-defined subtypes. Franz Boas made this
point in his essays on the proper use of anthropometry and metrical
analysis. Boas observed that the taxonomic confusion that characterized
physical anthropology's attempts to sort out human difference was due
in part to the subjective definitions of racial type, "an abstraction of the
striking peculiarities of the mass of individuals." What constituted "striking peculiarities" depended "largely upon the previous experience of the
observer, not upon the morphological value of the observed traits."[27] He
warned against arbitrarily segregating a particular group out of a larger
series simply on the basis of "certain metric values" of particular traits,
because, absent some reason to regard the traits as biologically significant, the segregation would merely introduce "a subjective element that
had no relation to the series itself."[28] A race, on the other hand, Boas
warned, must not be identified with some "subjectively established type,"
but "must be conceived as a biological unit, as a population derived from
a common ancestry and by virtue of its descent endowed with definite
biological characteristics."[29] Series composed for research and classification had to be carefully constructed to be as properly inclusive as
possible, deriving impressions of the characteristic qualities of a type

[27] Franz Boas, "The Relations between Physical and Social Anthropology," *Essays in Anthropology in Honor of Alfred Louis Kroeber* (Berkeley: University of California Press, 1936), reprinted in Boas, *Race, Language, and Culture*, pp. 172–175.

[28] Ibid., p. 178.

[29] Franz Boas, *Mind of Primitive Man*, revised ed. (New York: The Macmillan Company, 1948), p. 37.

from the series itself.[30] In theory, this method, which would define race after doing the statistical work, could lead to the kind of conclusion that Boas approached – that there are no statistically definable races. In practice, even Boas acknowledged this method was usually also grounded in assumptions – about the grouping of people along racial lines, the kinds of characteristics worth measuring, and the interpretation of the measurements and statistical analysis that frequently amounted to the same process of defining a priori typological categories.

Race, Race Mixing, and Environment: Boasian Racial Science

Franz Boas and the Anthropometrical Approach

From his first acquaintance with physical anthropology at Clark University in 1889, Franz Boas was perturbed by the raft of unexamined assumptions that riddled the study of racial questions. Despite mountains of data and a multitude of papers and pronouncements already produced, Boas recalled decades later how "shocked" he had been at the shoddy state of the science: "Nobody had tried to answer the questions why certain measurements were taken, why they were considered significant, whether they were subject to outer influences." These questions, which had still not been adequately answered in the intervening years, were the ones Boas had dedicated his physical anthropological work to addressing. "We talk all the time glibly of races," he noted, but "nobody can give us a definitive answer to the question what constitutes a race." He lamented to the end of his career the "lack of clarity in regard to what constitutes a type." Such confusion, he complained, was "the cause of the incredible amount of amateurish work produced for more than a century, but particularly by modern race enthusiasts."[31] Boas was convinced that "intelligent discussion" of racial questions could not proceed without better methods and definitions.[32]

Boas became increasingly frustrated with the simplistic and erroneous way that quantification was used to justify supposed racial classifications,

---

[30] Franz Boas, "The Measurement of Differences between Variable Quantities," *Quarterly Publication of the American Statistical Association* (Dec. 1922), pp. 425–445, reprinted in Boas, *Race, Language, and Culture*, pp. 189–190.

[31] Franz Boas, "Heredity and Environment," *Jewish Social Studies*, vol. 1. no. 1 (Jan. 1939), pp. 5–14.

[32] Franz Boas, "History and Science in Anthropology: A Reply," *American Anthropologist*, new series, vol. 38 (1936), pp. 137–141, reprinted in Franz Boas, *Race, Language and Culture*, pp. 309–310.

and the persistence of such methods long after he regarded them as having been soundly refuted. His criticism of Joseph Denicker in 1922 is telling. Boas, with typical understatement, bemoaned the imprecise way anthropologists used the concept of variation in their discussion of race and types. He was particularly appalled by the thoughtless use of averages in one or two characters to classify whole races. He singled out Denicker, who, in *The Races of Man*, assigned people to local types on the basis of certain group averages. "All those that have average statures, head indices, facial forms, nose forms and pigmentation falling within certain limits that may be expressed numerically were assigned by him to a certain subrace." But although this method yields a particular description of local types, Boas continued, "the biological significance of the observed differences remains undetermined." He noted that, lacking information on the significance of measured traits, the resulting groups were totally arbitrary:

If we call tall those populations whose average stature is more than 170 cm., their assignment to a subdivision will not be the same as the one obtained when we call tall those whose average stature is more than 172 cm. If no valid reason can be given for the choice of one or the other limit, then the subtype so established can have only a conventional descriptive meaning.[33]

Not only did investigators need to establish that such descriptive features were morphologically significant, they also had to distinguish between the environmental and hereditary influences.

Moreover, averages themselves were problematic, if used in a simplistic manner to characterize groups of people. The fundamental problem in delineating racial types, the one that made use of statistics essential, was that types were not in fact uniform, but variable (as they had to be to have any evolutionary significance). This meant that averages were a means of describing the variation found in whole sets of people (and hence, one of the main reasons that it made no sense to claim that an individual typified a race, as the typical traits were composites of all the people encompassed in the type). Thus, although averages provided a particular sort of information about a group of people, they also obscured information. Two groups of people might have different average measurements in a particular trait, but there might be individuals within both groups who have the same value. Boas used as his example *cephalic index* (the ratio of skull

---

[33] Boas, "The Measurement of Differences between Variable Quantities," in *Race, Language and Culture*, pp. 181–182.

length to breadth): Swedes might have an average of seventy-seven, Bavarians eighty-five, but both groups would contain individuals whose index was eighty-one. Simply on the basis of the cephalic index, the individuals who measured eighty-one could be placed in either group.[34] Conversely, groups with different ranges of variation might have the same average, and thus appear to be the same type. Relying simply on similar averages might obscure the existence of a mixed type. The extent of variation was thought to be an indicator of racial purity, although, Boas cautioned, not a straightforward one.

Relying on averages had another pitfall. Boas argued that, in fact, groups varied considerably among themselves and almost always overlapped with nearby groups. Moreover, human variation was actually more of a continuum than sets of discreetly bounded types – only "fundamentally different races," such as the "blond north European White and the dark Sudanese Negro" failed to overlap in any significant way. In reality, the north European and African were separated by a series of intermediate and overlapping steps, a picture of racial difference that arbitrary grouping and reliance on distinct averages obscured.[35] For Boas, the extent, kind, and direction of variation were more revealing than mere averages. And none of these numbers could be fully comprehended without consideration of environmental factors and the biological significance of the traits employed.

Boas was wary about reifying his statistics, acting as though a set of numbers or quantified relationships were biological realities, unless there was robust proof for such conclusions. In compiling measurements of the cephalic index for various Indian tribes, and plotting them on a curve, Boas discovered bimodal distributions (two peaks, rather than the single peak expected in a "normal" probability distribution) with such frequency that he felt sure they were not simply an artifact of small numbers, but due "to some actual reason."[36] Noting that the kinds of curves he was getting were often quite irregular – versus the kind of "normal" curves statistician Francis Galton had derived – Boas cautioned that his data on Indian-white mixes showed "conclusively that anthropometric curves are not always probability curves." The consequences for anthropological theorizing about races were significant. Much analysis in racial

---

[34] Ibid., pp. 182–183.

[35] Ibid., p. 41.

[36] Franz Boas "The Anthropology of the North American Indian," *Memoirs of the International Congress of Anthropology* (Chicago: Schulte, 1894), pp. 37–49, in Boas, *Race, Language and Culture*, pp. 191–201.

science relied on the comparison of "average" traits to define groups. But in populations characterized by a bimodal distribution, Boas argued, the average "would have no meaning whatever." According to Boas, based on the "biological law" that children tend to "revert" to their ancestral, or parental, "types," a bimodal distribution reflected the origins of a mixed race in two disparate parental stocks, and an "average" of those types in the offspring would not yield an accurate picture of the mixed race. However, if the parental types were not dramatically different, then the evidence of the mixed race might not appear as a bimodal distribution of two distinct variants within a group, but rather as an increased range of variation.

Boas himself was exceedingly cautious about how he constructed his studies, calculated his statistics, and interpreted his data. Always concerned with the quality of data collected, Boas noted that he strove to compile a sufficient quantity and type of measurements to assure that his calculations would show that the differences he found "were real, not accidental."[37] Even in his earliest projects, Boas was conscientious about minimizing errors in data collection by requiring multiple separate individuals to take each set of measurements and observations, at different times, and rejecting any returns whose accuracy he doubted. Like other anthropologists, Boas also noted that his subjects did not enjoy being measured and observed anthropometrically. In his 1892 study of Indian-white race mixing, Boas commented that the number of measures had to be kept to a minimum, focusing on "the most important ones," and only those that could be taken without forcing their subjects to disrobe. Boas does not elaborate on the circumstances, but his remark that "only by this restriction could a sufficient number of measurements be secured," itself suggests a degree of concerted resistance and refusal that reveals the complicated power and racial dynamics lurking behind Boas's text.[38]

The Science and Politics of Miscegenation

One of the earliest physical anthropological problems Franz Boas turned to in the 1890s was the study of Native Americans, both "full-blooded" and "half-blood," to delineate the "distribution of types of man in North America." The history of interest in "half-blood" racial hybrids dated to earlier nineteenth-century monogenist-polygenist debates about the

37 Ibid.
38 Ibid.

degree of difference between races of people. Boas was interested in both contemporary and historical populations, with the hope that through a study of language, culture, histories of migration and development, and bodies, both living and deceased, anthropologists could reconstruct "'The History of the American Race.'"[39]

Like much of Boas's work, his study of Indian-white mixed populations was spurred by frustration with inadequate data to answer persistently perplexing questions, along with a fortuitous opportunity to pursue a research program. By the 1890s, the United States had large collections of Indian skeletal material at the U.S. Army Medical Museum, the Peabody Museum at Harvard, and in Samuel Morton's large collection of skulls housed at the Academy of Natural Sciences in Philadelphia (dubbed the "American Golgotha" by some of the more ghoulish wags in the anatomical community). Boas found these collections unsatisfactory, the skeletal material rarely identified by tribe, sex, or degree of intermixing. Boas hoped to investigate "indispensable" living individuals. In 1891, the opportunity presented itself in the form of Frederic Ward Putnam, chief of the Harvard Department of Archaeology and Ethnology and director of the Peabody Museum, who was busy making preparations for the 1893 World's Columbian Fair. Putnam asked Boas, who was already preparing lectures at Harvard, to organize a study of "the physical characteristics of the North American Indians."[40] Coordinating a staff of trained young college students, Boas collected measurements on 17,000 "full-blooded and half-blooded Indians . . . distributed all over the North American continent."[41] Material gathered was used for an exhibition on physical characteristics of the North American Indian at the World's Fair, and for Boas's first study of race mixing.[42]

Despite his subsequent reputation as a particularist, rather than a theoretician, in his 1894 report on his study of Indian-white racial mixes, Boas laid out a number of methodological and theoretical considerations. Boas was fully conscious of the potential theoretical significance of his study. "There are few countries in which the effects of intermixture of races and of change of environment upon the physical characteristics of man can be studied as advantageously as in America," he wrote, "where a process of slow amalgamation between three distinct races is taking

---

[39] Ibid., p. 201.
[40] Boas, "Anthropology of the North American Indian," p. 192; Cole, *Franz Boas*, p. 152.
[41] Ibid.
[42] For more detail on the Indian exhibits, see Cole, "'All Our Ships Have Gone Aground,'" *Franz Boas*, pp. 152–166.

place." He likened the process to the "migration and intermarriage" that had been a "fruitful" source of intermixture in Europe, and which had "had the effect of creating strong contrasts in adjoining countries."[43] His analysis represented a fascinating mix of pre-Mendelian speculations on heredity; persistent concerns about the role of culture, history, and environment; and inklings of future positions. Like most anthropologists of his day, Boas was concerned about the state of Native Americans. He noted with concern that Indians were not only diminishing alarmingly in sheer numbers, but, from the perspective of physical anthropologists, many of those left were "mixed to a considerable extent with whites and negroes, so much so that in certain regions it is impossible to find a full-blooded individual."[44] But Boas viewed the "great frequency of half-breeds" as a particularly interesting phenomenon that afforded him a chance to study its hereditary effects, by comparing both full-blooded and mixed adults, as well as parents and children. He pronounced it "one of the most fruitful fields of investigation" in the entire project.

Methodologically, Boas did not attempt to divide the Indians into "races" or racial "types," but grouped the tribes in "geographical" areas, based on their present locations. He made two important caveats that reveal how, even by 1892, Boas was thinking about the populational mix of traits, living social units, and bodily types, as well as the effect of social group, history, and environment on the appearance of individual bodies and groups of people. Boas cautioned that the "types" discussed were living peoples, not representative of any "original types of the respective peoples" studied. He noted that, far from some straightforward descent from an original type, which could be read off living bodies, the living people very likely represented a process of mixing "over the course of the centuries with numerous other peoples," so much that any traces of an "original" type might have disappeared entirely. Moreover, Boas warned readers not to conflate or rashly correlate the "social and political organizations which we call tribes or nations" with "the physical unit which constitute the characteristics of the individuals in a certain region." In other words, the kinds of bodies found in a given region may or may not correspond to the social and political ways people have grouped themselves. Boas also noted that not only were bodily types not simply

---

[43] Franz Boas, "The Half-Blood Indian," originally published in *Popular Science Monthly* (Oct. 1894), reprinted in Boas, *Race, Language and Culture*, pp. 138–148.

[44] Ibid. He lists the Iroquis, Cherokees, Chickasaws, and Chocktaws as particularly mixed.

congruent with social forms, they were not necessarily stable over time and space, either. "The physical type," Boas noted, "is the result of the complex descent of a people and of the effect of the surroundings upon its physical development."[45]

Contrary to nineteenth-century expectations, Boas found that the mixed-race individuals showed considerable vigor and no signs of degeneration. He found that "half-breed" women had considerably higher fertility (that is they averaged more children), and that mixed-race individuals tended to be taller than "full-blooded" members of the same tribes, a pair of results that Boas concluded suggested that "mixture of races results in increased vitality." Indeed, he argued,

The difference in favor of the half-breed is so striking that no doubt can be entertained as to its actual existence. I believe the cause of this fact must be considered to be wholly the effects of intermixture, as the social surroundings of the half-breeds and of the Indians are so much alike that they cannot cause the existing differences.[46]

In addition to increased height and fecundity, Boas observed that mixed-race individuals tended to favor one or both parents. He noted, for example, that the color and texture of the hair, as well as eye color, tended to strongly favor the Indian parent. Even in breadth of face, which Boas contended was the characteristic in which white and Indian parents most differed, although the children were generally intermediate, most measurements were closer to either parent than to perfectly intermediate forms.[47] In other words, Boas concluded that "the effect of intermixture is not to produce a middle type," what would at that time have been understood as "blending" inheritance, "but that there is a tendency to reproduce ancestral traits," that is to "revert" to the forms of their parents, the supposedly "purer" lines.

Other than ruling out the effects of the environment and social setting, Boas was unsure just what kind of hereditary causes would have

---

[45] Ibid., p. 193.

[46] Ibid.

[47] Boas did note that Indians and American whites were particularly distinct in the breadth of their faces because American whites had "exceedingly narrow" faces, noting that in Europe, people in the eastern regions had considerably wider faces. Unlike many anthropologists at this time, who tended to see particular manifestations of traits as indicative of a racial type wherever they were found – a "Nordic" cephalic index indicated a Nordic individual anywhere in the world – Boas seems to have been suggesting that differences between distinguishable groups were notable only in local, historical context. Ibid., p. 194.

led to the observed effects. He conducted his study before the rediscovery in 1900 of Gregor Mendel's groundbreaking work on basic patterns of inheritance in pea plants that became the foundation of modern genetics. Lacking guidance from Mendelian genetics, and with the statistical innovations of biometrics still in their infancy, Boas used his own understanding of heredity and his own abilities in probabilistic thinking to conjecture about the hereditary process. He suggested, for example, that dark hair and a wide face might be "more primitive characteristics" than a narrow face and light eyes. Possibly, Indian characteristics were inherited "with great strength," he proposed, "because they are older." Or, he observed, perhaps, because in virtually all cases it was the mother who was Indian, perhaps this "had an influence," although he quickly noted that there was "no proof that children resemble their mothers more than their fathers." In some cases, early Mendelian theory would still have been little help, because characteristics such as face breadth or stature are not discrete, dominant or recessive, characters but continuously variable traits, something both Mendelians and biometricians struggled with for decades.

In 1915, three years after publishing his seminal paper on the influence of the environment on heredity, Boas addressed the International Congress of Americanists, on the topic "Modern Populations of America." There he argued for the utter lack of evidence underlying widespread claims of racial inferiority among mixed races, and in the course of doing so neatly reinscribed broadly foundational racial types. Boas was clear in his dismissal of the biological dangers of intermixing: "The claim has been made, and has constantly been repeated, that mixed races – like the American Mulattoes or the American Mestizos – are inferior in physical and mental qualities, that they inherit all the unfavorable traits of the parental races. So far as I can see, this bold proposition is not based on adequate evidence." Further, in border areas of the world where the "fundamental," markedly disparate races had mixed, such as "Japan, a country in which the Malay and the Mongol type come into contact; or the Arab types of North Africa, that are partly of Negro, partly of Mediterranean descent; or the nations of eastern Europe, that contain a considerable admixture of Mongoloid blood," no "careful and conscientious investigator" would "be willing to admit any deteriorating effect of the undoubted mixture of different races." Boas argued that "half-bloods" frequently lived under difficult conditions, often worse than those of "the pure parental races," and these conditions could well account for the "apparent weakness" critics "erroneously interpreted as due to effects of intermixture."[48]

But what initially sounds like a ringing denunciation of race mixing as a social and biological problem looks less clear-cut on further examination. Boas opened his address by noting that the United States was comprised of three "distinct types of populations": descendants of European immigrants (whom he likened to the population of the northern United States, Canada, and Argentina); populations "containing a large amount of Indian blood" ("Mestizos," whom he likened to populations in Mexico, Peru, and Bolivia); and populations "consisting essentially of mixtures of Negroes and other races" (which included the southern United States, the West Indies, and some areas of Central and South America).[49] Boas then explained that the development of mixed populations in North, Central, and South America "depended to a great extent upon the very fundamental difference in the relations between the Anglo–Saxon European immigrants and the Latin American immigrants." In other words, the American enforcement of hypodescent, or the "one drop rule," had shaped the racial character of New World nations. According to Boas, Latin American immigrants, men and women, were much more willing to form marriages and "unions" with what he referred to as "members of foreign races" (by which he meant indigenous people). Conversely, "Anglo-Saxon" women only rarely bore mixed-race children, preserving their "White blood"; "Anglo-Saxon" men, by fathering mixed-race children more frequently, provided "a considerable... infusion of White blood" to the "foreign" population. According to Boas, this process meant "the number of individuals with European blood is constantly increasing," because the children of white women fathered by white men

---

[48] Boas, "Modern Populations of America," Earnest Hooton later published remarks that largely supported Boas's and Shapiro's conclusions in "Development and Correlation of Research in Physical Anthropology at Harvard University," *American Philosophical Society, Philadelphia. Proceedings*, vol. LXXV, Philadelphia, 1935:

> In general these studies of race crossing seem to show no biological inferiority of racial hybrids, little or no heterosis or hybrid vigor, but a heightened fecundity, a segregation of inherited traits, the presence of multiple factors in their inheritance, and the impossibility of simple Mendelian interpretations. However, it is clear that the most important aspect of race mixture is the social selection of types which operates in favor of those showing a preponderance of the physical features of the superordinate race (p. 507).

[49] Franz Boas, "Modern Populations of America," originally in *Proceedings of the 19th International Congress of Americanists, Washington, December, 1915*, Washington, DC: 1917, pp. 569–575, reprinted in *Race, Language and Culture*, p. 18.

remained in the white population, while the children of "foreign women" fathered by white men moved into the "foreign" population and increased "the relative amount of non-European blood." Boas noted that in those cases where white women did marry indigenous or "half-blood" descendants, "a thorough penetration of the two races must occur; and if marriages in both directions [i.e., white men also married 'foreigners'] are equally frequent, the result must be complete permeation of the two types." Boas noted "there is very little doubt that the rapid disappearance of the American Indian in many parts of the United States is due to this particular kind of mixture."[50] I don't imagine he needed to tell his audience that this logic also applied to African Americans and other so-called foreign races.

In "Modern Populations of America," Boas displayed both his well-known criticisms of racial science and his distinctly less-familiar racialized perspectives. He harshly critiqued studies of race mixture as scientifically sloppy, basing bold conclusions on inadequate evidence and ignoring contradictory evidence, and he offered alternative explanations for apparent "weaknesses," including the importance of social environments. He also enumerated racial groups he regarded as "fundamental," groups that underlay the concept of mixed races, and depicted a central place for the role of miscegenation and white womanhood in the disappearance of indigenous peoples.

"Like a Bomb Dropped upon a Sewing Circle"[51]: Children of Immigrants and the Immutability of Race

Boas's caution stemmed in part from his recognition that little was understood about human heredity, its mechanisms, or their relationships to the various environments in which human beings lived. Boas retained this caution throughout his career, even while his ideas about races, types, and heredity shifted. In 1899, Boas published a review of William Ripley's book *The Races of Europe*, as well as an essay addressing recent "severe attacks" on the methods of physical anthropology.[52] In these

---

[50] Ibid., pp. 19–20.
[51] "Part of paper delivered at Anthropol. Club. Philadelphia 1935," p. 1, Box 4, Anthropology, Harry L. Shapiro Papers, Special Collections, Library, American Museum of Natural History, New York (hereafter AMNH-HLS).
[52] Franz Boas, "Review of William Z. Ripley, 'The Races of Europe,'" *Science*, new series, vol. 10 (Sep. 1, 1899) reprinted in *Race, Language and Culture*, pp. 155–159; Franz Boas, "Some Recent Criticisms of Physical Anthropology," *American Anthropologist*,

articles, Boas discussed race and type in a markedly static typological manner. Boas noted that one of the major criticisms leveled at physical anthropology regarded scientists' failure to identify reliable descriptive racial features in skeletons, and thus the apparent futility of using skeletons as the basis for good racial classifications. Critics seized on this failure, Boas argued, as evidence for "the belief, frequently expressed, that the characteristics features of each race are not stable," but rather that they were greatly influenced by the geographical and social environment. "It seems to me," Boas responded, "that these views are not borne out by the observations that are available.... While it may be impossible to classify any one individual satisfactorily, any local group existing at a certain given period can clearly be characterized by the distribution of forms occurring in that group."[53] Similarly, in his discussion of Ripley's terminology – Teutonic (instead of Joseph Deniker's Nordic), Alpine, and Mediterranean – Boas approved of his geographic labels for the European types because, "on the whole, human types are comparatively stable in given areas." He found the application of the term "race" to describe the varieties of Europeans unfortunate, however, as Europeans differed much less among themselves than they did as a group from "Africans and Mongols." Boas thought the term "race" should be reserved for "the largest divisions of mankind." To bolster his argument, he noted that there were a number of important similarities among the three varieties of Europeans, and that, in his opinion, Europeans were "a highly specialized form of the Mongoloid type" based on the "peculiar development of the nose and face, and decreased pigment."[54]

However, despite his broad concurrence with the most commonly employed racial terminology, Boas's surprising reply to physical anthropology's critics reveals a glimmer of the profound reversal he would make following his study of immigrants and their children. His argument continued:

The critics... will of course concede that a Negro child must be a Negro, and that an Indian child must be an Indian. Their criticism is directed against the permanence of types within the race; for instance, against the permanence of short or tall statures, or against the permanence of forms of the head.... The

new series, vol. 1 (Jan. 1899), reprinted in *Race, Language, and Culture*, pp. 165–171. George Stocking provides an excellent discussion of the development of Boas's ideas in physical anthropology, in "The Critique of Racial Formalism," in his volume *Race, Culture and Evolution*, pp. 161–194.
[53] Boas, "Some Recent Criticisms," *Race, Language and Culture*, p. 166.
[54] Boas, "Review of William Z. Ripley," *Race, Language and Culture*, p. 157.

insufficiency of the influence of environment appears in cases where populations of quite distinct types inhabit the same area and live under identical conditions. Such is the case on the North Pacific coast of our continent. . . . While this may be considered good evidence in favor of the theory of predominance of the effect of heredity, the actual proof must be looked for in *comparisons between parent and offspring.* If it can be shown that there is *a strong tendency on the part of the offspring to resemble the parent, we must assume that the effect of heredity is stronger than that of environment.*[55] [My emphasis.]

Boas noted that Karl Pearson and Francis Galton had developed a statistical method for making just such comparisons between parents and children, and furthermore, that "[w]herever this method has been applied, it has been shown that the effect of heredity is the strongest factor in determining the form of the descendant." But then Boas went on to make a crucial caveat for his future role in the physical anthropology of race and human diversity, and that of Harry Shapiro (who would not be born for three more years): Pearson's and Galton's methods had not been applied either to a series of generations or under conditions of considerable change in environment.

In 1899, Boas addressed various criticisms that had been leveled at physical anthropology for its failure to successfully describe and classify humanity, particularly a contingent of critics who were doubtful that a quantitative, metrical approach could succeed where a descriptive, morphological approach had failed. Boas regarded an organismal, holistic morphological approach, as well as cognizance of a group's history and culture, as crucial for attaining a full picture of racial types, but he also held out great promise for the metrical approaches developed by Galton and Pearson. Foreshadowing his seminal work a decade later, Boas actually argued for the likelihood that heredity was a more powerful determinant of bodily form than environment, but noted, in typically judicious fashion, that this was mostly a hopeful expectation that would make the origins and evolution of human difference intelligible to anthropology, not a proven fact. Proof, Boas argued, would come in "comparisons of parents and offspring," and the application of the metrical method to a series of generations, "under conditions in which a considerable change of environment has taken place." Such studies, Boas argued, offered the hope of a "definite solution of the problem of the effect of heredity and of environment."[56]

---

[55] Boas, "Some Recent Criticisms," *Race, Language and Culture,* p. 167.
[56] Boas, "Some Recent Criticisms," *Race, Language and Culture,* pp. 167–168.

In 1908, Franz Boas proposed a study that would apply Galton's and Pearson's metric method to a research problem that Boas had character-ized as a key question for the determination of the relative influences of heredity and environment on bodily form.[57] Boas wrote to Jeremiah W. Jenks, an economist heading the U.S. Immigration Committee, proposing to study immigrants to the United States and their children, both those born in their homeland and those born in the United States. Writing to a federal agency preoccupied with how to restrict immigration, Boas posed the problem as one of assessing the threat that the massive immigration of the late nineteenth century might pose to the vigor of the U.S. popula-tion, the solution as the new biometrical approach. "During the last ten years," Boas wrote,

attention has been drawn to the change in composition of our immigrant popu-lation. Instead of the tall blond north-western type of Europe, masses of people belonging to the east, central, and south European types are pouring into our country; and the question has justly been raised, whether this change in phys-ical type will influence the marvelous power of amalgamation that our nation has exhibited for so long a time. The importance of this question can hardly be overestimated, and the development of modern anthropological methods makes it perfectly feasible to give a definite answer to the problem that presents itself to us.[58]

Elsewhere, Boas discussed the virtues of this plan, not as a way to eval-uate the threat to America's Anglo-Saxon purity and superiority, but as a chance to study an ongoing human experiment in changing envi-ronments and their effects on supposedly stable and permanent physical features.

By the early 1910s, when Boas published the results of his immigrant studies, he was reluctant to claim he had solved so large a problem as the relative effects of heredity and environment, but he nonetheless felt

---

[57] See Stocking, "The Critique of Racial Formalism," *Race, Culture and Evolution*, pp. 175–180.

[58] "Changes in Immigrant Body Form," Letter from Franz Boas to J. W. Jenks, March 23, 1908, Stocking, ed., *The Shaping of American Anthropology, 1883–1911*, p. 202. For historiography on immigration and the discourse surrounding it in this period, see inter alia, John Higham, *Strangers in the Land: Patterns of American Nativism, 1860–1925* (New Brunswick, NJ: Rutgers University Press, 1955); Matthew Frye Jacobson, *Whiteness of a Different Color: European Immigrants and the Alchemy of Race*; Daniel J. Tichenor, *Dividing Lines: The Politics of Immigration Control in America* (Princeton, NJ: Princeton University Press, 2002); Mae Ngai, *Impossible Subjects: Illegal Aliens and the Making of Modern America* (Princeton, NJ: Princeton University Press, 2004); Aristide R. Zolberg, *A Nation by Design: Immigration Policy in the Fashioning of America* (Cambridge, MA: Harvard University Press, 2006).

that his study had produced some startling and significant results. In fact, his results were so stark and profound that his faith in the precedence of heredity over environment was permanently displaced in favor of a new conviction that physical and social environments were so powerful that even the most supposedly stable characteristics were capable of rapid and remarkable change. Boas had set up his study for maximum theoretical impact. Taking the large population of recent European immigrants and their children, Boas set out to study possible changes in the paradigmatic racial index: the skull. He and his assistant measured immigrants from various parts of Europe, corresponding roughly to the supposed European racial types (Nordic, Mediterranean, and Alpine, although Boas simply divided them into Northern, Eastern, Central, and Southern Europeans) who were living in New York City. Between 1908 and 1910, they measured almost 18,000 people, predominantly East European Jews, Bohemians, Neapolitans, and Sicilians, as well as smaller numbers of Poles, Hungarians, and Scots.

His results were "revolutionary."[59] Based on his statistical analysis of the data, Boas concluded that his study demonstrated that there were "decided changes in the rate of development" and the direction of change, as well as dramatic changes in apparently key bodily features "without change in descent," results which could "only be explained as due directly to the influence of environment." Moreover, he found that parents differed more from their American-born children than they did from their foreign-born offspring, and that this influence of the environment increased with the length of exposure.[60] "The influence of the American environment makes itself felt with increasing intensity," Boas asserted, "according to the time elapsed between the arrival of the mother and the birth of the child."[61] As to causes, Boas conceded that he had no idea how such changes were produced. He could only argue that none of the explanations proffered to that point to explain such differences were adequate. The "instability or plasticity of types," however, was clear.[62] His study had dealt a profound blow to the basis for so many typological classifications, such as those of Ripley and Deniker, that relied on

[59] Stocking, "The Critique of Racial Formalism," in *Race, Culture and Evolution*, p. 177.
[60] Ibid., pp. 177–178; Boas, "Changes in Bodily Form of Descendants of Immigrants," in Boas, *Race, Language, and Culture*, pp. 60–61, 74.
[61] Franz Boas, "Changes in the Bodily Form of Descendants of Immigrants," *American Anthropologist*, new series, vol. 14 (1912), pp. 530–562.
[62] Boas, "Changes in Bodily Form of Descendants of Immigrants," in Boas, *Race, Language, and Culture*, p. 72.

measures of the skull, a trait "which had always been considered one of the most stable and permanent characteristics of human races."[63] Boas had demonstrated that purportedly stable characters were "plastic," physical and social environments shaped human bodies, and equally important from his point of view, anthropometry and proper statistical analysis were invaluable tools for grasping subtle but crucial changes not apparent through mere morphological scrutiny or the simple calculation of averages.

For Boas, his study of immigrants shaped his views in new and significant ways. In contrast to his observations twelve years earlier in his critique of William Ripley's racial classification, Boas now regarded European types as "remarkable for their high variability." His understanding of what a racial type was had been transformed. In 1894, when he described his study of Indian-white mixes, Boas viewed Indian tribes as relatively stable entities existing in discernible geographical settings. Following his immigrant studies, Boas increasingly viewed racial types as populations that could only be characterized through statistical assessments of the range of traits and their averages, an "enumeration of the frequencies of individuals with distinctive forms."[64] Attempts to reduce some simple set of averages or set of stable "racial" characteristics obscured the complex webs of descent – in Boas's terms "family lines" and "fraternities" – the effects of constant migration and contact, and the consequent range of variation encompassed by any given population. Equally important for Boas, his study of immigrants renewed his faith in the value of anthropometry and the value of rigorously, quantitatively examining human diversity. About this he was explicit; the idea that his study would "destroy the whole value of anthropometry, in particular the study of the cephalic index[,] has been shown to have no importance. It seems to me, on the contrary, that our investigations, like many other previous ones, have merely demonstrated that results of great value can be obtained by anthropometrical studies." The point, for Boas, was not to establish stable morphological types and racial hierarchies, but rather to elucidate "the early history of mankind and the effect of social and geographical environment upon man," and the necessity of pursing researches that could demonstrate that the traits used to discriminate among races were "morphologically important" rather than arbitrarily

[63] Stocking, "The Critique of Racial Formalism," in *Race, Culture and Evolution*, p. 178.
[64] Franz Boas, *The Mind of Primitive Man*, p. 42.

chosen averages.[65] "A result of historical significance," Boas later argued, could only be obtained "by a study of the many genetic lines constituting a population," not based on some arbitrary conception of what was "typical," "but with due consideration of the variety of forms that occur, of their frequencies in succeeding generations, and of the response to varying environmental influences."[66]

Twenty-five years later, Boas continued to employ biometrics, statistics, and anthropometry to argue for the twin importance of heredity and environment in understanding the complexities of human bodies, their relations, and development. In "The Tempo of Growth of Fraternities," published in 1935, Boas turned to the growth of children to examine the relative effects of genetics and social conditions. Human growth and development had interested Boas since his earliest days at Clark University working with psychologist G. Stanley Hall in the 1890s.[67] Using data collected on children over a period of almost thirty years at the Hebrew Orphan Asylum in New York City, where "the conditions of nutrition, shelter and mode of life are as uniform as they can be obtained," Boas found that there were significant effects of both heredity and environment. Comparing brothers and sisters, Boas found that the tempo of growth, slow or rapid, seemed to be hereditary, a finding corroborated by Raymond Pearl's animal studies. Echoing the findings of Harvard geneticist William Castle, Boas noted that the genetic explanations for changes in characteristics such as length of body and tempo of development would be complex. Such traits "must be governed by many hereditary factors," and such phenomena were matters of the "general organization of the body" – growth of arms and legs were surely correlated, not independently varying hereditary traits. Boas also discerned evidence of the effect of the New York urban environment on the growth of children as well as the variable impact of the orphanage itself. Boas noted that in the early years of his study, when children were fed a poorly balanced diet and got little exercise their growth was retarded compared to their age peers, but when a new management after 1918 improved the food supply, provided medical care, and gave the children fresh air and exercise, development improved immediately.[68]

---

[65] Boas, "Changes in the Bodily Form of Descendants of Immigrants," *American Anthropologist*, p. 562; Boas, *Mind of Primitive Man*, p. 47.
[66] Boas, "The Relations between Physical and Social Anthropology," *Race, Language and Culture*, p. 173.
[67] Stocking, "Critique of Racial Formalism," p. 165.
[68] Franz Boas, "The Tempo of Growth of Fraternities," *Proceedings of the National Academy of Sciences*, vol. 21, no. 7 (Jul. 15, 1935), pp. 413–418.

Boas viewed humans as primarily products of a particular historical trajectory, cultural creatures who also happen to be biological ones who possess heritable characteristics, some of which can be modified through the influence of various sorts of environments. Boas stressed the cultural, environmental influences in studies of human heredity and appearance. Where Earnest Hooton enthusiastically embraced a biometrical approach to questions of human difference and tended to focus on the fixity of hereditary traits instead of their plasticity, Boas promoted rigorous statistical and anthropometric approaches, as adjuncts to morphological descriptions, in order to prevent erroneous conclusions about the hereditary nature of human differences. In general, Boas was extremely cautious about drawing conclusions about causation, correlation, and the genetic nature of various traits. But even a committed empiricist such as Franz Boas, who regarded compiling large amounts of data as preferable to advancing unsupported conclusions, was cheered by the possibility of eventually drawing some conclusions. He noted that one of the key virtues of his study of changes in head form among immigrants was its demonstration of the value of the anthropometric method, a result that he hoped would "strengthen our confidence in the possibility of putting [anthropometric methods] to good use for the advancement of anthropological science."[69]

Boas's faith in the efficacy of anthropometry properly applied to the problems of heredity and environment never waned. In 1941, not long before his death, Boas continued to pursue studies in physical anthropology that he hoped would clarify the processes that produce human diversity. Noting that Harry Shapiro's study of Japanese immigrants in Hawaii corroborated the conclusions of his 1911 study (see Chapter 4 in this book for a discussion of Shapiro's work), Boas constructed another study in human growth meant to distinguish between purely developmental environmental effect, such as improved diet, strictly hereditary causes immune to environmental influence, and the elusive third category, "actual physiologically determined changes," the sort of indices in which he and Shapiro detected the persistent effects of the environment on bodies. At the end of his career, he turned to the studies of growth and development he had conducted for fifty years, asserting that "Anthropometric investigations . . . will throw light upon our problem."[70]

[69] Boas, "Changes in the Bodily Form of Descendants of Immigrants," *American Anthropologist*, p. 562.
[70] Franz Boas, "The Relation between Physical and Mental Development," *Science*, new series, vol. 93, issue 2415 (Apr. 11, 1941), pp. 339–342.

The Science of Man and the Problem of Racism

Although Franz Boas is little known for his physical anthropology, the elements of that work that have been widely recognized are those that have been credited with undermining "scientific" racism. His study of changing head form among immigrants is routinely cited as an early, critical blow to scientific work purporting to demonstrate the reality of immutable racial types and hierarchies. Secondary scholarship is filled with accounts of Boas's lifelong efforts to combat racism and change scientific practice, in the schoolroom, civil rights, studies of human behavior, and amidst two world wars.[71] Much less often explored is the way his vigorous opposition to prejudice and discrimination coexisted with his ongoing efforts to practice a science of physical anthropology that could illuminate human physical variation. His fight against erroneous ideas, invidious distinctions, faulty logic, and poorly conducted investigations – in short, against bad racial science and popular fallacies – was married to his persistent efforts to use the methods and theories of physical anthropology and genetics to comprehend variation that he still understood in fundamentally racialized terms.

Boas's efforts to combat racism involved a two-pronged approach that would become the standard strategy employed by anthropologists in the mid-twentieth century. On one hand, following the revolutionary results of his head-form study, Boas consistently stressed what he viewed as the "facts" about race, facts that he argued undermined facile, subjective, hereditarian typologies. At the same time, Boas also inveighed against erroneous assumptions and false conclusions that he viewed as the basis for racism and discrimination. In this approach, a judicious, methodologically rigorous investigation of human physical and mental

---

[71] See, for example, George W. Stocking, Jr., "Franz Boas and the Culture Concept in Historical Perspective," *American Anthropologist*, new series, vol. 68, no. 4 (Aug. 1966), and "Critique of Racial Formalism;" pp. 867–882; Marshall Hyatt, *Franz Boas, Social Activist: The Dynamics of Ethnicity*, Contributions to the Study of Anthropology, number 6 (New York and Westport, CT: Greenwood Press, 1990); Smedley, *Race in North America*; Vernon J. Williams, *Rethinking Race: Franz Boas and His Contemporaries* (Lexington, KY: University of Kentucky Press, 1996); Lee D. Baker, *From Savage to Negro: Anthropology and the Construction of Race, 1896–1954* (Berkeley and Los Angeles: University of California Press, 1998); Julia Liss, "Diasporic Identities: The Science and Politics of Race in the Work of Franz Boas and W.E.B. Dubois, 1894–1919," *Cultural Anthropology*, vol. 13, no. 2 (May 1998), pp. 127–166. Zoë Burkholder, "'With Science as His Shield': Teaching Race and Culture in American Public Schools, 1900–1954" (PhD diss., New York University, 2008), and her subsequent monograph, *Color in the Classroom: How American Schools Taught Race, 1900–1954* (Oxford: Oxford University Press, 2011).

variation, along with a similarly serious study of culture, language, history, and environment, would lead to an informed understanding about the hereditary and social bases underlying human diversity in bodies and cultures. Race itself was only one part, albeit an important and interesting one, of this broader picture of humanity. Approached appropriately, with reasoned empiricism, race was not a source of invidious distinctions or oppressive action. Muddled, subjective racial studies were another matter, as were a range of folk notions about race and culture that fostered prejudice.

Boas spent more than forty years mounting arguments against misbegotten notions about race, culture, and human behavior. George Stocking has argued that prior to about 1909, Boas had not fully developed the notion of cultural determinism that would ultimately provide his, and others', counterargument to hereditarian racial determinism.[72] But already by the 1890s, he was marshalling a set of arguments that he would reiterate, adjust, and amplify as he continued over the decades to critique shoddy racial science and popular misapprehensions for professional and popular audiences. His views were not static over the course of his career – he became increasingly convinced that bodies were molded by environmental forces, and his critiques reflected the growth of evolutionary science and an increasingly Mendelian and populational perspective on heredity and change (although he never seemed to fully grasp the ramifications of population genetics for racial typologies).[73] Boas never abandoned the idea that large populations of human beings could be sensibly grouped into races on the basis of shared hereditary somatic, and possibly mental, features. But by the end of his career, which coincided with the rise of the Nazis in Germany and advent of the Second World War, he increasingly emphasized the critical determinative effects of culture on human beings whose physical form he continued to view in racial terms.

Already by 1894, when he delivered an address to fellow anthropologists at a meeting of the American Association for the Advancement of Science (AAAS), Boas was arguing against the idea of immutable racial purity and superiority, and challenging white supremacist philosophies, with evidence of cultural diffusion and racial mixing. People, and the cultural traditions and languages they bore, had intermingled since the

---

[72] George W. Stocking, Jr., *A Franz Boas Reader*, p. 220.
[73] John S. Allen, "Franz Boas's Physical Anthropology: The Critique of Racial Formalism Revisited," *Current Anthropology*, vol. 30, no. 1 (Feb. 1989), pp. 79–84.

earliest recorded histories, he argued. History demonstrated that a variety of peoples had attained the "highest type of culture," before "sinking back into obscurity." Against claims of contemporary members of "the white race," who, on the basis of their "wonderful achievements," considered themselves beings "of a higher order as compared to primitive man," Boas noted that among the great ancient races were Hamites, Semites, Aryans, and Mongols.[74] "The Arabs," Boas noted, "who were the carriers of civilization were by no means members of the same race as Europeans, but nobody will dispute their high merits."[75] The ability to develop advanced societies was more a matter of historical contingencies – "favorable conditions" such as amenable "habitat" and commonalities among subject peoples, instead of "striking racial differences" that forestalled "amalgamation" and assimilation, or devastating epidemics.[76]

Boas also sounded another of his recurring themes in 1894. Throughout his career, Boas found arguments that distinctive differences in bodily traits provided evidence of racial superiority utter nonsense. His rejoinder to these fallacious arguments, was, in the first instance to highlight the exceedingly selective catalog of traits offered as proof of superior standing. Boas noted, for example, that although the "color of the skin, the form of the hair and the configuration of the lips and nose distinguish the African negro clearly from most other races . . . it would be easy to find among members of the [native] American race, for instance, lips and nose which might be mistaken for those of a negro." His point, which he would reiterate and elaborate for the next four decades, was that human beings possessed enormously varied physiques, so diverse that what at first appeared to be easily bounded racial types turned out to grade into each other in "innumerable transitions between the races of man." Groups overlapped in their variations such "that a number of characteristics may be common to individuals of both races."[77] No race stood in possession of an exclusive set of superior features. In 1894, Boas countered supremacist arguments with a litany of anatomical and physiological evidence. He had "no doubt that great differences exist in physical characteristics of the races of man" and enumerated a number of them, including a discussion of notable differences between European,

---

74 Franz Boas, "Human Faculty as Determined by Race," as published in the American Associations for the Advancement of Science, *Proceedings* 43 (1894): pp. 301–327, in Stocking, ed., *A Franz Boas Reader*, pp. 221–242 (pp. 221–223).

75 Ibid., p. 225.

76 Ibid., p. 226.

77 Ibid., p. 227.

African, and Asian heads and faces. He argued that although elements of African physiognomy put them "slightly nearer the animal than the European type," noting in particular features of the head and face that "remind us of the higher apes," many of the same features were "not entirely absent among the white races," and "variations belonging to both races overlap."[78] Whether such differences extended to cognitive capacities were unclear to Boas, although he allowed they might. Written just two years before *Plessy v. Ferguson* would enshrine "separate but equal" in federal law, Boas concluded his address with a prediction that only partially undercut white supremacy but held out hope for more racial parity in American society:

[T]he average faculty of the white race is found to the same degree in a large proportion of individuals of all other races, and although it is probable that some of these races may not produce as large a population of great men as our own race, there is no reason to suppose that they are unable to reach the level of civilization represented by the bulk of our own people.[79]

By the 1930s, Boas found the idea of racial superiority grounded in some set of characteristic hereditary traits nonsensical. Although he remained convinced that the major racial divisions – Mongoloid, Negroid, and Caucasoid – were both taxonomically useful and biologically valid, and argued that some smaller divisions and locally distinct groups might also qualify as races, much of his discussion of heritable physical variation was couched in terms of changing and overlapping constellations of characteristics. Although Boas never fully abandoned race, he became increasingly skeptical of it as any sort of determinative force. In 1894, he had thought it entirely possible that evidence would one day demonstrate mental differences among races corresponding to their physical variations.[80] By the 1930s, following his own studies of the effects of environments, as well as studies by psychologists such as Otto Klineberg of differential racial performance on intelligence and other mental tests, Boas argued that the overwhelming determinative forces in human life were environmental. Family, work, cultural practices, social institutions, and the physical environment were the predominant forces shaping not only human lives and societies, but bodies as well. Speaking again to the AAAS in 1931, Boas argued that "ethnological evidence is all in favor of the assumption

---

[78] Ibid., p. 230.
[79] Ibid., p. 242.
[80] Ibid., pp. 230–242.

that hereditary racial traits are unimportant as compared to cultural conditions." In defining race, Boas still enumerated the traditional diagnostic
traits, although he also consistently noted that not all people fit well into
the widely accepted groups. "Whites, with their light skin, straight or
wavy hair and high nose, are a race set off clearly from the Negroes, with
their dark skin, frizzly hair and flat nose," he noted. But "not quite so
definite is the distinction between East Asiatics and European types," he
said, "because transitional forms do occur among normal White individuals, such as flat faces, straight black hair and eye forms resembling the
East Asiatic types; and conversely European-like traits are found among
East Asiatics."[81] Although Boas contended that racial differences were
real enough, by 1931, he was also arguing that they were small.

One argument Boas consistently repudiated was the hereditarian idea
that culture or nation was an outgrowth of race. He rejected any easy
causality, or even correlation, between particular physical constitutions
and distinctive cultural formations. For example, concern about events
in Germany led Boas to give a speech on the perils of conflating hereditary and environmental sources of behavior to the World Congress on
Population in Paris in 1937.[82] Although Boas never mentioned the Nazi
regime and its racial doctrines, the fact that the speech appeared two years
later in the first issue of *Jewish Social Studies* suggests that people at the
time perceived the broader social and political import of his arguments.[83]
At the Congress, Boas argued that there was no anthropological evidence
that racially discrete groups of people, "different human types," displayed
"distinct innate personalities." Studies of human anatomy, physiology,
mind, and behavior had not produced any evidence that "habits of life
and cultural activities are to any considerable extent determined by racial
descent."[84] The idea that a particular racial group bore a distinctive,
uniform "personality" was founded, Boas argued, on a set of misconceptions about both race and culture. First, he argued, the concept of race
had to be clearly understood. Racial types could not be "naively" based

---

[81] Franz Boas, "Race and Progress," in *Race, Language and Culture*, pp. 3–17 (p. 4).
[82] Franz Boas, "Heredity and Environment," *Jewish Social Studies*, vol. 1, no. 1 (Jan.
1939), pp. 5–14.
[83] He did note, however, in the course of discussing the range of variability found in
distinct populations, that "among German groups...the headform of family lines at
each end of the series are so different that they never overlap but behave like two distinct
races," a remark surely calculated to undermine the notion of a pure Aryan German
race. Boas, Ibid., p. 6.
[84] Ibid., p. 13.

on "subjective attitudes" and contingent "impressions," often governed by "irrational" impulses and "individual experiences."[85] Rather, a race "must be conceived as a biological unit, as a population derived from a common ancestry and by virtue of its descent endowed with definite biological characteristics." These "units," however, were highly variable and unstable, "subject to a multitude of outer influences . . . under the varying conditions of life."[86] Moreover, human history demonstrated that changes in race and culture were not linked. Recent research had proven that "the same race shows great differences in different environment[s], while . . . different races react alike in the same environment."[87] Citing the work of psychologists, Boas also argued that "intelligence, emotions, and personality are expressions of both innate characteristics and experience based on . . . social life."[88] Cognitive skills varied greatly depending on an individual's social location, Boas argued, noting dramatic differences in performance on intelligence tests depending on how long a test subject had lived in the United States, or whether the person grew up in an urban or rural setting.[89] Boas argued that there was a strong desire, "an emotional drive," to see "the life of a people in its whole setting," and to leap to "the unproved opinion" that "not only in individuals, not only in hereditary [family] lines, but in whole populations bodily build determines cultural personality." Just as there was no "unity of body build" in even the most homogenous population, so, too, there was no singular "cultural personality." Such an idea was not only mistaken, it was "a poetic and dangerous fiction."[90]

Conclusion

Franz Boas was a central figure in the development of American anthropology and American racial science. But while Boas has been widely known for his opposition to racism, he is much less widely known as an active racial scientist, someone who studied, published, and lectured on human racial variation until the 1940s.

Boas was indeed an opponent of "scientific" racism, as he vigorously opposed science he viewed as misguided, ill conceived, or that promoted

[85] Ibid., pp. 5–6.
[86] Ibid., p. 6.
[87] Ibid., p. 7.
[88] Ibid.
[89] Ibid., p. 8.
[90] Ibid., pp. 13–14.

discrimination. Throughout his career he policed his profession for shoddy practice, faulty logic, and inadequate theorizing. He objected to much of what passed for a science of human variation on three principal grounds. First, key questions about racial traits and types had been either woefully neglected or existing studies were so riddled with grave methodological problems that their conclusions were highly suspect. Worse, two critical theoretical problems bedeviled an effort to sort out human variation: investigators had no idea which physical traits had any evolutionary or hereditary significance, and little was understood about the relative influence of heredity and environment on "racial" characteristics, even if they knew with more confidence what those were. Despite these very serious theoretical and methodological problems, Boas believed that carefully crafted studies, employing sophisticated statistical analyses and the best genetic theory, might begin to make sense of the heterogeneity of humanity.

But for Boas, racial science was not merely a matter of measuring carefully selected bodies, sifting the numbers, and applying theories of heredity. Boas saw bodies and cultures locked into a pattern of change and adaptation in response to local conditions (including not only the physical environment, but also the historical and current social context). Boas was not an advocate of somatic essentialism or determinism. Bodies did not dictate cultural practices or individual character. Rather, Boas thought physical anthropology could describe differences among populations, offer clues to the history of migrations, and possibly illuminate the still murky question of whether the very evident variations in human appearance had any biological significance or not. Far from divorcing biology from culture, Boas strove to join the study of race, culture, and language in his effort to capture the diversity and plasticity of human appearance and behavior.

Franz Boas's physical anthropology theorized human diversity as shifting in response to the gamut of environmental factors and human interactions, but his work nonetheless retained remnants of fixed racial types characteristic of earlier racial science. Other anthropologists in the first decades of the twentieth century clung to those older, romantic typological traditions that imagined races as fixed natural kinds associated with essential physical and psychic qualities. Physical anthropology in the 1920s and 1930s struggled with how to conceptualize and investigate the complexities of human diversity, and grappled with how to situate racialized peoples and racial science for American audiences. At the Field

Museum of Natural History in Chicago, Henry Field imagined a typological vision of race and evolution that offered a quite different account of bodies and culture from the historicized, environmentally grounded vision of race Franz Boas constructed. In Chapter 3, I explore the tensions inherent in crafting an exhibition about race for the American public in the interwar period.

3

Order for a Disordered World

The Races of Mankind *at the Field Museum of Natural History*

Introduction: The Races of Mankind

Filling the entire Chauncey Keep Memorial Hall on the first floor of the Field Museum of Natural History, the *Races of Mankind* presented the American public with racial taxonomy on an unprecedented scale[1] (Figure 3.1). Meant as one of two marquee halls displaying the fruits of the field of physical anthropology, the *Races of Mankind* offered a depiction of race rooted in a nineteenth-century romantic typological tradition. Most of the exhibition space was given over to the display of nearly 100 life-size bronze sculptures of the "principal" human racial types, each created by artist Malvina Hoffman specifically for the Field Museum. Hoffman, a renowned sculptor with studios in New York and Paris, and a former pupil of Auguste Rodin, created the figures over a period of several years, in some cases traveling across the world to find her models. The exhibition, mounted in 1933 to coincide with the Century of

---

[1] Other scholarship examining the Field Museum's *Races of Mankind* hall includes: Linda Nochlin, "Malvina Hoffman: A Life in Sculpture," *Arts Magazine* (Nov. 1984), pp. 106–110; P. H. Decoteau, "Malvina Hoffman and the 'Races of Mankind,'" *Art Journal* (Fall 1989/Winter 1990), pp. 7–12; Jeff Rosen, "Of Monsters and Fossils: The Making of Racial Difference in Malvina Hoffman's Hall of the Races of Mankind," *History and Anthropology*, 2001, vol. 12, no. 2, pp. 101–158; Gregory Foster-Rice, "The Visuality of Race: 'The Old Americans,' 'The New Negro' and American Art, c. 1925" (PhD diss., Northwestern University, 2003); Linda Kim, "Malvina Hoffman's *Races of Mankind* and the Materiality of Race in Early Twentieth-Century Sculpture and Photography" (PhD diss., University of California, Berkeley, 2006); Marianne Kinkel, *Races of Mankind: The Sculptures of Malvina Hoffman* (Urbana, IL: University of Illinois Press, 2011); Rebecca Peabody, "Race and Literary Sculpture in Malvina Hoffman's 'Heads and Tales,'" *Getty Research Journal*, no. 5 (2013), pp. 119–132.

FIGURE 3.1. View of the *Races of Mankind* hall, Oceanic races on the left, African on the right. Courtesy of The Field Museum, #A78455.

Progress World's Fair in Chicago, was initially a huge success, attracting 3 million visitors in the first year alone,[2] and garnering wide coverage in the press.[3] By the time the exhibit was complete, the Field Museum had paid Hoffman $125,000, on top of expenses incurred sending her and her husband, Samuel Grimson, and two assistants throughout Asia for eight months in search of models.[4]

[2] Stanley Field to Malvina Hoffman, Oct. 30, 1933, Box 16, Field Stanley, President of Chicago Natural History Museum, Correspondence (c. 1932–1935), Malvina Hoffman Collection, 850042-1, Special Collections, The Getty Center for the History of Art and the Humanities, Los Angeles, California (hereafter MHC); *Annual Report of the Director to the Board of Trustees, 1933*, vol. X, no. 1, publication 328 (Chicago: Field Museum of Natural History, Jan. 1934), pp. 91–92.

[3] *Annual Report of the Director to the Board of Trustees, 1933*.

[4] Berthold Laufer, "The Projected Hall of the Races of Mankind (Chauncey Keep Memorial Hall)," *Field Museum News* (Dec. 1931) vol. 2, no. 12; Henry Field to Marshall Field, Feb. 22, 1930, Races of Mankind Correspondence, vol. 12, 1920–1950, Henry Field Papers, Museum Archives, The Field Museum (hereafter FM-HF); *Annual Report of the Director to the Board of Trustees, 1933*, pp. 15–16, 35.

From the outset, physical anthropologist Henry Field's plan for the *Races of Mankind* was typological. Given his training and long association with physical anthropologists Arthur Keith, Dudley Buxton, and Earnest Hooton, it is not surprising that Field conceived the *Races of Mankind* in the long-standing tradition of what Donna Haraway has called the "romantic typological approach" of nineteenth-century racial scientists such as Paul Broca, Paul Topinard, Joseph Deniker, and William Ripley.[5] The Field Museum exhibition was one of the last incarnations of the popular race galleries commonly found at late nineteenth and early twentieth century world's fairs in their "villages" filled with "indigenous" people performing their lives for visitors, and in dozens of racial atlases published on both sides of the Atlantic, designed to titillate middle-class armchair adventurers and "anthropologists" with illustrations of exotic peoples from around the world.[6] In touting the Field Museum exhibit as an opportunity for visitors to encounter a virtual collection of living representatives from around the world, as Field Museum Anthropology Curator Berthold Laufer did, the museum was wittingly or unwittingly proffering the same rationale for partaking in their spectacular conflation of race, nation, and primitivism as had barkers at the 1893 Columbian World's Fair Midway and authors of tomes such as *Living Races of Mankind: A Popular Illustrated Account of the Customs, Habits, Pursuits, Feasts, and Ceremonies of the Races of Mankind throughout the*

[5] Donna Haraway, "Remodeling the Human Way of Life: Sherwood Washburn and the New Physical Anthropology, 1950–1980," in George W. Stocking Jr., ed., *Bones, Bodies, Behavior: Essays on Biological Anthropology*, History of Anthropology, vol. 5 (Madison: University of Wisconsin Press, 1988), pp. 217–218. For an excellent discussion of nineteenth- and twentieth-century theories of race, including Broca, Topinard, Ripley, Deniker, and especially Franz Boas, see Stocking, ed. *Race, Culture and Evolution* (Chicago: University of Chicago Press, 1968). Also, other essays in Stocking, ed., *Bones, Bodies, Behavior*; Elazar Barkan, *The Retreat of Scientific Racism: Changing Concepts of Race in Britain and United States between the World Wars* (Cambridge: Cambridge University Press, 1992); Stocking, "Ideas and Institutions in American Anthropology: Toward a History of the Interwar Period," in Stocking, ed., *Selected Papers from the American Anthropologist, 1921–1945* (Washington, DC: American Anthropological Association, 1976), pp. 1–44; and Frank Spencer, ed., *A History of American Physical Anthropology, 1930–1980* (New York: Academic Press, 1982).

[6] Andrew Zimmerman argues that the development of German racial anthropology was crucially shaped by its participation in the public display of supposedly "natural peoples" whose actual complexities challenged and undermined anthropologists' rigid categories and neat primitive/civilized dichotomies. *Anthropology & Antihumanism in Imperial Germany* (Chicago and London: University of Chicago Press, 2001).

World.[7] Although Henry Field acknowledged that pure races no longer existed, as did virtually all anthropologists in the early twentieth century, he nonetheless believed that various groups approached pure types, and moreover, that within definable human racial populations, individuals could be found who typified the set of racial characters that defined their group. Henry Field envisioned his gallery of racial types as an exhibit that would "be the most popular, not only in the museum, but on the continent," an exhibition that would, with its companion hall devoted to European prehistory, tell nothing less than the "Story of Man."[8]

Once complete, the *Races of Mankind* might more properly be described as a typological pastiche, rather than the thoroughly romantic typological exercise initially envisioned by Henry Field. As a pastiche, the exhibit reflected an uneasy combination of older and newer approaches to the problem of racial types, and the ongoing contest among anthropologists over the most basic disciplinary questions: What is the nature of human variation and "race"? How is race best studied? What is the import of our findings? The exhibit promoted the older, "romantic" typological notions championed by Henry Field and Arthur Keith that reified cultural,

---

[7] H[enry] N[eville] Hutchinson, *Living Races of Mankind: A Popular Illustrated Account of the Customs, Habits, Pursuits, Feasts, and Ceremonies of the Races of Mankind throughout the World* (London: Hutchinson, 1902). The most (in)famous of these volumes is probably Josiah Clark Nott and George R. Gliddon, *Types of Mankind; Or, Ethnological Researches, Based upon the Ancient Monuments, Paintings, Sculptures, and Crania of Races, and upon Their Natural, Geographical, Philosophical, and Biblical History. Illustrated by Selections from the Unedited Papers of Samuel George Morton, and by Additional Contributions from L. Agassiz, W. Usher and H. S. Patterson* (Philadelphia: Lippincott, Grambo, 1854). Other popular titles include: Cephas Broadluck, *Races of Mankind; With Travels in Grubland* (Cincinnati: Longley, 1856); Robert Brown, *The Races of Mankind, Being a Popular Description of the Characteristics, Manners and Customs of the Principal Varieties of the Human Family* (London, New York (etc.): Cassell, Peter & Galpin, 1873–1876); Walter Graham Blackie, *The Comprehensive Atlas & Geography of the World; Comprising an Extensive Series of Maps, a Description, Physical and Political, of All the Countries of the Earth; A Pronouncing Vocabulary of Geographical Names, and a Copious Index of Geographical Positions: Also Numerous Illustrations Printed in the Text, and a Series of Coloured Engravings Representing the Principal Races of Mankind* (London: Blackie & Son, 1882); H[einrich] Leutemann, A[dolf] Kirchoff, *Graphic Pictures of Native Life in Distant Lands: Illustrating the Typical Races of Mankind* (London: George Philip & Son, 1988); H[enry] J[ohn] Fleure, *The Races of Mankind* (London: E. Benn, and Garden City, NY: Doubleday, Doran: 1928).

[8] Henry Field, *The Track of Man: Adventures of an Anthropologist* (New York: Doubleday & Company, Inc., 1953), p. 132.

especially national or ethnic, identities and physical variation as "race."
But Berthold Laufer also shaped the final exhibition by his attempts to
minimize its ethnographic excesses, to emphasize strictly physical manifestations of human variation, put human behavior and culture into social
and historical context, and inject a message of human unity amidst an
otherwise relentless focus on difference and hierarchy by insisting that
an imposing "Unity of Man" sculpture be placed at the center of the
hall.[9]

Despite methodological urgings of biometricians such as Earnest
Hooton, who urged Field to abandon his predetermined typology in
favor of a more statistically valid one, and the efforts of Berthold Laufer
to restrain the exhibit's stereotypic excesses and inject a message of
unity into an otherwise hierarchical presentation, Henry Field's *Races
of Mankind* remained an overwhelmingly traditional typological presentation. It reflected nineteenth-century racial logic in its insistence on
organizing human variation into dozens of narrowly defined racial types,
each rooted in one or more of three foundational racial "stocks": Negro,
White, and Mongoloid. Ultimately, Malvina Hoffman's sculptures were
presented in the scientific arena of a natural history museum in a way
that, rather than problematizing race, presented race as a natural category, purportedly objectively discovered and elucidated by the science
of physical anthropology. The exhibit explicitly, through texts referring
to the sculptures, and implicitly in its conception and presence in a natural history museum, promulgated the idea that races were bounded,
natural entities that could be objectively and unambiguously recorded
and understood. The exhibit suggested that, armed with incontrovertible
facts about the races, anyone could go out into the world, distinguish one
race from another, and select an individual (who necessarily existed) who
embodied the typical characters of that race.

Of Fields, Fairs, and Museums

The Field Museum of Natural History's association with World's Fairs
and the Field family reached all the way back to its origins in the collections gathered for the 1893 Columbian Exposition. In a city founded in

---

9 Henry Field, "The Story of the Hall of Physical Anthropology," Mar. 12, 1930, Folder:
   "Found Original in V. II 1916–1930, plans, proposals, notes," Malvina Hoffman, Races
   of Mankind, Department of Anthropology Archives, The Field Museum (hereafter FM-
   ROM).

commerce, agriculture, and transportation, a city associated with scrappy trade and unpolished immigrants, Chicago's elites in the late nineteenth century looked for ways to raise the city's cultural status and "elevate the tone of the public" that had reached 1 million souls.[10] To this end, the wealthy elite organized and promoted not only the Columbian Exposition, but also a set of prominent high-culture institutions, including the Art Institute, Symphony Orchestra, University of Chicago, and Newberry and Crerar Libraries. Organized to retain and display the scientific materials collected for the 1893 Fair, the Columbian Museum, as it was first named, joined these august institutions as a venue for the city's burgeoning population that had doubled in every decade since 1840.[11] By 1933, the Field Museum of Natural History was one of Chicago's signal cultural institutions, a site of scientific study and civilized leisure. It marked Chicago as one of the nation's and the world's leading cities.

Chicago's natural history museum was named for its most generous founder, Marshall Field, the department store magnate, and Henry Field's great-uncle. Marshall Field was one of the wealthy elite, one of the "Protestant, native-born sons of predominantly British descent" who viewed the cultural development of Chicago as their responsibility and prerogative.[12] The museum was founded as the Columbian Museum of Chicago in 1893, reflecting its origins in the collections of the World's Columbian Exposition. In 1905, it was first named the Field Museum of Natural History to honor Marshall Field's critical, and enormous, financial support of the museum. The new name also more explicitly positioned the museum as a science museum. In 1943, the fiftieth anniversary of its founding, the museum was renamed the Chicago Museum of Natural History. After long-serving museum President Stanley Field's death in

---

[10] Stephen Becker, *Marshall Field III: A Biography* (New York: Simon and Schuster, 1964), p. 43.

[11] Helen Lefkowitz Horowitz, *Culture and the City: Cultural Philanthropy in Chicago from the 1880s to 1917* (Chicago and London: University of Chicago Press, 1989), pp. 29, 41. Horowitz notes that the 1870s, the decade following the Great Fire, were an exception to this otherwise steady growth.

There is no single, definitive historical account of the Field Museum. For a fuller history of the Field Museum, particularly regarding anthropology, see Stephen E. Nash and Gary M. Feinman, eds., *Curators, Collections and Contexts: Anthropology at the Field Museum, 1893–2002*, Fieldiana: Anthropology, New Series, no. 36 (Chicago: Field Museum of Natural History, 2003); Warren Haskin, *Anthropology at the Field Museum, 1894–2000*, manuscript, Museum Archives, The Field Museum, Chicago; Steven Conn, *Museums and American Intellectual Life, 1876–1926* (Chicago: University of Chicago Press, 1998).

[12] Horowitz, p. 45.

1964, the museum returned to its original name, the Field Museum of Natural History, in recognition of Field's devoted service to the museum and the integral role of the Field family in its creation and continuation.[13]

Marshall Field was one of Chicago's most generous philanthropists. He was a founding member of the Chicago Academy of Fine Arts, which established the Art Institute, he donated funds for the Chicago Symphony, and he was the largest single investor in 1893 World's Fair.[14] Field was sometimes a reluctant or entrepreneurial philanthropist. He rejected a request to support the Newberry and Crerar Libraries and required matching funds from the University of Chicago (as did John D. Rockefeller, its principal benefactor). Ultimately, Field donated $361,000 to help reestablish the foundering university and donated or sold at low rates some of his land holdings, including the Midway Plaisance site of the current university and a plot used for athletics called the "Marshall Field."[15]

Field also initially declined to support creation of a natural history museum following the Columbian Exposition.[16] Edward Ayer, a Chicago lumber magnate with a passion for Indian artifacts, persuaded Field to change his mind. Ayer's devotion to the notion of a natural history museum and his persistence convinced Field to undertake what would become his largest philanthropic effort. Again, Field demanded matching funds, initially donating $1 million, if another $500,000 could be raised. Before his death in 1906, Field donated another $1 million; at his death, he left the Museum a further $8 million.[17]

By the time Henry Field was hired in 1926, his association with the museum was practically a foregone conclusion (Figure 3.2). Not only was his great-uncle Marshall Field the namesake and founder of the museum, but his second cousin Marshall Field III (whom Henry called "Marshie"), was a museum trustee who frequently funded expeditions and exhibitions.

---

[13] "Marshall Field, 1893–1956," *Chicago Natural History Museum Bulletin*, vol. 26, no. 12 (Dec. 1956), p. 2.

[14] Becker, pp. 43–45.

[15] Ibid., pp. 44–46.

[16] Horowitz, p. 46.

[17] Becker, p. 46; Edward Everett Ayer, "In Re: Founding of the Field Museum," in Nash and Feinman, eds. *Curators, Collections, and Contexts*, pp. 49–52; Thomas Wakefield Goodspeed, "Marshall Field," *The University of Chicago Biographical Sketches*, vol. 1 (Chicago: University of Chicago Press, 1922), pp. 1–34. Marshall Field was thought to be the fifth wealthiest American at his death, with an estimated estate of $120 million. David R. Wilcox, "Creating Field Anthropology: Why Remembering Matters," in Nash and Feinman, *Curators, Collections, and Contexts*, p. 35.

FIGURE 3.2. University of California African Expedition. Henry Field, measuring a worker in the Southwest Sinai, ca. 1947. The original caption for this photo in Henry Field's autobiography read: "Bedouin being measured by author in southwestern Sinai during University of California African Expedition. The Bedouins of Sinai belong to the basic Mediterranean stock. They were friendly but shy about being examined or photographed." *Source*: *The Track of Man: Adventures of an Anthropologist*, Greenwood Press, Publishers (New York, reprinted with permission of Doubleday & Co., [1953] 1969). Courtesy of Juliana Lathrop Field.

Marshall Field III would eventually prove a critical figure in the development of Henry's anthropology halls, through his connections to sculptor
Malvina Hoffman and by providing at least $100,000 for the *Races
of Mankind*.[18] Another second cousin, Stanley, was museum president,
and a notable benefactor in his own right. In the course of his fifty-
three years as president, Stanley Field gave a total of some $2 million to
the museum, frequently making up deficits in the operating budget and
donating money for various projects.[19] Henry's arrival as the museum
anticipated the opening of the next World's Fair presented a striking historical symmetry. Once again, members of the elite Field family would
work at the confluence of an international exposition and a landmark
civic institution to elevate the standing of the city and their museum in a
way that would reach millions of Americans.

"Whither Now Mankind?"[20] Order for a Disordered World

Whatever Henry Field's limitations as an anthropologist, he, more than
Berthold Laufer, or arguably anyone else in museum anthropology in
America at that time, had a knack for innovation and synthesis in museum
display that resulted in two of the most widely praised and copied exhibitions in anthropology for several decades: the *Hall of the of the Stone Age
of the Old World*, featuring Frederick Blaschke's Neanderthals, and the
Races of Mankind hall. Because city leaders founded the Field Museum,
anthropologists and museum officers justified displaying the fruits of science as not only the dissemination of accumulated knowledge about the
world, but as a civic duty. The Field Museum was not only a platform for
anthropologists to propound a vision of their discipline and the world it

---

[18] See Footnote 53 for further details regarding benefactors' funding of the *Races of Mankind*.

[19] Stanley Field was independently wealthy; in 1917, he earned more than $800,000 ("Sues to Recover Refund," *New York Times*, Jun. 21, 1931, p. 29). He started his career in 1893 as an employee of Marshall Field & Company, serving as a vice president from 1906 to 1918, and remained associated with the firm throughout his life. His work in the banking industry began in 1913; he briefly served as chairman of the Continental Illinois National Bank and Trust, one of several firms with which he was associated ("Stanley Field Heads Continental Illinois, Nation's Largest Bank Outside New York," *New York Times*, Jan. 14, 1933, p. 19). He also served on the Boards of Commonwealth Edison Company, regional railroads, Children's Memorial Hospital, and the Brookfield Zoo. See also "University of Chicago Honors Stanley Field," *Field Museum News*, vol. 2, no. 1 (Jan. 1931), p. 2; and "Stanley Field, 1875–1964," *Chicago Natural History Museum Bulletin*, vol. 35, no. 12 (Dec. 1964), pp. 2–3, 8.

[20] Field, *Track of Man*, p. 226.

described, but also a venue through which museum trustees and officers could participate in a broader cultural discourse about the nature and direction of society.[21]

The basic plan for the *Races of Mankind* hall that opened in 1933 was largely conceived by Henry Field, and first presented to Berthold Laufer in 1927. Originally, Field and Laufer planned to use the same medium in the *Races of Mankind* that was ultimately used in the *Stone Age* hall: newly developed flesh-colored wax and plaster figures decked out with real human hair and glass eyes – what Field called "true realism."[22] The eventual switch to bronze is surprising given its obvious limitations in conveying the bodily characteristics once thought crucial to discriminating among supposed racial types. But bronze, despite its deficiencies, actually suited Field's racial and exhibitionary scheme remarkably well. Malvina Hoffman preferred bronze for aesthetic and artistic reasons; its primary appeal to the anthropologists lay elsewhere. Certainly, the elegance of the medium, as well as the spectacle likely to be created by the exhibition of bronze sculptures on an unprecedented scale, motivated the museum.[23] But even more critical than the notice 101 bronze sculptures would bring to the museum was the metaphorical and epistemological significance of the medium. Bronze symbolized permanence, stability, and the static, embodied nature of race, as Henry Field, Stanley Field, and to some degree, Berthold Laufer, understood it, while simultaneously engaging visitors in an imaginative act of racial reinscription.

Field, Laufer, and museum literature repeatedly stressed the value of Hoffman's statues as a permanent representation of the world's races, making them forever available for study, and in particular as a way to preserve vanishing primitive races. Indeed, Laufer claimed that owing to

[21] On the civic and sociopolitical roles of the modern museum, see Tony Bennett, *The Birth of the Museum: History, Theory, Politics* (London and New York: Routledge, 1995).

[22] Field, *Track of Man*, p. 197. Shortly after Hoffman was hired, Henry Field wrote her at her studio in Paris. "I am sure," he wrote, "that you will agree with me that the material to be employed must be such as to render possible variations in skin color and the application of hair." She did not. Henry Field to Malvina Hoffman, Feb. 5, 1930, Races of Mankind Correspondence, Henry Field, vol. 12 1920–1950, FM-HF.

[23] Herbert Ward had done bronzes of a few African natives for the Smithsonian Museum, which were on Henry Field's mind when he sought an artist for the *Races of Mankind* exhibit, but no museum had tried to present an entire physical anthropological exhibit including the number of figures in the Field Museum exhibit, nor had it been attempted in bronze. Mary Jo Arnoldi, "Herbert Ward's Ethnographic Sculptures of Africans" in *Exhibiting Dilemmas, Issues of Representation at the Smithsonian*, ed. Amy Henderson and Adrienne L. Kaeppler (Washington, DC: Smithsonian Institution Press, 1997), pp. 70–91.

"the rapid extinction of primitive man due to white man's expansion over the globe many a vanishing race will continue to live only in the sculptures displayed in this hall."[24] Arthur Keith called the exhibit "a permanent abode... for representative members of the human family."[25] According to Henry Field, in the *Races of Mankind* exhibit, "modern man met with his own image... portrayed in imperishable bronze."[26] Laufer and Field believed their exhibit was more than a compendium of current anthropological theory, a neutral exposition on race. Rather, they intended it to address social instability by emphasizing the rationally ordered and determined nature of life, in which vanishing primitive races could be preserved and better understood, despite their inevitable demise at the hands of more civilized and progressive human races. Under the heading of "Social Anthropology," in the technical portion of the exhibit, displays were initially devoted to delineating racial problems in the United States, immigration issues, and antimiscegenation laws, promoting the efficacious application of racial science to national and world problems. In contrast to a threatening perception of society imperiled by decay and degeneration, the Field Museum offered a stable and familiar world order, frozen in bronze, at once humanitarian in its theme of unity and its preservation of endangered "primitives," and reassuring in its reinforcement of the natural place of Europeans and Americans at the top of the racial heap.

Hoffman's bronze sculptures, the *Races of Mankind* exhibit as a whole, indeed the entire physical anthropological project of the Field Museum, were intimately tied to the reinforcement of order in a world that seemed increasingly and frighteningly chaotic. The pervasive faith in American scientific, commercial, and social progress characteristic of the late nineteenth and early twentieth century had been severely shaken in the 1910s and 1920s. The horror and scale of World War I fundamentally undermined belief that civilization was marching ever forward. The subsequent failure of the League of Nations did nothing to improve Americans' sense of security in a contentious world. At home, fear for the fate of society took ugly forms: nativism, red scares, race riots. Widespread union strikes, particularly in the coal and steel industries, convulsed communities and heightened growing fear of communism. The years between the Great War and the Great Depression were marked by virulent

---

[24] Laufer, "Hall of the Races of Mankind," 1931, p. 3.
[25] Henry Field, *The Races of Mankind*, Popular Series, Anthropology Leaflet 30, Fourth Edition (Chicago: Field Museum of Natural History, 1942), p. 11.
[26] Field, *Track of Man*, p. 226.

antiradicalism and efforts to hunt down radicals thought to endanger civil society, particularly communists and anarchists. Fear of moral and intellectual decline spurred eugenic theories and culminated in the nativist anti-immigration National Origins Act of 1924, whose restrictions were aimed at the influx of southern and eastern Europeans, as well as Latin America. In the South, the Ku Klux Klan hit its peak membership of 5 million in 1925. In the North, the rapid influx of African Americans from the South during the Great Migration dramatically changed the complexion of Chicago and other cities and sparked social tensions; in Chicago, the black population more than doubled between 1920 and 1940.[27] The Prohibition Amendment of 1919 was a boon to organized crime; in Chicago, Al Capone created a $60 million empire in the 1920s through the sale of liquor and drugs, gambling, and prostitution. By 1929, when Field first contacted Hoffman, the world economy was falling apart; in 1933, when the exhibit opened, the country was entrenched in the Depression.[28]

One response of American science and business to domestic and foreign threats was to reinvigorate Americans' faith in progress, hence the theme of the 1933 World's Fair – the Century of Progress. At the fair, "scientific idealism" – what Robert Rydell has called "a deification of the scientific method and glorification of anticipated scientific solutions to social problems" – was promoted to 32 million visitors in Chicago through a wide range of industrial and scientific exhibits, including a fountain of a man, woman, and robot entitled *Science Advancing Mankind*.[29] Visitors to the

[27] See "The Second Ghetto and the Dynamics of Neighborhood Change," in Arnold R. Hirsch, *Making the Second Ghetto: Race and Housing in Chicago, 1940–1960* (Chicago: University of Chicago, 1998), pp. 1–39, for a brief account of the development of the "first" and "second" ghettos in Chicago. See also Allan H. Spear, *Black Chicago: The Making of a Negro Ghetto, 1890–1920* (Chicago: University of Chicago Press, 1967); St. Clair Drake and Horace R. Cayton, *Black Metropolis: A Study of Negro Life in a Northern City* (Chicago: University of Chicago Press, 1993).

[28] See inter alia, Isabel Leighton, *The Aspirin Age, 1919–1941* (New York: Simon and Schuster, 1949); William Tuttle, *Race Riot: Chicago in the Red Summer of 1919* (New York: Athenum, 1970); Matthew Frye Jacobson, *Whiteness of a Different Color: European Immigrants and the Alchemy of Race* (Cambridge, MA: Harvard University Press, 1998); David M. Kennedy, *Freedom from Fear: The American People in Depression and War, 1929–1945* (Oxford: Oxford University Press, 1999); Gary Gerstle, *American Crucible: Race and Nation in the Twentieth Century* (Princeton: Princeton University Press, 2001); David Paul McDaniel, "A Century of Progress?: Cultural Change and the Rise of Modern Chicago, 1893–1933," 2 vols. (PhD diss., University of Wisconsin-Madison, 1999).

[29] Robert Rydell, "The Fan Dance of Science, America's World's Fairs in the Great Depression," *Isis* 76 (1985), pp. 529, 533.

fair could encounter Indian "villages," created as ethnological exhibits under the guidance of University of Chicago anthropologist Fay Cooper-Cole, as well as a "Darkest Africa" concession on the midway that offered fairgoers evocative depictions of race, whether or not they were intended as such. Physical anthropology and the science of biometrics was represented at the fair by the Harvard University Anthropology Lab, a joint venture between Harvard, the Peabody Museum, Smithsonian Institution, and corporations that donated equipment, housed in the Hall of the Social Sciences.[30] Anthropologist C. W. Dupertuis and his staff measured 6,000 fairgoers anthropometrically, yielding what Earnest Hooton called "the most valuable sample of the native American population of moderate means heretofore gathered." By "native American" Hooton did not mean the indigenous peoples of the United States, but rather used common anthropological parlance for white Americans whose ancestry could be traced back in the United States for at least two generations. The concept of "native" or "old" American distinguished this population from recent immigrants, as well as from indigenous peoples, African Americans, and other non-white populations, regardless of their ancestry. Hooton did not indicate whether Americans of color, particularly African Americans, given Chicago's large black population, were among those measured. But based on Robert Rydell's discussion of African American participation in the fair, it seems unlikely that a large proportion, if any, of the laboratory's subjects were non-white.[31]

Henry Field and Berthold Laufer planned the opening of their anthropological exhibits to coincide with the Century of Progress Fair, hoping to attract some of the fair's enormous attendance, and more crucially, to participate in the reinforcement and preservation of American social order by demonstrating to museumgoers that man had traveled an illustrious and progressive path. The 1933 fair commemorated the 100-year anniversary of the incorporation of Chicago on the swampy banks of Lake Michigan. It was also the 40-year anniversary of the 1893 Columbian Exposition in Chicago, itself a popular commemoration of the 400-year anniversary of Columbus's arrival in the New World. According to exposition President Rufus C. Dawes, the 1933 fair was "the first attempt in an enterprise of this nature to explain to a world-wide public the accomplishments of

---

[30] Robert Rydell, *World of Fairs: The Century of Progress Expositions* (Chicago: University of Chicago Press, 1993), pp. 82–84, 104–105, 165–171.

[31] Aleš Hrdlička, *The Old Americans* (Baltimore: Williams and Wilkins, 1925); Earnest A. Hooton, "Development and Correlation of Research in Physical Anthropology at Harvard University," *American Philosophical Society, Philadelphia. Proceedings*, vol. LXXV (Philadelphia, 1935), pp. 510–511; Rydell, *World of Fairs*, pp. 104, 166.

those explorers of nature's laws who, working in the field of pure science, have opened up new realms for man's enjoyment." In the summer of 1933, as the Depression ground on, Americans flocked to Chicago for a bit of escape and a dose of hope, presented in a story about the steady progress of American society, and especially American science. If capitalism was proving to be an uncertain foundation for the future, human ingenuity and scientific know-how had offered a "century of progress," and promised more.[32]

The Field Museum's "Story of Man"

In 1927, when Berthold Laufer asked Henry Field, as a new assistant curator, what he would like to do at the museum, Field outlined his "dream" for two physical anthropological exhibits telling the "Story of Man" in adjoining halls, a plan that superceded Laufer's own longstanding efforts. One of Field's exhibits became the *Hall of the Stone Age of the Old World*, depicting prehistoric man in Europe from 250,000

---

[32] "Chicago Quickens World's Fair Work," *New York Times* (Apr. 13, 1931), p. 5. For other contemporary accounts of the Century of Progress Fair, see R. L. Duffus, "The Fair: A World of Tomorrow," *New York Times* (May 28, 1933), p. SM1, 3 pgs.; William A. Robson, "Chicago's 'Century of Progress,'" *The New Statesman and Nation* (Jul. 29, 1933), pp. 132–133. Robert Rydell has written extensively on the world's fairs in the United States, and in particular on the scientism displayed at the Century of Progress Fair in Chicago. For the latter, see Rydell, *World of Fairs*. Also, his first book, *All the World's a Fair, Visions of Empire at American International Expositions, 1876–1916* (Chicago and London: The University of Chicago Press, 1984). There is a large literature, following Rydell's watershed work, on the world's fairs and expositions. Notable in their attention to anthropology and race are: Nancy J. Parezo and Don D. Fowler, *Anthropology Goes to the Fair: The 1904 Louisiana Purchase Exposition* (Lincoln: Nebraska University Press, 2007); Matthew F. Bokovoy, *The San Diego World's Fairs and Southwestern Memory, 1880–1940* (Albuquerque: University of New Mexico Press, 2005); Christopher Robert Reed, "*All the World is Here!": The Black Presence at White City* (Bloomington: Indiana University Press, 2000); Patricia P. Morton, *Hybrid Modernities: Architecture and Representation at the 1931 Colonial Exposition* (Cambridge, MA: MIT Press, 2000); Burton Benedict, *The Anthropology of World's Fairs: San Francisco's Panama Pacific International Exposition of 1915* (Berkeley, CA: Lowie Museum of Anthropology; London and Berkeley: Scolar Press, 1983); Alison Griffiths, *Wondrous Difference: Cinema, Anthropology & Turn-of-the-Century Visual Culture* (New York: Columbia University Press, 2002); Penelope Harvey, *Hybrids of Modernity: Anthropology, the Nation State and the Universal Exhibition* (New York: Routledge, 1996); Pascal Blanchard, et al., *Human Zoos: Science and Spectacle in the Age of Colonial Empires* (Liverpool University Press, 2008); Theda Purdue, *Race and the Atlanta Cotton States Exposition of 1895* (Athens: University of Georgia Press, 2010); Sadiah Qureshi, *Peoples on Parade: Exhibitions, Empire, and Anthropology in Nineteenth-Century Britain* (Chicago: University of Chicago Press, 2011).

to 6,000 BC, and the other became the *Hall of the Races of Mankind*. These two halls would portray, according to Field, the "complex history of mankind . . . exactly as it developed, link by link."[33]

The ostensible goal of these exhibits was educational and humanitarian, implicitly exhorting museumgoers to accept the reassuring vision of a progressive, unified path of humanity. In his preface to the Field Museum *Races of Mankind* leaflet, Laufer articulated this hope:

Anthropology is essentially a science of human understanding and conciliation based on profound human sympathies that extend alike to all Races of Mankind. . . . If the visitors to the hall will receive the impression that race prejudice is merely the outcome of ignorance and will leave it with their sympathy for mankind deepened and strengthened and with their interest in the study of mankind stimulated and intensified, our efforts will not have been futile and will have fulfilled their purpose.[34]

Laufer also carefully tried to limit the scope of the exhibit, criticizing "the general confusion of the terms race, nationality, language, and culture" in the study of race, reminding visitors that race was an exclusively biological concept. Foreshadowing distinctions Ruth Benedict would make several years later, Laufer stressed that race "means breed and refers to the physical traits acquired by heredity" and not "the total complex of habits and thoughts acquired from the group to which we belong . . . the social heritage called culture." Laufer applied this distinction between biological and cultural origins to the behavior of nations as well as individuals. It was cultural traditions, he argued, not the biological origins of its members that determined national behavior.[35]

Henry Field was less modest in his goals. According to Field, visitors to the *Hall of the Stone Age* viewed an iconography of human progress. A visitor to the *Stone Age* hall could, "within the space of a half hour, walking past the eight dramatic and colorful dioramas . . . read in true-to-life chapters the past quarter of a million years of Man's history."[36] These chapters included: weapons and fire, "the first step toward supremacy over the beasts"; two men, a woman and child as the "first family"; Cro-Magnon, the "first artist"; and the development of art, agriculture, domesticated animals, pottery, tools, homes, and ritual burial.[37] The

[33] Ibid., p. 131.
[34] Field, *The Races of Mankind*, 1942, p. 8.
[35] Ibid.
[36] Field, *Track of Man*, p. 211.
[37] Ibid., pp. 212–214. This narrative is reminiscent of nineteenth-century social evolutionary schemes promulgated by anthropologists such as Lewis Henry Morgan in *Ancient Society; or Researches in the Lines of Human Progress from Savagery through Barbarism*

allegory of prehistoric man provided visitors not just facts about the development of humanity but a moral for modern existence:

Prehistoric man shows the fluctuating history of man. He shows us that there is hope – there is always hope – for the future of mankind.... Man's struggles and victories began several million years ago – and those struggles were against greater odds, those victories more inspiring, than any man has known since the time that records were first inscribed on pictographic tablets.[38]

Lurking in Field's morality tale and behind the construction of his exhibits was a hierarchical vision of race that positioned European races and their supposed ancestors as the key actors in human history. Understanding the relationship of the two halls as a single story of man's evolution leaves non-European humans in a decidedly marginal position, one physically instantiated in the design of the *Races of Mankind*. If man's story amounted to the cultural and racial evolution from Neanderthal to Nordic, by implication the rest of the world's races either had failed to reach the peak or were degenerate branches. This implication existed in tension with the other main thrust of the *Races of Mankind* exhibit – the fundamental unity of mankind. In contrast to Laufer's hopeful vision, Field's views echoed those of his tutor and exhibit advisor Arthur Keith, who propounded a divisive view of race as the source and result of evolutionary competition. In Keith's view, vanishing races were not merely a tragic result of cultural conflict; they were evidence of the evolutionary struggle among races for survival of the fittest. This tension between a crudely ranked typology rooted in morphology and heredity, and a vision of human unity underlying diversity based in environmental and cultural differences, reflected conceptual tensions inherent in a typological, evolutionary system of classification, global differences among anthropologists on these questions, as well as local differences between Laufer, who was primarily an ethnologist with Boasian sympathies, and Henry Field, trained as a physical anthropologist by Arthur Keith, L. H. Dudley Buxton, and Earnest Hooton.

Henry Field and the Riddle of Man

Henry Field's ideas about anthropology and race had first developed at Oxford University, under the tutelage of Henry Balfour, Robert Ranulph

---

to Civilization (New York: Henry Holt, 1877); and Edward B. Tylor, in *Researches into the Early History of Mankind and the Development of Civilization* (London: John Murray, 1865).

[38] Field, *Track of Man*, pp. 215–216.

Marett, and Leonard Halford Dudley Buxton. Of his tutors at Oxford, Buxton had the most influence on Field's physical anthropology. Buxton was a member of the Department of Human Anatomy and accompanied Field on his first field trip to Kish in Iraq (Figure 3.2). Field acquired a deep fondness for Buxton in his years at Oxford and on their expeditions and turned to him repeatedly over the years for advice about the *Races of Mankind* hall, his fieldwork, and questions of physical anthropology, until Buxton's death in 1939. Field attributed his enthusiasm for physical anthropology to Buxton and found him an excellent teacher.[39] In his principal work, *Peoples of Asia*, Buxton demonstrated a thorough knowledge of the field, both theories propounded by past authorities as well as contemporary ones, and the problems of measurement and description faced by physical anthropologists, as well as a commitment to a typological view of race. Buxton was especially keen on statistical work, arguing that the unsophisticated mathematical work of early anthropologists contributed little to a true understanding of racial populations. He strongly advocated the use of moderately sophisticated statistical computations, such as standard deviations, measures of dispersion, probable error, correlations, and coefficients of variation. He considered it possible to construct something such as a "Coefficient of Racial Likeness" that would give a numerical value to the combined characters of a race and offer a definitive method of distinguishing racial classes.[40] Buxton was actively involved in efforts within physical anthropology to standardize anthropometric methods, a persistent complaint among practitioners who bemoaned the profusion of studies that could not be reliably compared because of wide variation in methods of measurement, observation, and analysis, including such basic issues as which characteristics to study, which instruments to use and how to use them, and which indices were most revealing of racial associations.[41]

Buxton's faith in biometrics was one enthusiasm that apparently was not passed along to Field, although Field did retain an enthusiasm for collecting data of all kinds and taking measurements of fossil and

---

[39] Field, *Track of Man*, pp. 36–40; and Gabrielle H. Lyon, "The Forgotten Files of a Marginal Man: Henry Field, Anthropology and Franklin D. Roosevelt's 'M' Project for Migration and Settlement," (MA thesis, University of Chicago, 1994).

[40] L. H. Dudley Buxton, *The Peoples of Asia* (New York: Alfred Knopf, 1925), pp. 7–31.

[41] G. M. Morant, M. L. Tildesley, and L. H. Dudley Buxton, "Standardization of the Technique of Physical Anthropology," *Man*, vol. 32, no. 193, 1932; L. H. Dudley Buxton, "The Essential Craniological Technique," *Journal of the Royal Anthropological Institute of Great Britain and Ireland*, vol. 63 (Jan.–Jun. 1933), pp. 19–47.

living peoples on his Mesopotamian expeditions. Field did go so far as to submit his data for analysis in Earnest Hooton's Statistical Laboratory, a move that necessitated reconfiguring his material to conform to Hooton's analytical categories. In the same way, Field also manipulated his data to conform to a scheme devised by Arthur Keith, which required creating indices and categories of analysis that had not informed collection of the data. But, in marked contrast to Harry Shapiro, who also worked closely with Hooton, Field seems to have had no ideas about statistical analysis of his own. He relied very heavily on the advice of Hooton, Keith, and Buxton for guidance in how to manipulate and characterize the voluminous data he eagerly collected.[42] His career was marked by many descriptive publications of data, often collected in somewhat haphazard fashion by an array of associates with questionable anthropological credentials, but little in the way of original synthetic or analytical work. Instead, Field's true gift seems to have been imagining and producing compelling visual human narratives – his Story of Man. Although he could not take sole credit for their success, Henry Field was instrumental in conceiving and seeing to completion two of the most striking, popular, and culturally resonant exhibitions mounted by any natural history museum in modern America.

Field's fixation on the "story of man" broadly conceived in space and time, and in the racial story of man in particular, predated his work for the Field Museum and guided his research program before, during, and after his tenure at the museum.[43] Field's enthusiasm for the broad problem of the origins, migrations, culture, and physical changes in the inhabitants of Kish was born of his work with Buxton and his discussions of the "riddle of ancient man in Mesopotamia," a problem he pursued for three decades.[44] In the 1920s, anthropologists regarded Kish as "the seat of the

[42] Henry Field, *The Anthropology of Iraq, The Upper Euphrates,* Anthropological Series, Field Museum of Natural History, vol. 30, part I, no. 1, May 31, 1940, Publication 469, pp. 8, 10, 32, passim.

[43] Field resigned from the museum in October 1941, ostensibly to begin work for the federal government on the wartime "M" Project. Field glides over his resignation in his autobiography, *Track of Man* – it is not mentioned. Field continued to work in anthropology after leaving the Field Museum, making expeditions to Russia, India, the Middle East, and Arab regions, especially Iran and Iraq. Much of his work was privately published or published under the rubric of Field Research Projects. During WWII, he worked for the federal government on the "M" Project, a study of the problems of refugees, migration, and resettlement. *Track of Man,* pp. 321–380; *Field Museum News,* Nov. 1941, p. 6. On Field's career, particularly his years on the "M" Project, see Lyon, "The Forgotten Files of a Marginal Man.

[44] Field, *Track of Man,* pp. 83–85; and Lyon, "Forgotten Files."

world's earliest civilization," and as such it was the site of numerous joint expeditions between the Field Museum and Oxford University.[45] In 1925, while Field was still a student at Oxford University, his uncle, Barbour Lathrop, gave him $1,000, with which he and Buxton traveled to Kish, where they joined the Joint Expedition as "volunteer physical anthropologists." In 1927, after Field had been hired as an assistant curator at the Field Museum but before work on the physical anthropology halls had begun, he joined the expedition in an official capacity as a physical anthropologist. In 1934, after the *Races of Mankind* opened, Marshall Field funded another excursion, this time with Henry in charge of the expedition.[46]

Although much of the work at Kish primarily addressed cultural and archaeological questions, Field was particularly interested in racial questions. The "riddle" that provoked Field, and Arthur Keith, who advised him on his studies and researches, was the origin of the white race and its (superior) civilization. Field's research, undertaken in part as dissertation work in 1927 to earn a doctorate in physical anthropology with Earnest Hooton at Harvard,[47] continued the collection of measurements of living people and excavations of skeletal material begun with Buxton in 1925. In one of his monographs, *The Anthropology of Iraq*, Field enumerated the racial questions that he hoped his research would illuminate: Were the ancient inhabitants related to the current Arab population? Had the population changed in the last 6,000 years? And how were the Arab inhabitants of Mesopotamia related to their neighbors and to

---

[45] "Expedition at Kish Resumes Operations," *Field Museum News*, Dec. 1931, p. 3.

[46] Field discusses these trips at length in his autobiographical account of his anthropological work, *The Track of Man*. See also, Henry Field, *The Anthropology of Iraq, The Upper Euphrates*, Anthropological Series, Field Museum of Natural History, vol. 30, part I, no. 1, May 31, 1940, Publication 469, pp. 7–8; also Henry Field, *Folklore and Customs of Southwestern Asia*, Prepared for Publication by Henry Field, Curator of Physical Anthropology, Field Museum of Natural History, Anthropological series, Field Museum of Natural History, vol. 33, no. 1, Dec. 30, 1940, Publication 484. Apparently this volume was printed but never distributed, and the volume number was reused. Personal Communication, Ben Williams, The Field Museum Library, Oct. 9, 1995.

[47] Field apparently never received this degree, and instead was awarded a Doctor of Science degree from Oxford University in 1937, on the basis of his work at Kish as well as his work on his two physical anthropology halls for the Field Museum (he submitted a monograph on Kish, which included a 40,000-word introduction by Keith, and "a large album of photographs" from the two halls). He had previously earned a Diploma in Anthropology at Oxford. Henry Field to Earnest Hooton, April 1, 1927; Earnest Hooton to Henry Field, April 13, 1927, vol. 13, FM-HF; Field *Track of Man*, pp. 318–320.

people in Europe, Asia, and Africa?[48] Arthur Keith encouraged Field in this endeavor, telling him he was "very interested in your observations on the Mesopotamians." In their correspondence, Keith stated the ultimate goal of racial researches in the Middle East more plainly than Field did in print. "You are opening a new chapter for us students of races," he wrote Field. "For understanding the white man, a knowledge of Arabs is a fundamental necessity. I suspect these modern Arabs to be the degenerate descendants of the pioneers of civilization."[49]

Whereas most physical anthropologists searched for human origins in Africa or Asia in the interwar period, Field looked to the canonical – indeed biblical – site of origin.[50] Field's search looked back not 6 million, 600,000, or even 60,000 years – periods in which fossil evidence placed human ancestors in Africa, Asia, and Europe – but 6,000 years, in a quest for an originary moment for historic man, a chapter to fill the gap in his story, between the development of prehistoric man and the basic elements of human existence, and the wide diversity of races, led by the "white man" in Europe and America. Field searched for the narrative that would tell the tale of how the white race progressed through time from the prehistoric epoch to its Nordic supremacy.[51] By 1935, Field had begun to construct his tale. The story began with the earliest evidence of writing – a pictographic tablet unearthed at Kish – and the earliest technology – a wheeled chariot found among sacred buildings. To this evidence of the substrate of great civilization Field added evidence confirming the events of "the greatest written word": discovery of the biblical deluge, the "Flood of Noah," and Nebuchadnezzar's temple. In the linear, progressive narrative Field favored, the archaeological evidence from Kish linked Mesopotamian culture to the key moments in Western history: the Greek and Roman civilizations, followed by the birth of Christ, after which "the story takes definite shape, in which certain makers of history stand out in bold relief." These "makers of history," so easily recognized that he had no need to enumerate them for his audience, were contrasted to "different" forms of civilization, developing "along

[48] Henry Field, *The Anthropology of Iraq, The Upper Euphrates*, Anthropological Series, Field Museum of Natural History, vol. 30, part I, no. 1, May 31, 1940, Publication 469, p. 7; also, pp. 16, 89–90.
[49] Arthur Keith to Henry Field, March 3, 1931, Box 5, Hoffman, Malvina, Correspondence, Keith, Sir Arthur, 850042–79, MHC; Field, *The Anthropology of Iraq*, pp. 89–90.
[50] Henry Field, "The Story of Man," Science Service Radio Talks Presented Over the Columbia Broadcasting System, *The Scientific Monthly* (Jul. 1935), pp. 61–65.
[51] Ibid., pp. 64–65.

different lines" in Egypt, China, India, and Persia at the same time. This
story of historic man, missing from the halls of the Field Museum, casts
light on the implicit story embedded in the *Races of Mankind*. In this
light, Malvina Hoffman's Nordic man evokes not only Greek "civilization" and white supremacy, but also Christianity, making his superiority
not only biologically and culturally inevitable, but divinely ordered as well
(see Chapter 4, Figure 4.9 in this book). Nordic man epitomized the conceptual slipperiness of racial typing that makes it at once fundamentally
corrupt and pervasively powerful – he symbolized the conflated meanings
of whiteness that were at the root of its construction and dissemination,
and that Field, Keith, and many others were at pains to undergird with
scientific evidence.

Unity or Diversity: The Competing Philosophies of Berthold Laufer and Arthur Keith

*The Races of Mankind, An Introduction to Chauncey Keep Memorial
Hall* was more than just a handy guide for the Field Museum visitor
who sought more information about the sculptures and displays found
in the hall of physical anthropology. The authors of the volume were
also the key authors of the exhibition, despite the fact that one of them –
Arthur Keith – was not employed by or formally associated with the Field
Museum. In their separate and distinctive contributions to the leaflet, each
revealed through form and content their conceptual roles in the planning
and mounting of the *Races of Mankind*, and thereby also revealed some
of the tensions embedded within the exhibit.

Henry Field was responsible, under Laufer's editorial supervision, for
the bulk of the leaflet: a discussion of "race biology"; descriptions of the
races by geographic area – Africa, Europe, Asia, America, Oceania; and a
brief enumeration and description of the "Special Scientific Exhibits."[52]
Unlike Laufer's and Keith's contributions, Field produced a text more
akin to the conventional, seemingly dispassionate and objective, descriptive account that reads as a synthetic compilation of an uncontested
body of knowledge about race. No doubt, this was deliberate, providing

---

[52] Ibid., pp. 13–40. The leaflet also included a brief bibliography and a list of Hoffman's
sculptures. The bibliography included work by Franz Boas, Alexander M. Carr-Saunders,
L. H. Dudley Buxton, Joseph Deniker, Alfred C. Haddon, Melville Herskovits, Earnest
Hooton, Aleš Hrdlička, Waldemar Jochelson, Alfred Kroeber, Rudolf Martin, William
Ripley, Herbert H. Risley, Charles G. Seligman, William Sollas, Edward P. Stibbe, Louis
Sullivan, and W. D. Wallis.

another avenue for conveying the serious and scientific character of the
Field Museum's undertaking. It also reflected the character of Henry
Field's contributions to the content of the exhibit. His major contribution – and it was a critical one – was the narrative shape of the two
physical anthropology halls. When it came to the details – the selection
of types to display, the accuracy of Hoffman's figures, the content of
the technical exhibits – he deferred to Laufer and other anthropologists,
especially Arthur Keith, Earnest Hooton, and Dudley Buxton.[53]

Berthold Laufer, curator of anthropology, wrote the preface.[54] Befitting his supervisory role, Laufer began with descriptive information that
served to legitimize the seriousness, importance, and scientific authority of the exhibition. In the first of three brief paragraphs that open the
preface, Laufer, noting his own responsibility for the hall, explained that
it had been planned since 1915, finally opening after "long and mature
deliberation and thorough study of every detail." He next implied the
significance and seriousness of the exhibition by listing the benefactors
and their sizable contributions, leaving the names Field, Schweppe, and
Keep, prominent local benefactors, to signal the provenance of the hall.[55]
Following the financial benefactors, Laufer introduced Arthur Keith,

[53] The list of anthropologists and anatomists Henry Field, Berthold Laufer, and Malvina
Hoffman consulted in the course of planning and mounting the *Races of Mankind* is
long. It is headed by a handful, beyond Keith, who were regularly sought out: Bux-
ton, Hooton, Henry Wellcome, T. Wingate Todd, Wilton M. Krogman, Henry Balfour,
Adolph Schultz, Alfred C. Haddon. Others were Franz Boas, Aleš Hrdlička, Roland
Dixon, Davidson Black, Paul Stevenson, Fay Cooper-Cole, A. R. Radcliffe-Brown,
Clark Wissler, Harry Shapiro, William Lessa, F. Wood Jones, Eugen Fischer, Eigon
von Eickstedt, Stephen Langdon, A. J. H. Goodwin, Eric Thompson, George Montan-
don, Joseph Shellshear, Josef Weninger, Harry Laughlin, Charles Davenport, Morris
Steggerda, Melville Herskovits, Hella Poch, Charles Merriam, Charles Wellington Fur-
long. In addition, there were a number of anthropologists consulted in the countries
where Hoffman traveled, including B. S. Guha. These lists are not comprehensive; nor
do they include the variety of printed references utilized (including, e.g., work by William
Ripley and photographs of Native Americans by Edward Curtis), as well as the many
local government and police officials, and other acquaintances, who procured models
for Hoffman.
[54] Field, *The Races of Mankind*, 1934, pp. 3–6.
[55] Of the total cost, $50,000 was a bequest by Chauncey Keep, a former trustee, to the Field
Museum that the trustees used to fund the exhibit. The $18,000 "Unity of Mankind"
sculpture was paid for by Laura Schweppe, wife of financier Charles H. Schweppe,
and daughter of John G. Shedd, founder of the Shedd Aquarium next door to the Field
Museum on the Chicago lakefront. Marshall Field III and Mrs. Stanley Field contributed
the balance of more than $100,000. Berthold Laufer, "The Projected Hall of the Races
of Mankind (Chauncey Keep Memorial Hall)," *Field Museum News* (Dec. 1931) vol. 2,
no. 12; Henry Field to Marshall Field, Feb. 22, 1930, Races of Mankind Correspondence,

noting his prestigious offices and authoritative publications on a range of subjects bearing on human biology, evolution, and race. Remarkably, after establishing his exhibition's pedigree, Laufer then proceeded to call into question the very ground on which it was organized by providing a cautionary meditation on the pitfalls of racial classification that vexed physical anthropology, including, notably for an exhibition opening in 1933, the illegitimacy of the term "Aryan." He concluded his portion of the leaflet by stressing the fundamental unity of human kind and his hope that visitors would leave with their "sympathy for mankind" deepened.

Arthur Keith authored the introduction.[56] He constructed in his few pages both a paean to Malvina Hoffman's artistic intuition for race and a relentless accounting of the evolution of difference and the "aberrant forms" it produced. Where Laufer's preface destabilized and problematized physical anthropology's object, Keith persistently naturalized and reified the hall's racial types as biologically occurring kinds easily sensible to "the man in the street."[57] Thus, perhaps remarkably, the brief pamphlet constructed to guide the visitor untutored in anthropology through the exhibition, managed in forty-four small pages to encapsulate, in all their tensions and contradictions, the visions of race at the heart of the *Races of Mankind*.

Berthold Laufer: Historicizing Racialized Others

For Henry Field and anthropological advisors such as Arthur Keith and Earnest Hooton, diverse lifeways and behavior were evidence of profound hereditary racial differences. They saw divergent cultures as secondary to, and in important respects as the result of, divergent bodies. Berthold Laufer viewed the relationship differently, and his perspective informed Hoffman's sculptures, attendant technical displays, and the literature and promotional materials that accompanied the exhibit.

Berthold Laufer, like Franz Boas, grew up in Germany. He studied at the University of Berlin, the Seminar for Oriental Languages in Berlin, and the University of Leipzig, where he received his doctorate in 1897. In 1898, Franz Boas hired Laufer to lead the Jesup North Pacific Expedition to Saghalin Island and the Amur River region to document the traditional culture of China and collect material for the American Museum of

vol. 12 1920–1950, FM-HF; *Annual Report of the Director to the Board of Trustees,* 1933, pp. 15–16, p. 35.

[56] Ibid., pp. 7–12.

[57] Ibid., p. 9.

Natural History. From 1901 to 1904, Laufer led the Jacob H. Schiff Expedition to China, also for the American Museum. While in New York, he was a lecturer in anthropology and Asian languages, and a colleague of Franz Boas at Columbia University. In 1907, Laufer joined the Field Museum, becoming Curator of Anthropology in 1915 (Figure 3.3). His specialization was Eastern Asia, particularly archaeology and ethnology, although he was regarded as competent in all fields of anthropology. He became the leading sinologist in the United States, known for his wide interests and knowledge, as well as his mastery of languages of India, Central Asia, and the Far East. The Field Museum acquired a large collection of Chinese art and antiquities as a result of his expeditions; the Newberry and John Crerar Libraries in Chicago benefited from his collection of 40,000 Chinese books.[58] Although the majority of his writings were confined to Chinese arts and material culture and Asian languages, a few of his more popular writings and especially his published and private thoughts on the *Races of Mankind* exhibit offer insight into his conception of human race and physical anthropology.[59]

Fundamentally, Laufer believed that mind, body, and behavior were primarily rooted in cultural traditions, environmental conditions, longstanding habits of mind shared by groups, and a common human psyche rather than in somatic, hereditary factors. Like Franz Boas, Laufer's perspective was shaped by a German historical, philosophical, and anthropological tradition that emphasized a historicized understanding of human culture, combined with a belief in the psychic unity of mankind. Differences, even seemingly profound ones, could be explained as the result of a

---

[58] "Obituary," *Nature*, Oct. 13, 1934, p. 562; Arthur W. Hummel, "Berthold Laufer: 1874–1934," *American Anthropologist* N.S., 38, 1936, pp. 101–111; "The Death of Dr. Laufer, Curator of Anthropology," *Field Museum News*, Oct. 1934, p. 2; Walter Hough, "Dr. Berthold Laufer: An Appreciation," *Scientific Monthly*, vol. 39, no. 5 (Nov. 1934), pp. 478–480; Roberta H. Stalberg, "Berthold Laufer's China Campaign," *Natural History*, Feb. 1983, vol. 92, pp. 34, 36. Laufer was born in Cologne, Germany, in 1874. He committed suicide in Chicago on September 13, 1934, apparently after having been diagnosed with terminal cancer. Stanley Field to Malvina Hoffman, Sep. 24, 1934, Box 30, Field Museum of Natural History, Miscellaneous Correspondence, 850042-3, MHC.

[59] Laufer had a wide range of interests. His publications included "History of the Finger-Print System," "The American Plant Migration," "Tobacco and Its Uses in Africa," and "The Giraffe in History and Art." Among his papers at the Field Museum are notes on the origin and distribution of varieties of potato. This was in addition to his authoritative works on Chinese archaeology and ethnology, particularly work on ancient pottery, bronzes, jades, and precious stones, as well as Buddhist and Tibetan literature, "Dr. Berthold Laufer," p. 479; "Obituary," *Nature*; Berthold Laufer Papers, Department of Anthropology, Museum Archives, The Field Museum (hereafter FM-BL).

FIGURE 3.3. Berthold Laufer (right) in Hankow, ca. 1904. Courtesy of The Field Museum, #A98299.

complex combination of human mental capacities, physical environment, and the particular historical development of a people. Bodies themselves were not irrelevant, but they were not the most fundamental unit of analysis for anthropologists trying to understand human diversity. Bodies were markers, embedded in a sociohistorical complex. They provided evidence of human history, much in the way language or kinship systems did. In their form, bodies reflected the cultural traditions, human migrations, and local conditions that shaped groups of people over time. As such, simply measuring, cataloguing, and classifying them made no sense without also understanding their cultural history, broadly understood to include mind, language, behavior, and environment. Mind, bodies, and behavior were the result and shapers of culture. Laufer firmly believed that all humans were essentially alike in mental capacities and that major differences, even to the level of seemingly physical capabilities, were shaped by differing, longstanding cultural backgrounds.[60]

Laufer offered an explicit analysis of this kind in a foreword to a series of monographs on Oriental society published in *The Open Court* in 1933. His subject was the misinterpretations and excesses of "grandiloquent orators" who favored broad contrasts between Eastern spiritualism and Western materialism. He discussed the ostensible essential differences between peoples of the East and West – a subject on which "an avalanche of platitude and blah has fallen" – and dismissed a hereditarian, racialist interpretation of the widely noted differences between Asians and people of European descent. In the course of a few sentences he revealed some of the basic themes that guided his anthropological work.

We must not take East and West in the sense of abstract ideas, which will inevitably lead to vague idealizations, but must sense them as living realities in their proper setting and perspectives. In the first place, all orientals taken individually are not radically different from ourselves, they are just as human as you and I, subject to the same human emotions and passions; all shades and grades of character are found among them. Armenians, Arabs, Persians, Indians, Chinese, and Japanese are as shrewd, keen and enterprising industrialists and merchants as any in the Western world.... The fundamental divergencies are not between

---

[60] For an excellent treatment of the German origins of Boasian anthropology, see essays in George W. Stocking, Jr., ed., *Volkgeist as Method and Ethic, Essays on Boasian Ethnography and the German Anthropological Tradition*, History of Anthropology, vol. 8 (Madison: University of Wisconsin Press, 1996), including Stocking, "Boasian Ethnography and the German Anthropological Tradition," pp. 3–8; Matti Bunzl, "Franz Boas and the Humboldtian Tradition: From *Volkgeist* and *Nationalcharakter* to an Anthropological Concept of Culture," pp. 17–78; Benoit Massin, "From Virchow to Fischer: Physical Anthropology and 'Modern Race Theories' in Wilhelmine Germany," pp. 79–154; and Julia Liss, "German Culture and German Science in the *Bildung* of Franz Boas," pp. 155–184. See also James Whitman, "From Philology to Anthropology in Mid-Nineteenth-Century Germany," George W. Stocking Jr., ed., *Functionalism Historicized: Essays on British Social Anthropology*, History of Anthropology, vol. 2 (Madison: University of Wisconsin Press, 1984), pp. 214–229.

individuals or classes of people, but are deeply sunk in the thoughts of the folk
mind fostered by a different background of civilization. There are only two such
fundamental differences between the East and the West, which may be tersely
formulated as the difference between the ego and the non-ego and the difference
between the definite and the indefinite article.[61]

Laufer argued for the unity of humanity in which people of both the East
and West were capable of the attributes assigned to both groups – it was
true that India had been "a dreamland of mysticism," but that had not
prevented Indians from "cornering the world market in precious stones
for two thousand years."[62] The basic differences that anthropologists
studied Laufer ascribed to a "folk mind" created through long cultural
traditions, what he here called the "background of civilization."[63] Rather
than turning to a biological explanation for differences between large
groups of people, Laufer offered a psychological explanation. Divergent
social traditions created "ego" and "non-ego" cultures and their resultant
"mental complexes." According to Laufer, "egocentric" Western culture
was rooted in a heritage received from classical antiquity in which individuals were glorified and overvalued, leading to the codification of individual rights. Eastern "non-ego" cultures were focused on the individual's
duties to family and the state. The Chinese "mental complex" had always
been "focused on the cosmos, the joy and deification of nature, striving
for the beyond and reveling in dreams of eternity and immortality."[64]

   To illustrate the difference between the ego and non-ego mental complexes, Laufer discussed the aspect of culture in which he thought the
distinction was most pronounced: art. Published only a few months after
the *Races of Mankind* opened, and while a number of its figures remained
to be completed, Laufer could not help but have had the hall, with its
gallery of bronze sculptures, each signed at the base by Malvina Hoffman, in his mind as he wrote. In his essay, Laufer claimed most Western
artists were motivated principally by "vanity, ambition, self–love and an
inordinate craving for fame and notoriety." Sculptors, he said, "ostenta-
tiously . . . carve their names in huge letters on their . . . sculptures and are

---

[61] Berthold Laufer, "East and West," *The Open Court*, vol. 47 (no. 8), number 927, Dec.
   1933, p. 473.
[62] Ibid.
[63] Laufer made the same point in another essay, this time devoted to the differences between
   the Chinese and Americans. He argued that apparent psychological differences "do not
   spring from a basically divergent mentality or psyche but are merely the upshot of a
   distinct set of traditions and education based upon the latter." "Sino-American Points
   of Contact," *The Scientific Monthly*, Mar. 1932, vol. 34, reprinted in *The Open Court*,
   p. 499.
[64] Laufer, "East and West," p. 474.

prone to ascribe their work to their own merit and genius," forgetting their predecessors, teachers, and the tradition within which they worked. Conversely, although the Chinese had produced some of the most skillful pottery and bronzes ever created, none were signed, the artists being "too modest and too sensible to mar their productions" with their signatures. None, even "superlatively great" artists, were "fool enough to believe," or to "flatter" themselves into believing, that they were personally responsible for their creations, attributing them instead to a higher power, the influence of ancestors, or the will of Heaven. Whereas for Western artists, work was produced to make a living or please contemporaries, for Eastern artists it was a sacred process of honor and salvation.[65] Thus, a seemingly simple act – to sign or not to sign a piece of artwork – took on broad significance for Laufer, who saw in this gesture evidence of the underlying "folk mind" that distinguished one group of people from another.

Not only did Laufer find many aspects of Asian culture appealing, and in some cases superior, to Western culture, his argument also militated against viewing differences as inevitable, or shifts in the East toward Western practices as necessarily evidence of progress. Laufer used the term "civilization" in a manner that retained its Arnoldian sense of perfected, superior attainments, particularly in arts and letters, but it was complemented by his more Boasian use of "culture" to refer to the set of beliefs and practices that characterized the lifeways of groups of people. Moreover, Laufer's understanding of civilization and culture was not the nineteenth-century evolutionary one, in which peoples progressed through definite stages, some reaching the level of "civilization," while others remained stuck at lower levels of cultural sophistication and capabilities. Although he clearly adhered to a cultural hierarchy in which some cultures sat at the top of the scale as "civilized" peoples and others inhabited the "primitive" bottom, his placement of peoples on that scale was not a simple reflection of evolutionary stages or an equation of civilization with whiteness because, for Laufer, the existence of definable races was not simply a function of the laws of heredity and biology, but a contingent result of the complex of mind, tradition, environment, bodies, and time.[66]

[65] Laufer, "East and West," p. 474.

[66] Stocking, "The Ethnographic Sensibility of the 1920s" in *Ethnographer's Magic*, especially pp. 284–290. For the classic nineteenth-century formulation of cultural evolutionary theory, see Lewis Henry Morgan, *Ancient Society*. Other historical work that examines the concept and implications of cultural evolution in anthropology and American society include Gail Bederman, *Manliness and Civilization: A Cultural History of*

Thus, Laufer did not see in Hoffman's actively posed figures either a hereditarian racial destiny or an inappropriate conflation of culture and physical variation. Nor was he eager to embark on an elaborate biometric project to exhaustively quantify the racial differences they sought to portray in the halls of the Field Museum. Instead, Laufer, like Field and Keith, embraced Hoffman's artwork. He praised Hoffman's figures for the insights they offered visitors, enabling them to study "the physical functions which are more important for evaluation of a race than bodily measurements."[67] In other words, Laufer believed that if one looked to bodies for evidence of race, then movement, behaviors, and gestures revealed more than mere morphology, because what made groups of people most racially distinct was not primarily skin color, head form, nose shape, or any of a hundred other features (although these had their place in delineating race), but the long-accumulated effects of a particular historical, cultural trajectory, effects that were evident not only in artifacts, language, or social relations, but literally in the bodies and minds of a people. It is this philosophy that explains the significance of the informational display case full of disembodied hands and feet cast in writing, painting, and other supposedly racially characteristic behaviors that Laufer directed Henry Field to include in the *Races of Mankind* (Figure 3.4).

Arthur Keith and the Place of Prejudice

Given Laufer's entreaty to museum visitors, he and Henry Field could not have picked an anthropologist with more diametrically opposed views to write the Introduction for the *Races of Mankind* pamphlet.[68] Keith had been a professor of comparative anatomy at the Royal Institution and was both professor and conservator of the museum at the Royal College of Surgeons in London. In addition, he was a one-time president of the Anatomical Society of Great Britain, Royal Anthropological Institute, and British Association. He was born in Scotland and received his bachelor's degree from Aberdeen. His work began with primate anatomy,

---

Gender and Race in the United States (Chicago: University of Chicago Press, 1995); and Griffiths, *Wondrous Difference*.

[67] Laufer, "Hall of the Races of Mankind," p. 3.

[68] Keith also wrote a promotional review of the *Races of Mankind* for *The New York Times*, lauding the exhibit's value to anthropology and Malvina Hoffman's accomplishment. The review included twelve photos of Hoffman's figures. "Races of the World: A Gallery in Bronze," *New York Times*, May 21, 1933, p. SM10, 2 pages.

FIGURE 3.4. Case of hands and feet, *Races of Mankind* hall (1933). Courtesy of
The Field Museum, #A78448.

moved on to comparative anatomy, and finally included human evolution
(Figure 3.5). He was noted for his argument, based on paleontological
evidence, that big brains were the key criterion distinguishing hominids
from primates, that primate and *Homo* lines diverged at an early date,
and that the main human races split at an early date from a single original human ancestor ("human" being defined by Keith as bearing a skull
capacity at the lower limits of modern humans – 1000 cc.).[69] His theories of human evolution were intimately connected to theories of racial

[69] Arthur Keith, *The Antiquity of Man* (London: Williams and Norgate, Ltd., seventh
impression, Jan. 1929, vol. 2), p. 732. For other discussions of Keith's racial and evolutionary theories, as well as his professional activities and involvement with the Piltdown skull, see Elazar Barkan, *The Retreat of Scientific Racism, Changing Concepts of
Race in Britain and the United States between the World Wars* (Cambridge: Cambridge
University Press, 1992), especially pp. 46–53; George W. Stocking Jr., ed., *Bones, Bodies, Behavior. Essays on Biological Anthropology*, History of Anthropology, volume 5
(Madison: University of Wisconsin Press, 1988), especially Frank Spencer, "Prologue
to a Scientific Forgery: The British Eolithic Movement from Abbeville to Piltdown,"
pp. 84–116, and Michael Hammond, "The Shadow Man Paradigm in Paleoanthro-
pology, 1911–1945," pp. 117–137; Elazar Barkan, "Mobilizing Scientists against Nazi
Racism, 1933–1939," pp. 180–205; and Frank Spencer, ed., *A History of Physical
Anthropology 1930–1980* (New York: Academic Press, 1982).

FIGURE 3.5. Malvina Hoffman and Arthur Keith at Downe House, Charles Darwin's home in England, 1936. Courtesy of Research Library, The Getty Research Institute, Los Angeles (850042), #r34636_850042_b82_001, and Derek Ostergard.

evolution and classification, because, in his own nationalistic version of natural selection, he believed that the engine of human evolution was competition between races, which he often equated with nations.

In perhaps his most important work, *The Antiquity of Man*, first published in 1915, revised and reprinted seven times, Keith argued that the competitive mechanism of evolution worked in the past as well as in the present. In the last edition of *Antiquity*, published in 1929, Keith estimated that the modern human ancestor of all human races dated back at least 500,000 years, an early date by contemporary standards.[70] Such an early date was necessary, according to Keith, because, after the emergence of a modern human type, sufficient time had to be allowed for differentiation into existing racial types.[71] After excluding most known hominoid fossil remains from the main line leading to modern humans in his genealogical tree, he offered the "true significance" of all these dead-end, extinct types.[72] "When we look at the world of men as it exists now, we see that certain races are becoming dominant; others are disappearing. The competition is world-wide and lies between varieties of the same species of man." By analogy, the various species of fossil man competed locally. "Out of that welter of fossil forms only one type has survived – that which gives us the modern races of men."[73]

The implications of Keith's evolutionary view of race were made explicit in his public presentations. On June 6, 1931, Keith, newly elected as Lord Rector of the University of Aberdeen, gave an acceptance address to the students entitled *The Place of Prejudice in Modern Civilization*, in which he offered an evolutionary explanation and justification for racial prejudice.[74] Keith's evolutionary mechanism had four parts: the action of hormones produced "new characters"; these new characters were preserved and augmented through physical isolation from other such groups of people; isolation was brought about and enforced by "tribal spirit," which was the machinery of race prejudice; through race prejudice, groups of people competed, some succeeding, dominating, and extinguishing the others. In Keith's view, "a nation was an incipient race." In the modern world, "a conglomeration of massed communities – nations, states, vast empires" had superseded naturally produced tribes, creating unrest between populations thrown together for economic and political reasons,

---

[70] Keith, *Antiquity*, 713. Keith put the split of ape and human lineages at 1 million years in the past, p. 732.

[71] Ibid., pp. 714–715, 720. Keith listed four principal racial types in his text – black, brown, yellow, and white – but five in an evolutionary tree accompanying the text – Negro, Negroid, Australoid, Mongoloid, and Caucasoid.

[72] The excluded types, in addition to the Neanderthal, included Rhodesian man, *Eoanthropus* (Piltdown man; Keith suggested that although likely not a direct ancestor, Piltdown most closely resembled a common ancestor for Neanderthal and modern man), *Pithecanthropus* (Java man), and *Australopithecus* (Keith marked this fossil as an ape ancestor, rather than as a human ancestor following Raymond Dart; Keith also placed it at the end of Micoene rather than at the beginning).

[73] Ibid., p. 725.

[74] Sir Arthur Keith, *The Place of Prejudice in Modern Civilization*, QUEST, 1973.

rather than through the natural organizing principle of race. The source of the unrest was the inherent tribal spirit that still pulsed with race prejudice.[75] Race prejudice was the motor of human progress: "The human heart, with its prejudices, its instinctive tendencies, its likes and dislikes, its passions and desires, its spiritual aspirations and its idealism, is an essential part of the great scheme of human evolution – the scheme whereby Nature, through the eons of the past, has sought to bring into the world ever better and higher races of mankind."[76] In modern nations, according to Keith, this racial prejudice took the form, among others, of patriotism and a love of independence, binding nations together. Keith argued that the pacifist goals of the League of Nations and others who sought to bring all humanity together into "a single harmonious united tribe" would be achieved only by sacrificing everyone's "racial birthright" through a common pooling of blood, each race distributing to a common progeny the inherited traits that their ancestors had struggled to preserve and enhance. Such pacifist commingling was anathema to Keith. In his view, it would undo thousands of years of improvement in the most successful races, reducing all people to a middling melding of the healthy and unhealthy, strong and weak, in an unnatural, antievolutionary intervention in the normal course of human development. Conversely, the cost of continued race prejudice was one he justified; competition was the price of continued progress: "[R]ace prejudice and, what is the same thing, national antagonism, have to be purchased, not with gold, but with life.... Nature keeps her human orchard healthy by pruning; war is her pruning-hook."[77] That some races were disappearing might be lamentable, but it was merely the result of the cold logic of evolutionary progress. The best that anthropology could offer was to understand this mechanism and the resulting intricate racial hierarchy it produced, and preserve as best it could, for science and posterity, examples of the races of men that were inevitably vanishing into history.

Keith also publicized his views in the United States.[78] In essays published in the *New York Times*, he applied his theories to race in America, a place that "riveted" the attention of anthropologists on "the unfolding of

[75] Ibid., pp. 7–14.

[76] Ibid., p. 26.

[77] Ibid., pp. 45, 47, 48–49.

[78] Sir Arthur Keith, "Creating a New American Race," *The New York Times*, Jun. 2, 1929, p. SM1. Between 1926 and 1932, Keith wrote a series of eighteen essays for the *Times* Sunday magazine concerning human evolution and race, including "The Evidence for Darwin Is Summed Up" Sep. 14, 1927, p. XX1, and "Whence Came the White Race?" Oct. 12, 1930, p. SM1.

the greatest experiment in race building the world has ever seen." Despite the diversity in the United States, Keith was interested only in a subset of American variety. "Will the European racial elements," Keith asked, "so blend that out of the amalgamation there will emerge a new and uniform race of mankind which will be 100 percent American?" His answer was yes. Keith even went so far as to predict that this new race would resemble the current president – Herbert Hoover, a type "permeated with a living, active virility."[79] Indeed, he regarded "the establishment of a white republic in North America" as "the most momentous anthropological event of historical times." But the path to a nation of Hoovers was complicated by the presence of 12 million Americans of non-European descent. These races also were engaged in the ongoing evolutionary struggle to "creep up the scale of racial differentiation," but unlike the white Europeans, their prospect was the "hard and ugly fact" of extinction, according to Keith. Indeed, the anthropologist could only explain the existence of such inferior varieties of modern human by recourse to a theory of evolutionary competition through which nature sought "a more perfect type of human being." As to the question of what the new white race would "do" with millions of African Americans, Keith dismissed the prospect of "fusion" (he elsewhere warned that to mingle "strains" that were too disparate would "ruin" nature's experiment), and instead suggested that time would tell whether white and black Americans could "resolve to remain apart" to work out their "separate evolutionary destinies." With the advent of the United States, the "white stock" was given a "dominant position in the racial rivalry of the world." In Europe and America, Keith argued, "statesmen are awakening to the fact that breeding a race is as important – nay, is more important – than the accumulation of national wealth." The evidence? Movements in Europe for "'national self-determination' – which is a claim for freedom and independence so that a people or nation may work out its racial destiny," immigration restrictions in the United States, and "strong schools of eugenics" that provided Keith "clear proof that white peoples are resolved that economics is not the mistress of their destiny but their slave." For Keith, "all the auguries for the future are in favor of a world domination of the white type."[80]

---

[79] In 1932, Keith estimated that the new white race in America was 65 percent Nordic, 23 percent Mediterranean, and 12 percent Alpine, "America: The Greatest Race Experiment," *The New York Times*, Jan. 24, 1932, p. SM6.

[80] Keith, "Creating a New American Race"; Keith, "America: The Greatest Race Experiment"; Keith, "The Greatest Test for Mankind," *The New York Times*, Feb. 8, 1931,

In these popular essays, Keith aligned himself with a "new school" of anthropology that had gone beyond the "old school's" limited and strict adherence to a "zoological" approach confined to measurement and classification. According to Keith, members of the new school felt the zoological method failed to take the evolutionary nature of race formation seriously and was inadequate to address the kinds of social and political problems with which anthropologists were confronted. "Indeed," he argued, "it is becoming evident to even the most highbrowed anthropologist that the despised politician has formed a truer conception of race than had the professional student of mankind." According to Keith, nationalities were evolutionary units, races in the first stages of development, and politicians were able to recognize incipient races within their populace although they were "not distinguishable by the methods of the anthropologist." The fact that old-school anthropologists had had difficulty sorting people into discreet categories, finding that many people appeared to display a blend of characters, was to Keith evidence, not of increasing mixing and dilution of racial types, much less a failure of the race concept altogether, but instead evidence that incipient races were in the early stages of differentiation. In Keith's view, difference was increasing under the pressure of evolutionary competition, with the inevitable result that some groups would survive and others perish.[81]

The arguments put forward in *The Place of Prejudice* provoked reaction in Britain and the United States. *Nature* ran an editorial endorsing his views. *The New York Times* ran three articles and an editorial reporting on Keith's views and reactions to it, including criticisms from Franz Boas. H. H. Horne, professor at the New York University School of Education, was quoted as arguing in response to Keith that the purpose of prejudice was repression, and that "man's intelligent pruning hook is birth control"; the unintelligent pruning hook was "death by accident in our machine age." Boas used his platform as president of the American Association for the Advancement of Science, to attempt to refute Keith's argument. In the first evening's general address at the meeting in Pasadena, California, Boas argued that Keith's argument was fallacious. "War eliminates the physically strong; war increases all the devastating scourges of mankind such as tuberculosis and genital diseases; war weakens the growing generation,"

p. 77; Keith, "Science Ponders Man's Future," *The New York Times*, May 2, 1926, p. SM1.

[81] Keith, "Creating a New American Race"; Keith, "America: The Greatest Race Experiment."

he argued. Boas reminded his audience that there had been much mixing between African Americans and whites, as well as among Native Americans and whites, in American history with no ill effects, and that no race was inherently inferior to any other. Moreover, he argued that if racial antipathy had been "implanted by nature" as Keith proposed, it would have been expressed through "interracial sexual aversion," which plainly was not the case.[82]

In contrast to Berthold Laufer's call for unity and sympathy in the preface to the Field Museum booklet on the *Races of Mankind*, in his introduction Arthur Keith called on the visitor to attend to human difference. Echoing his remarks about politicians, Keith argued that all people were amateur anthropologists by virtue of their ability from birth to discern subtle and gross differences in the appearance of people they encounter, from family members to members of their community to "stray people from distant lands." Arguing that ordinary people had no need of measurements and statistics to recognize race, Keith also argued that such an intuitive and unerring recognition of race was proof of its existence. Professional anthropologists only compiled data in order to quantify and describe race, rather than prove its existence. Yet this state of humanity, so evident to the casual observer, required voluminous measurements and statistical treatment for the professional to ascertain, and was "so shifting and indeterminate in nature, that scientific measurement can never rival the accuracy and completeness of the rule of thumb method practiced by the man in the street." Despite the (commonly held) contradiction inherent in the notion that something so self-evident required scrupulous and myriad measurement to define, this argument comported with Keith's overriding belief in the inherent racial prejudice in all people and its instinctual operation. Thus, he could go on to argue that "true artists" in particular had a highly developed intuition for racial qualities. For Keith, race prejudice was not the outcome of ignorance, as it was for Laufer; rather, it was the source of insight. Confronted with the *Unity of Mankind* group, Keith noted not the links between races but rather highlighted differences among humans. "It is right that this group should hold a central position in the hall, for no one can look at those

---

[82] "Progress and Prejudice," *Nature*, vol. 127, no. 3216, Jun. 20, 1931, pp. 917–918; "Sir Arthur Keith Holds War Helpful," *The New York Times*, Jun. 7, 1931, p. 12; "Disputes Keith View as to Race Prejudice," *The New York Times*, Jun. 8, 1931, p. 13; "War and Prejudices Called Ill for Man," *The New York Times*, Jun. 16, 1931, p. 5; "Spears and Pruning Hooks," *The New York Times*, Jun. 17, 1931, p. 19.

three figures without asking the question, 'Why this diversity of racial type?'"[83]

Earnest Hooton: Populations, not Types

The Field Museum's choice to focus their anthropological exhibit of human variation on a large series of striking sculptures depicting individuals, each of whom typified one of the principal living races, presented a fundamental decision that focused the entire exhibition. But it reflected basic assumptions about the nature of race and how to study it that not all of Henry Field's colleagues shared. Prominent among the dissenting voices was Earnest Hooton, one of the most influential physical anthropologists in the United States in the interwar years.[84] Certainly, it would be hard to exaggerate his influence on the profession via his students; as a professor of anthropology at Harvard University, he produced nearly an entire generation of physical anthropologists. In addition, Hooton presided over the Peabody Museum's Anthropological Laboratory and the analyses, articles, monographs, and instruments it produced. And he was unmatched among his peers in the output of widely read popular tomes on human race and evolution – among them, *Up from the Ape* and *Apes, Men and*

---

[83] Sir Arthur Keith, "Introduction," in Henry Field, *The Races of Mankind*, Popular Series, Anthropology Leaflet 30, Fourth Edition (Chicago: Field Museum of Natural History, 1942), pp. 10–11.

[84] There is no full biography of Hooton; see "Hooton, Earnest Albert," *Current Biography*, 1940, pp. 397–400; "Memorium: Earnest Albert Hooton," *American Journal of Physical Anthropology* 12 (1954), pp. 445–453; Stanley Garn and Eugene Giles, "Earnest Albert Hooton, November 20, 1887–May 3, 1954," National Academy of Sciences, *Biographical Memoirs* 68 (1995), pp. 167–179; Harry L. Shapiro, "Earnest A. Hooton, 1887–1954 in Memoriam cum Amore," *American Journal of Physical Anthropology*, 56 (1981), pp. 431–434. For scholars' assessments of him, see, inter alia: Melissa Banta and Curtis Hinsley, eds., *From Site to Sight: Anthropology, Photography, and the Power of Imagery* (Cambridge, MA: Peabody Museum Press, 1986); Elazar Barkan, *The Retreat of Scientific Racism*; Haraway, "Remodeling the Human Way of Life"; Jonathan Marks, *What It Means to Be 98% Chimpanzee: Apes, People and Their Genes* (Berkeley and Los Angeles: University of California Press, 2002); Nicole Rafter, "Earnest A. Hooton and the Biological Tradition in American Criminology," *Criminology* 42, no. 3 (Aug. 2004), pp. 753–772; Nicole Rafter, "Apes, Men and Teeth: Earnest A. Hooton and Eugenic Decay," in Susan Currell and Christina Codgell, eds., *Popular Eugenics: National Efficiency and American Mass Culture in the 1930s* (Columbus: Ohio University Press, 2006), pp. 249–268; John P. Jackson, Jr., Nadine M. Weidman, eds., *Race, Racism, and Science: Social Impact and Interaction*, Science and Society Series (New Brunswick: Rutgers University Press, 2006); Jonathan Marks, *Human Biodiversity: Genes, Race and History* (New Brunswick, NJ: Transaction Publishers, 2009).

Morons. Like Henry Field, Hooton had studied physical anthropology at Oxford University under R. R. Marett and Arthur Keith. By training and orientation Hooton was sympathetic to Field's project. Like Field, Hooton embraced a hereditarian view of race, believing that people could be grouped meaningfully into types, and that much else besides physical conformation was attributable to a basic racial essence. But the advice he offered Field on how to acquire and display racial types did not conform to Henry Field's kind of typology, one fundamentally rooted in nineteenth-century notions of comparative anatomy, morphology, and a rudimentary understanding of heredity. Unlike Field, Hooton had developed a statistical, populational vision of race that demanded a different set of methods. Had Field followed his advice, it would have resulted in a quite different exhibition, dispensing a quite different rhetoric of race.

Henry Field approached the problem of how to represent types as a question of locating appropriate individuals, not as a question of how to depict an entire group. Field expected to seek out typical individuals for Malvina Hoffman to model by mining collections of anthropological photographs and traveling to far-flung locales in search of suitable subjects. Hooton rejected this approach and in fact saw the idea of a series of individuated figures – of real people – as highly problematic. Writing Field only a month after Hoffman was hired, Hooton discussed problems associated with sculpting life-size figures, and suggested that Hoffman create figures that were an embodiment of appropriate average characteristics, and as a concession to Field's insistence on representing real people, suggested that Hoffman then "give it a head copied from some photograph."[85] He later advised Field that Hoffman's sculptures should be based on a study of each type, as defined by measurements, observations, and photographs of "at least a hundred adult males in each area in which she is to work." This data, he wrote, "should represent an adequate sample of each race studied, which could be statistically elaborated and graphs and charts constructed to supplement" the bronzes.[86] In addition, he said samples of hair should be collected and skin pigmentation measured.

In Hooton's view, race, as a set of bodily characteristics, was something that existed at the level of populations, rather than inhering in individuals. This was especially true because no pure races existed. If living individuals

[85] Earnest Hooton to Henry Field, March 10, 1930, Races of Mankind, vol. 13, FM-HF.
[86] Earnest Hooton to Henry Field, September 4, 1931, Box 3, MHC.

were products of racial mixing, such that any individual exhibited traits that were characteristic to a greater or lesser degree of more than one race, then there would have been few, if any, living individuals who were strictly typical of one pure race. In a letter discussing races in India with Field, Hooton concurred with his Harvard colleague Roland Dixon that "one can't expect to get much in the way of pure types racially. What one can get is a series of more or less mixed examples, showing characteristic blends in different areas."[87] To find a "typical" individual required statistical studies of racial traits in a population to determine what was typical ("averages," in Hooton's terms) before attempting to find a reasonably representative individual. Thus, Hooton initially advised Henry Field to abandon the idea of depicting actual individuals in favor of a more truly representative aggregate "individual."[88]

Strictly speaking, the abstract, statistical notion of race as a set of characteristics spread in varying frequencies across a population is incompatible with representation in a single human body. "Race" in this sense exists only at the level of the population, so a "type" could only be described in aggregate and statistical terms. Given the long-standing intermixing of human populations and the complex distribution of "racial" traits, Hooton thought an investigator could only rigorously identify typical traits, and possibly typical individuals, by a biometrical study of a sufficiently large population to make results statistically valid. Hooton's suggestion to Field that he could embody racial averages in an individual figure that would be typical shows the tenacity of romantic typologies and the persistent belief that racial categories represented something real. Hooton, like Field, expected to be able to predetermine a set of racial types

---

[87] Roland Dixon to Earnest Hooton, January 17, 1932, in Earnest Hooton to Henry Field, January 19, 1932, Races of Mankind, vol. 13, FM-HF.

[88] Interestingly, this is reminiscent of Francis Galton's photographic experiments with "types." Galton employed composites in an attempt to distill characteristic types for criminals and the chronically ill from the 1870s to the early 1900s. The individuality of Galton's subjects dissolved behind the accumulation of common features in his composites, but practitioners were frustrated by the generality of the resulting types. See Francis Galton, "Composite Portraits, Made by Combining Those of Many Different Persons into a Single Resultant Figure," *The Journal of the Anthropological Institute of Great Britain and Ireland*, vol. 8 (1879), pp. 132–144. For scholars' assessments of Galton's practice, see Allan Sekula, "The Body and the Archive," *October*, vol. 39 (Winter 1986), pp. 3–64; Daniel Akiva Novak, *Realism, Photography, and Nineteenth Century Fiction* (Cambridge and New York: Cambridge University Press, 2008); M. G. Bulmer, *Francis Galton: Pioneer of Heredity and Biometry* (Baltimore: The Johns Hopkins University Press, 2003); Sharrona Pearl, *About Faces: Physiognomy in Nineteenth-Century Britain* (Cambridge, MA: Harvard University Press, 2010).

that would be investigated in the field. The result of Hooton's method still would have been a racial type, but not one that reified race as a natural kind instantiated in every human being. Hooton's proposed biometrical project did not envision an artist, or even an expert anthropologist, easily plucking a typical individual out of the mass of humanity simply by visually sizing them up. Hooton's biometrical, populational notion of race could support the existence of racial types, but it didn't reinforce the "common sense" romantic notion that race was real because it was plainly evident to "the man on the street."

Although anthropologists at the Field Museum acknowledged that pure races did not exist, a program of measurement such as Hooton urged was rejected. Instead the races exhibited were preselected from various extant classifications, before any measurements in the field were taken. Armed with a roster of these types, Hoffman and local anthropologists were expected to identify representative individuals based on a set of trained eyes, not anthropometric verification of their subjects' true representativeness. It was only after Hoffman's subjects had been selected for modeling that they were measured and photographed.[89] Indeed, Arthur Keith supported Field in his plan, arguing that "no more measurements were needed, as the world is full of them."[90] The repeated theme informing Field Museum replies to critics and instructions to Hoffman was the underlying conceptualization of races as discrete, identifiable physical entities. Field seemed relatively unconcerned about the ability of Hoffman or anthropologists advising her to select properly representative individuals because he believed that such individuals existed to be found, whereas Hooton's concern stemmed from his belief that race was a more elusive statistical concept that could not be simply read off of (or onto) any particular individual, and instead could be "seen" only in the aggregate.

Conclusion

In the interwar period, scientists and scientific institutions, including the Field Museum, continued to promote for the public a vision of themselves

---

[89] In some cases, Henry Field gave Malvina Hoffman photographs and provided her with sets of measurements to use in constructing her figures. Hoffman herself also selected photographs to model some of her African subjects, taken from Georges-Marie Haardt's photographs of African people he encountered on his 1926 expedition. See Georges-Marie Haardt and Louis Audouin-Dubreuil, *The Black Journey: Across Central Africa with the Citroën Expedition* (New York: Cosmopolitan Book Corporation, 1927).

[90] To Malvina Hoffman from Stanley Field, September 9, 1931, Box 3, MHC.

as dispassionate purveyors of uncontested facts about the natural world. The *Races of Mankind* presented to the public a simple exposition of scientific data and consensus about natural racial kinds, obscuring the actual nature of the scientific process and the process of racial construction. Contrary to the account commonly offered to the public, science does not proceed simply by scientists observing truths, which the natural world offers for their scrutiny. Exhibition planning is an opportune place to glimpse the much more contested, messy reality of scientific theorizing and consensus building. Crafting displays forced Field Museum planners to finally commit themselves to statements and representations of "fact" about race, even when they and other experts in the field could not agree. Facts about race emerged not as self-evident truths literally embodied in human diversity, but rather through a complex, socially situated process. Despite assertions that racial types were stable biological entities, their actual philosophical, sociological, and material construction renders them acutely unstable, subject to the vicissitudes of societal and disciplinary formations. The Field Museum literally cast in bronze an anthropological compromise, presenting a rigid racial typology, in individual racial figures, along with a racialized peon to the idea of unity within diversity.

Chapter 4 examines in detail the process through which race, as a product of contending anthropological, social, and political ideologies of bodies and cultures, was constructed in the *Races of Mankind* at the Field Museum in the late 1920s and early 1930s.

4

Mounting the *Races of Mankind*

Anthropology and Art, Race and Culture

Introduction

The overtly typological racial scheme, presented as an unproblematic scientific consensus in the *Races of Mankind*, not only masked the profound theoretical and methodological differences among anthropologists planning the exhibition but also obscured the tensions and contingencies that formed a critical part of the process of racial formation at the Field Museum of Natural History. Selecting, finding, and casting an array of human beings to typify principal racial types were far from a straightforward empirical, representational process. The contested, contingent, socially embedded process revealed in the creation of the *Races of Mankind* demonstrates quite clearly how deeply entangled race and culture was in interwar anthropology, even in an exhibition explicitly devoted solely to the form of human bodies.

The *Races of Mankind* was conceived as an explicitly typological exhibit. In the late 1920s, as Henry Field composed and revised a list of racial types for his hall of physical anthropology, he drew upon longstanding as well as contemporary racial classifications that reflected the way race had been defined in Europe and the United States since the eighteenth century. From their earliest instance, in the categories of naturalists such as Carl Linneaus and Johann Friedrich Blumenbach, racial taxonomies reflected the conflation of language, geography, nationality, culture, and physical appearance in zoological classifications that reified "race" as a series of distinct, stratified types, culminating in Europeans.[1]

[1] Carl von Linné (or Carolus Linneaus) divided human beings into *homo europaeus albescens*, *homo americanus rubescens* (later *rufus*), *homo asiaticus fuscus*, and *homo*

The seventy-two races depicted in the *Races of Mankind* hewed to this imperialist vision of racial typology.[2] Each of Hoffman's figures was meant to portray both a general racial type and its particular instantiation in some individual. Thus, Hoffman's "Kashmiri Man," seated cross-legged, peacefully meditating, was meant to represent both the Kashmir racial type from the Indian subcontinent as well as its particular instantiation in the person of Prakash Narain Haksar, a twenty-five-year-old student Hoffman encountered in Paris in the 1930s[3] (Figures 4.1 and 4.2).

But visitors to the *Races of Mankind* never knew who Prakash Haksar was. They were never given the kind of personal information that would have transformed the racial types and their evocative specificity into actual individuality. No names or personal histories were associated with Hoffman's sculptures. Instead, the exhibit confronted museumgoers with a panoply of figures displaying a racial typology that included categories such as "Tamil Climber," "Shilluk Warrior," "Chinese Scholar," and "Nordic Man." Dramatic individuation was not envisioned as a means to evoke the complex realities of the interwar world, in which a young man from Kashmir and a woman from America might both find themselves in Paris on the eve of the French Colonial Exposition, one to pursue his education, another her artistic career.[4]

Instead, at the *Races of Mankind*, Hoffman's dramatic sculptures invigorated old racial typologies, subsuming the particularities of individuals

---

africanus niger. In addition to various supposed defining physical variations in features such as hair, skin tone, and facial structures, Linneaus also distinguished the races by character. Linneaus described *Americanus*, for example, as stubborn, prone to anger, and governed by traditions. He described *Africanus* as cunning, inattentive, and ruled by impulse. *Europeanus* was described, not surprisingly, as clever, inventive, and governed by laws. Carl von Linné, *Systema Naturae, 1735* (Nieuwkoop: B. de Graff, 1964). Johann Friedrich Blumenbach divided humans into Caucasian, Mongoloid, Ethiopian, American, and Malay types. He attributed the cause of human variation to the effects of climate, custom, diet, and other non-hereditary factors. Johann Friedrich Blumenbach, *On the Natural Varieties of Mankind. De Generis Humani Varietate Nativa, 1775,* translated and edited from the Latin, German and French originals by Thomas Bendyshe (New York: Bergman Publishers, 1969).

[2] There were 101 individual sculptures in the exhibition, but a number of these represented the same racial type (i.e., the Ituri Pygmy family has three separate figures, a man and a woman holding an infant; the Australian aboriginals are represented by two men, a woman, and a boy; etc.).

[3] *Field Museum Bulletin*, v. 54 (Feb. 1983), p. 7; "Complete Set of Photographs of Sculptures by Malvina Hoffman," Anthropological Archives, vol. I, 401A, marginalia signed "B. Bronson, 17 Nov. 1980."

[4] Ibid. For secondary literature on race, anthropology, and world's fairs and expositions, see p. 87 n 32.

FIGURE 4.1. Malvina Hoffman's "Kashmiri" figure, *Races of Mankind* (1933). Courtesy of The Field Museum #MH34.

to a taxonomic racial hierarchy and a narrative of human progress and "civilization." The Field Museum constructed a form of racial essentialism rooted in bodies and cultures that resonated with earlier typologies, as well as long-standing racial and ethnic stereotypes, repackaged for a new generation under the guise of scientific authority and artistic genius for mass consumption and edification. In *Races of Mankind*, science was combined with art to evoke the "true" essence of racial types in a way no set of bones, charts of nose shapes, comparative scales, or plaster casts could. Hoffman's art and the individuals she modeled were deployed as part of an authoritative compendium of current anthropological knowledge intended to educate a wide public about the modern

FIGURE 4.2. Prakash Narain Haksar (1906–1969), a student in Paris in the early 1930s. According to his son, Aditya N. Haksar, former ambassador of India in Washington, DC, the family came from Srinaga, Kashmir, India. Courtesy of The Field Museum, #A80798. Samuel B. Grimson.

scientific discipline of physical anthropology. The typological pastiche presented in *Races of Mankind* also reflected not only a long history of anthropological racial constructions but also Malvina Hoffman's humanist ethos, rooted in the folk wisdom of racial stereotypes and assumptions about the superiority of modern Western "civilization."

The *Races of Mankind* was promoted as an unproblematic, uncontested venture, one in which an artist, herself specially selected, could proceed to locate, measure, and re-create in bronze individuals who bore on their physiques the tell-tale marks of a race, one of a finite set of universally recognized living races. This chapter explores the contested process through which the *Races of Mankind* exhibition was mounted at the Field Museum of Natural History in the 1920s and 1930s. The expressive, aesthetic appeal of Malvina Hoffman's sculptures, and their detailed, ethnographic naturalism, combined with the purported empirical rigor of the physical anthropology and scientific authority of the natural history museum, created a powerful vehicle for delivering to the American public a vision of race that promoted both division and unity. The *Races of Mankind* offered visitors a vision of race that was rooted simultaneously in the biological essentialism of racial typology, the cultural currency of ethnic stereotypes, and the idealism of human unity. A tangled mix of racial and cultural theorizing, philosophical and methodological disagreement, compromise and convenience, lay behind the supposedly straightforward science of race presented to the American public by the Field Museum.

Displaying Humans: An Object for the Natural History Museum

Malvina Hoffman's life-size bronze sculptures, many of them full-length figures captured in active poses brandishing various artifacts, were a remarkable departure from typical interwar natural history museum practice, particularly for physical anthropology. Early planning for a hall of physical anthropology had hewed to much more traditional museological practices. Prior to Henry Field's arrival in 1926, Berthold Laufer spent a decade outlining a relatively modest technical exposition of anthropological racial science, presenting man's "place in nature."[5]

Laufer, an expert in Chinese ethnology and language, had taken on the task of planning an exhibition in physical anthropology as head of the department and in the absence of someone with more appropriate expertise. Not surprisingly, then, his plans for a hall reflected the type of content and displays found at other museums around the country. Laufer's earliest proposals featured didactic displays of bones, plaster casts, charts, and photographs, revealing physical anthropology's roots in comparative anatomy, craniometry, and anthropometry. Ultimately, elements of this sort of material and type of display became part of the *Races of Mankind* "special scientific" exhibits that supplemented Hoffman's sculptures in a separate room, rather than the core of the exhibition, as Laufer had once imagined.

By the twentieth century, natural history museums were virtual charnel houses, stuffed with skulls and bones collected for decades. Having

---

[5] Berthold Laufer to F. J. V. Skiff, April 18, 1916, Correspondence of Henry Field, 1916–1933, Malvina Hoffman, History of the Hall of Races of Mankind 1930–1934, Races of Mankind, Department of Anthropology Archives, The Field Museum (hereafter FM-ROM); Berthold Laufer, "The Projected Hall of the Races of Mankind (Chauncey Keep Memorial Hall)," *Field Museum News* (Dec. 1931), vol. 2, no. 12.

centered their practice and classifications for a century on skulls, physical anthropologists looked first to crania and reconstructions of the head to instruct the public about race. Skulls showing the various shapes used for classifying races – dolichocephalic, mesocephalic, and brachycephalic – were commonly displayed in anthropological exhibits, along with other skeletal material and comparative charts of anatomical elements critical to typological racial taxonomies (hair color and form, for example, and shapes of noses, eyes, and lips). Exhibits also often presented demographic and statistical analyses of various "racial" populations. At the American Museum of Natural History in New York, J. Howard McGregor created life-size sculpted busts of human races, showing morphological features important to scientists, adding "as a concession to popular taste" hair and a slight beard.[6] The American Museum's racial reconstructions prior to the 1950s never included full-size figures, however. Ethnological dioramas had long used fully clothed plaster or wax figures posed engaging in social activities or standing still, primarily to illustrate elements of culture (clothing, textiles, ornament), but which suggested race, as well. Some investigators interested in human types created life-size composite figures to illustrate "average" Americans, including a figure based on measurements of WWI U.S. Army draftees created in the 1920s, as well as "Normman" and "Norma," figures created by physician Robert Latou Dickinson and sculptor Abram Belskie, first displayed in 1939 at the New York World's Fair.[7] Human evolution, the other province of physical anthropology, employed illustrations of early hominids, including Neanderthals by Arthur Keith and Marcellin Boule as well as Charles Knight's paintings for American Museum murals. At the Field Museum, Frederic Blaschke created full-size reconstructions of Neanderthals for the *Stone Age* hall, based on Boule's drawings and theories, in a novel plaster-and-wax medium.

After Henry Field arrived as a new assistant curator of physical anthropology in 1926, Laufer offered him the opportunity to craft a new plan for a hall of physical anthropology, anticipating expanded exhibition space in the new museum building in Grant Park. Ultimately, Field's grander plans for two halls, depicting the Stone Age in one and racial types in

6 Quoted in Ronald Rainger, *An Agenda for Antiquity: Henry Fairfield Osborn and Vertebrate Paleontology at the American Museum of Natural History, 1890–1935* (Tuscaloosa and London: The University of Alabama Press, 1991), p. 170.
7 For a discussion of the creation and meaning of these figures, see Julian B. Carter, *The Heart of Whiteness: Normal Sexuality and Race in America, 1890–1940* (Durham and London: University of North Carolina Press, 2007), pp. 1–3, 161 (n. 2).

another, largely supplanted Laufer's earlier, more modest designs. Field Museum officers, trustees, and anthropologists all wanted to mount a dramatic exhibit that would attract visitors, enhance the museum's reputation among its peers, and compete favorably with the World's Fair when it opened across the street. Henry Field sought an artist who could produce representations of humans as racial types in a manner that would provide both the requisite scientific precision and the artistic drama that would "draw the greatest crowds for decades to come."[8] In their quest for an artist who could produce "realistic portraits with an artistic flair,"[9] Field Museum staff were led to Malvina Hoffman at the suggestion of Marshall Field III, a museum trustee and Henry Field's second cousin. By the late 1920s, Hoffman was highly regarded for her portraiture and realistic sculpture, important qualifications to museum staff and trustees, who wanted figures that "anyone could recognize and feel to be authentic without being repulsive."[10] When Assistant Curator Henry Field, visiting Hoffman in her New York studio to discuss the project, saw marble busts of Africans she had done after a trip to Africa in 1926, he was convinced she could attain the standard of realism and drama they sought.

In Malvina Hoffman Field Museum staff hoped they had found a sculptor whose artistic sensibility coincided with their scientific and dramatic needs, one who could skillfully produce elegant and mimetic figures without also introducing a subjective vision. Realist artists such as Hoffman saw their artistic project as the expression of natural forms through detailed life-like renderings, catching their subject in a moment that revealed something essential, and evoking it through the artistic blending of meticulous detail and classical forms. Reflecting on her role in the *Races of Mankind* project, Hoffman described herself as a technician or conduit. "I had to efface my own personality completely," she wrote, "and let the image flow through me directly from the model to the clay, without impediment of any subjective mood, or conscious art mannerism on my part."[11] In the context of the Field Museum sculptures, this kind of realism was the artistic counterpart of empiricism and

[8] Henry Field, *The Track of Man: Adventures of an Anthropologist* (New York: Doubleday & Company, Inc., 1953), p. 132.

[9] Ibid., pp. 189–190.

[10] Malvina Hoffman, *Heads and Tales* (New York: Charles Scribner and Sons, 1936), p. 3.

[11] Hoffman, *Heads and Tales* 1936, p. 12; also Malvina Hoffman, *Sculpture Inside and Out* (New York: Bonanza Books), 1939, and Malvina Hoffman, *Yesterday Is Tomorrow: A Personal History* (New York: Crown Publishers), 1965.

scientific method in its faith in and reliance on observation of presumably stable natural entities. Naturalistic realist art implied a belief that the world could be sensibly reduced to essentials, accurately reproduced, and ordered. Hoffman's bronze sculptures were intended to be permanent, accurate reproductions of human individuals that Henry Field and Berthold Laufer could arrange in a museum hall to represent the natural order of mankind. Hoffman's realistic style suited scientific exhibition because it did not challenge the existence or form of the scientists' natural order or Western social order, as some forms of modernism sought to do through distorted shapes and spaces, but rather replicated them by working within Western traditions of form and aesthetics to evoke "natural" beauty. Hoffman's bronzes implied an acceptance and reinforcement of prevailing paradigms of natural and social order that was conveyed to the viewer through life-like detail, frozen in metal, catalogued, and arrayed in the space of a science museum.

To ensure the utmost authenticity in the figures, Hoffman was instructed to travel across the globe to discover some of her subjects in situ. Characterizing her travels as a scientific expedition to find and bring nature to the public, museum literature never failed to highlight the lengths to which the staff and Hoffman had gone to bring representatives of the world's people to the museum for public edification. Her quest was cast as if she were a zoologist, traveling across the world to find her subjects in the wild and appropriate their life (in clay, rather than through taxidermy) for display in a museum setting.[12] Her travels had the aura of scientific authenticity and authority, of directly observing the truth of naturally occurring racial types in their habitats. Rather than shooting, stuffing, and mounting her subjects in dioramas, Hoffman copied them in clay, to be transformed into a "skin" of bronze, mounted on pedestals in a gallery of humanity. Like Carl Akeley's lions and gorillas, mounted for the African hall at the American Museum of Natural History, Hoffman's figures were promoted to the public as authentic. Naturalistic styles were not presented as harboring a point of view, as a collection of objects tainted with purposive intervention, but as a peephole into reality.[13] That Hoffman actually modeled many of her figures from photographs and workers at the French Colonial Exposition in Paris, expediently selected friends and acquaintances for models, and manipulated her subjects'

---

[12] Donna Haraway, "Teddy Bear Patriarchy," *Primate Visions. Gender, Race, and Nature in the World of Modern Science* (New York: Routledge), 1989, p. 38.

[13] Ibid.

clothing and accoutrements to present more "primitive" types, were facts hidden from visitors to the museum, who were offered a narrative of empirical, objective science cast in bronze.

In sending Malvina Hoffman around the world to consult with anthropologists, observe local people, and then return to the west to cast her impressions into bronze for display in a Chicago museum, the Field Museum was participating in a legacy of institutional imperialism in which artifacts, and occasionally people, were appropriated, via trade or conquest, for study and display. In the nineteenth century, it was not uncommon for individuals representing "exotic" peoples to be brought to Europe or America for display at fairs, museums, and scientific gatherings to educate and titillate. Although the early twentieth century remained an imperial era in which Western depictions of colonial locales often included genuine or improvised "natives," such as those at the 1931 French Exposition Coloniale, museums, despite their need for authenticity, generally eschewed the spectacle of live humans. In the *Race of Mankind* exhibition, the elegance of Hoffman's art and the museum setting removed the sideshow aura from the observation of native peoples, by removing the authentic people and the unscientific carnival atmosphere. Museum patrons could examine exotic people unable to respond or return the look. The exhibit encouraged a kind of voyeurism, sanitized of shame in the culturally approved setting of a museum in which such seeing is given scientific sanction as the pursuit of knowledge and cultural sophistication. The bronzes were exhibited in a setting reminiscent of an art museum, without their cultural context, save for the few objects Hoffman depicted in her sculptures.

In their reliance on objects to convey information and authenticate scientific claims, natural history museums also rely on an epistemology of realism in which sight is the privileged sense, the sense through which we receive knowledge about the world.[14] The *Races of Mankind* exhibit was clearly based on this epistemology, in which the bronze sculptures, as well as the other displays, relied exclusively on the vision of viewers to convey their information, whether textual, schematic, or sculptural. Yet visitors could not resist the inherently tactile nature of sculpture. Despite the dual prohibition, as both a natural history exhibit and an art gallery, visitors encountered the *Races of Mankind* with their hands

[14] Nelia Dias, "The Visibility of Difference: Nineteenth Century French Anthropological Collections," Sharon Macdonald, ed., *The Politics of Display: Museums, Science, Culture* (New York and London: Routledge, 1998), pp. 36–52.

as well as their eyes. Assistant Curator Henry Field noted that visitors were particularly taken with the elongated lip of the Ubangi woman and the Afghan man's navel.[15] The museum counteracted this infraction by reapplying the patina as visitors' touching rubbed it off, maintaining both the illusion that the sculptures were untouchable and the integrity of the sculptures as integrated individuals, rather than artificial creations mimicking life.

The museum setting allowed a kind of contact with exotic peoples impermissible in exhibitions of living people. Just as visitors could gawk at the statues without shame or fear that the figures would respond, so, too, could they give in to their desire to touch the exotic, to possess him or her, without fear. In the ultimate objectification, real people had been transformed into "durable monuments" to racial hierarchy, arrayed for public consumption and scientific study. It was the perfect anthropological collection: living racial types, culled from all over the world, transported to Chicago and arranged in a single hall, permanently preserved, in every detail, and available to sight, touch, and measurement in a way no genuine living subject ever was. It was a gallery of race calculated to be the envy of every physical anthropologist.

The Exhibition Hall: Visualizing Race and Hierarchy

The typological, hierarchical racial taxonomy displayed in the *Races of Mankind* was embodied not only in Malvina Hoffman's dramatic, life-size bronze sculptures but also in the very arrangement of the hall itself. While Hoffman worked in New York and Paris to craft her sculptures, Henry Field and Berthold Laufer worked and reworked blueprints for what started as a *Hall of Physical Anthropology* and became the *Races of Mankind*. The shift in title reflected more than the evolution from working draft to fully realized exhibition. The embodiment of racial types in Hoffman's lifelike sculptures and their arrangement in the museum hall reflected an increasingly hierarchical, narrative presentation. Plans for the hall demonstrate how the exhibition became less a vehicle for

---

[15] Henry Field, *Track of Man*, p. 227. In the 1990s, the sculptures on display around the museum corridors were explicitly labeled, asking visitors not to touch them because it rubs off the patina and destroys the effect intended by the artist, casting the figures as pieces in an art museum rather than as scientific objects. In addition to the Afghan man's navel, the Mangbetu woman's breasts, the Chinese rickshaw puller's poles, the Indian Brahmin's knees and parts of other figures, such as noses and genitals, were shiny registers of museumgoers' behavior.

demonstrating the scientific methods and conclusions of anthropological racial science, as Berthold Laufer had once imagined it, and more and more a vehicle for promoting a vision of hierarchy and racial essence.

Initially, in 1930, Field and Laufer came up with a list of 164 racial types to be depicted in the *Races of Mankind*, a scheme that was progressively whittled down, first to 147 types, and then finally to 100, after President Stanley Field directed them to "cut out some of the less important types," once he realized how expensive a hall full of bronze sculptures would actually be. According to Henry Field, when Stanley Field received Hoffman's initial estimate, he nearly scrapped the project, saying she "proposed a six-figure sum of such proportions that it sounded like a national debt."[16]

When the *Races of Mankind* opened, visitors first encountered three successive galleries of sculptures, followed by a smaller room devoted to the methods, data, and conclusions of physical anthropology. In the main galleries, visitors viewed ninety-four life-size bronze and four stone sculptures set on pedestals. Each sculpture was labeled with a descriptive name, a small map locating the race geographically, and a notation indicating which of the three principal human racial stocks the figure represented: Caucasoid, Negroid, Mongoloid, or some mix of these. Thirty of the sculptures were full-length figures meant to represent the principal races; an additional sixty-eight life-size heads and busts represented important subdivisions associated with those primary racial types.[17] The

---

[16] Field, *Track of Man*, pp. 134, 193; Henry Field, "The Story of the Hall of Physical Anthropology," March 12, 1930, Folder: "Found original in V. II 1916–1930, plans, proposals, notes," Malvina Hoffman, FM-ROM.

[17] When complete, the exhibition contained 101 figures. Of these, three were heroic-size figures included in the "Unity of Mankind" sculpture at the center of the exhibit. These three figures represented what the Field Museum regarded as the main human racial divisions – Caucasoid, Negroid, and Mongoloid – not particular racial types. They were each six feet, eight inches tall. Of the remaining ninety-eight life-size figures, sixty-eight were heads or busts and thirty were full-length figures. When the exhibit opened June 6, 1933, nine of the life-size figures had not yet been completed by Malvina Hoffman (a full-length Navaho, a head of a Bedouin, and busts of a Turkish man, Igorot man from the Philippines, Pueblo woman and Jicarilla Apache man from New Mexico, Carib man from South America, Berber man from Morocco, and Toda man from India). All but the Bedouin head were added to the exhibition in October 1934. In addition, four of the heads were carved in stone. Thus, there were a total of ninety-seven bronze figures (ninety-four life-size) ultimately exhibited in the *Races of Mankind*. "Eight New Sculptures of Racial Types Added to Chauncey Keep Memorial Hall," *Field Museum News*, vol. 5, no. 11 (November 1934), p. 3; Field Museum of Natural History; *The Races of Mankind*, Popular Series, Anthropology Leaflet 30, Second Edition (Chicago: Field Museum of Natural History, 1934), pp. 42–44; Hoffman, *Heads and Tales*, p. 339.

figures were arranged by geographic region within a hall comprised of two long rectangular galleries on either side of a central octagonal room. Full-length figures dominated the hall; heads and busts were shown in alcoves near the figure representing the principal race to which they were related (Figure 4.3).

The center of the exhibit was an octagonal room dominated by a monumental statue entitled the "Unity of Mankind." The fifteen-foot statue featured a pillar, surmounted by a terrestrial globe, and surrounded by three "muscular" men representing the main human races – black, white, and yellow – positioned under the continent associated with his "racial stock" and each grasping "the weapon by which it has defended its own boundaries"[18] (Figure 4.4). Laufer, who had advocated for the sculpture, intended it to emphasize the unity of mankind as one species, and mark the heart of the exhibit. In promotional literature he described the statue as a symbolic triumvirate emphasizing humanity as a "well-defined, fundamentally uniform species, which has spread all over the surface of the earth and conquered almost every inhabitable spot."[19] Surrounding this "imposing central monument" were European and indigenous American races: the full-length Nordic, Sicilian, Navaho, and Blackfoot figures, along with heads and busts of other European and American races.[20] Leading up to and away from the central room were what Laufer termed "avenues of primitives" – the African and Oceanic races as visitors entered the hall, Asians as they headed toward the technical displays at the back.[21]

Following the exhibit through the galleries of bronzes the visitor came to the last section of the hall, a room devoted to what museum literature

18 Hoffman, *Heads and Tales*, p. 335.
19 Ibid.
20 Berthold Laufer, "Projected Hall," When the exhibit opened, a number of figures remained to be completed. (See Footnote 17). Between the opening in June 1933 and the completion of the Native American figures in October 1934, the central room contained some of the Asian figures. See Laufer, "Hall of Races" and "Eight New Sculptures." The central room, when complete, contained four full-length figures: the Sicilian Man and Nordic Man on the right and the Blackfoot Man and Navaho Man on the left. The remaining busts and heads were, from the right, the Italian Man, French Man, Lapp Man, Englishman, Breton Woman, Basque Man, Austrian Man, Russian Man (formerly Caucasian), Eskimo Woman, Eskimo Man, San Ildefonso Woman, Jacarilla Apache Man, Tehuelche Man, Sioux Man, Carib Man, and Maya Man. Box 3, Malvina Hoffman Collection, 850042, Special Collections, The Getty Center for the History of Art and the Humanities, Los Angeles, California (hereafter MHC).
21 Ibid.

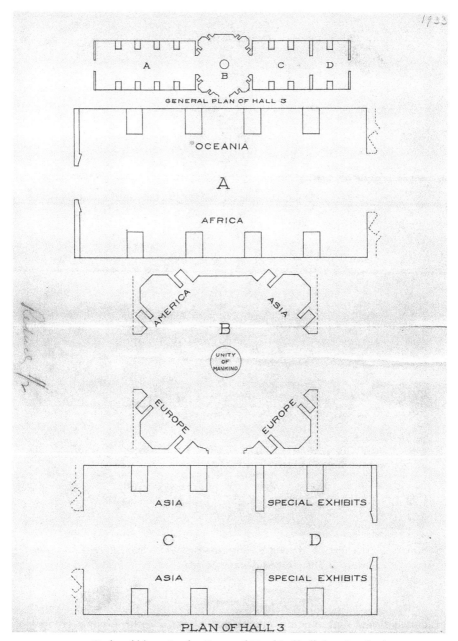

FIGURE 4.3. Undated blueprint for *Races of Mankind* hall showing final arrangement and layout. Courtesy of The Field Museum, GN91846d.

FIGURE 4.4. "Unity of Mankind," Showing the White man, *Races of Mankind* (1933). Courtesy of The Field Museum, #MH89.

called "special scientific exhibits." These exhibits, conforming to more conventional museological practices in physical anthropology, compensated for the inability of Hoffman's bronze sculptures to illuminate the actual methods and criteria used to define races, and addressed – or suggestively avoided – the evolutionary and sociological context for the museum's racial taxonomy. Approximately a quarter of the hall was devoted to these quantitative and comparative displays, including examples of actual hair types, charts on eye color, skeletal displays, and photographic transparencies depicting racial types in color.[22] The technical exhibits outlined some of the methods of physical anthropology, provided statistical and geographical information, displayed comparative anatomy, and illustrated cultural and behavioral attributes associated with race. The existence of the special scientific exhibits indicated the extent to which the curators themselves found Hoffman's realistic figures insufficiently informative.

As plans for the hall went through successive modifications between 1927 and 1933, the arrangement of sculptures became increasingly hierarchical, while the theoretical, technical, and methodological elements were increasingly marginalized. While Henry Field and Berthold Laufer were preoccupied with conforming their hall to their own and other anthropologists' (not always compatible) expectations for a hall that would represent the world's "principal" races, and secondarily, the practice

[22] Malvina Hoffman, Stanley Field, Berthold Laufer, and Henry Field discussed the use and placement of transparencies repeatedly throughout the years of planning for the hall. Hoffman and her husband, Sam Grimson, who was responsible for the film and photographs taken on their "world tour," selected photographs to be displayed and supervised the coloring of the transparencies. Some sketches of the hall show transparencies in the hall alongside Hoffman's sculptures, and correspondence reveals Hoffman's development of a variety of schemes for their display, arguing that they would provide "local color and life-like interest to the bronze subjects and [add] a great deal of supplementary value to the collection." Her attention was focused on positioning them with her sculptures, but in a manner that would not detract from her figures by casting unflattering light or colors on them, attracting disproportionate attention, or disrupting the streamlined gallery design. Her ultimate, preferred scheme involved setting the transparencies inside walls adjoining the sculptures, lit from the inside when activated by visitors using buttons in the display (an early example of interactive exhibition), so that light from the inset boxes would not be continuously cast on the sculptures. It appears that this scheme was never put into place. Shortly before the exhibit opened in 1933, Stanley Field wrote Hoffman that they would have to put off displaying her transparencies. Photos of the exhibition either do not offer vantage points that shed light on this question, or seem to suggest that these transparency insets were not built. Moreover, a diagram of the technical portion of the exhibit indicates two cases devoted to revolving sets of transparencies, as does the pamphlet printed to accompany the exhibition. Malvina Hoffman to Stephen Simms, April 15, 1933, Box 16, Simms, S. C., Director of Field Museum, Correspondence, 1930–1935, 850042-2, MHC; Stanley Field to Malvina Hoffman, April 15, 1933, and Malvina Hoffman to Stanley Field, May 11, 1933, Box 16, Field Stanley, President of Chicago Natural History Museum, Correspondence (c. 1932–1935), 850042-1, MHC; "1932 orig. Vol. I" Floor Plans, Blueprints, Hall of Physical Anthropology, From vols. I & II, Malvina Hoffman, FM-ROM; *The Races of Mankind*, 1934, p. 40; Henry Field, *The Races of Mankind*, Popular Series, Anthropology Leaflet 30, Fourth Edition (Chicago: Field Museum of Natural History, 1942), p. 47.

of physical anthropology, the visual and experiential impact of the hall was never far from their minds.[23] The basic plan to arrange Hoffman's sculptures in geographic groups remained the same from the beginning, reflecting long-standing assumptions about the regional and continental basis of racial types. But significant changes were made over time to the shape and appearance of the hall, the positioning of European and non-European races, the location of the "Unity of Mankind" statue, and the placement and arrangement of the "technical" exhibits.

Initially, the schematic plans for the *Races of Mankind* hall suggest no obvious racial hierarchy. The earliest plans, drafted in 1927, depict various races grouped by geographic region ("Europe," "Africa," "Mongol," "American," "Australia," "South Seas"), arrayed around the perimeter of a single rectangular hall, interspersed with display cases offering anthropological data and sociological information (including "comparative anatomy anthropoids & man," and "Negro Problem" positioned between "Africa" and "Australia").[24] By 1929, most of the technical displays had been relegated to the far end of the hall (including an entire wall devoted to "U.S. Racial Problems"). The blueprint showed more racial types, and now indicated precise spatial arrangements in a series of alcoves, but the arrangement still did not imply a full-blown racial ranking. Indigenous Americans were positioned between Indonesians and North Asians, and across the hall from East, West, and Central Asians, suggesting anthropologists' theory that Native Americans were descended from Asians who had migrated across the Bering Strait, but the arrangement implied no particular relationship or rank. Europeans were set in the middle of the gallery, on the south wall, between Africans and Asians, and directly opposite Polynesians. The "Unity of Mankind" figure appears for the first time on this 1929 diagram, positioned toward the end of the rectangular hall, centered across from indigenous American types on the north wall and Asian types on the south. With the dominating "Unity" statue at the end of the hall of races, the Europeans in the middle occupied no special place – indeed, given their small number and the correspondingly small amount of space devoted to them, they seemed rather insignificant in the story of human variety. Although the hall opened with

[23] For example, anthropologists were concerned that visitors have an uninterrupted view of each full-length figure from across the room. A 1928 blueprint noted "feet of figure visible to a 5′ line of sight at distance of 20′." Document in "1916–1930 plans, proposals, notes, found orig. in vol. II," FM-ROM.

[24] Plans from 1927, Folder: "Found Original in V. II 1916–1930, 1916–1930, plans, proposals, notes," Malvina Hoffman, FM-ROM.

Africans on one side, Tasmanians, Australians, and "Negritos"[25] on the other (along with a display on humans and apes), implying that the visitor was encountering the most primitive races first, Europeans did not hold a correspondingly exalted position, but were simply one set of races among many arranged around the remainder of the hall.[26]

Significant changes to exhibition designs took shape in 1930 and 1931. Toward the end of 1930, while Henry Field was traveling in Europe visiting anthropologists and museums, researching collections, and purchasing items for his halls, Stanley Field cabled him that a decision had been made to move the *Races of Mankind* and the *Stone Age* exhibits to more prominent, larger halls on the first floor of the museum. This meant that the space for Hoffman's bronzes increased from 75 feet to 160 feet, a move that provided room to arrange the sculptures in a way that would give her figures, each stationed on a platform or nestled in a lighted niche, its own distinctive space, instead of crowding her works together like so many skulls in a display case.[27] Then, in June of 1931, while visiting Hoffman at her Paris studio, Stanley Field and Marshall Field "agreed that the idea of dividing the entire space into two Halls would be advisable." The Fields also discussed various details regarding furnishing, finishes, and lighting, agreeing to model the museum hall in Chicago on the modernist style of the Dutch Pavilion at the French Colonial Exposition.[28] In addition, they "decided to do away with as many glass cases as possible."[29] The cumulative effect of these changes was to transform the hall from a traditional natural history exhibit filled with glass cases full of specimens, instruments, and charts into a space more

---

[25] "Negrito" is a term coined by European explorers and used by anthropologists in this period to refer to indigenous groups, generally of short stature and dark skin, in the Malay Peninsula and Philippines.

[26] Diagram from 1929, Malvina Hoffman, FM-ROM.

[27] Stanley Field to Henry Field, Nov. 12, 1930, Races of Mankind Correspondence, vol. 12, 1920–1950, Henry Field Papers, Museum Archives, The Field Museum (hereafter FM-HF).

[28] "Paris, Consultations with Anthropologists, June 1931," Box 16, MHC. In her notes on the visit, Hoffman remarked that both Fields "heartily agreed" that the natural-wood background and "severe plain lines for cases, benches, stands, etc would be the best scheme to follow in the Hall of Man" and that "the center main group" should be lit to create a "circular effect."

[29] "Paris, Consultations with Anthropologists, June 1931," Box 16, MHC. An early, undated sketch for the "Hall of Physical Anthropology" show a 175-foot hall filled with glass cases displaying the same racial types and technical displays as the 1927 and 1929 plans. Document in Folder: "Found Original in V. II 1916–1930, plans, proposals, notes," Malvina Hoffman, FM-ROM.

like an art museum, whose architecture, materials, lighting, and layout highlighted the sculptural works on display.

By 1933, the final hierarchical plan for the hall had been worked out in detail. In the pamphlet that accompanied the exhibition, Henry Field offered visitors an implicit itinerary that seemingly belies the racial hierarchy evident in the arrangement of Hoffman's sculptures and the space itself. Field told visitors that the races "are exhibited in the following order: Africa, Europe, Asia, America and Oceania," and his guidebook proceeded to discuss the races in that order. But to follow Field's path through the exhibition hall would have taken visitors on a journey to and fro through the halls, passing by the "special" displays (discussed in a separate, very brief section). Museumgoers following Field's advice would have entered the hall, attending to Africans on their right while ignoring Oceanic figures on their left; proceeded through the central room, noting only the Europeans; moved on into the hall on the far side, viewing all the Asians, skipping the "special" exhibits, instead turning back through the octagonal center to study the American races; and ended up where they started, to finally examine the Oceanic races. A much more plausible itinerary would have reflected the nature of the space and the figures in it. It seems much more likely that visitors, seeing the heroic "Unity" figure framed at the end of a long hall, would have moved toward it, down Laufer's "avenue of primitives," linger in the central room at the peak of the exhibition, and then continue into the succeeding gallery, ending in the room of technical displays.

Between 1927 and 1933, plans for how to showcase race within the halls of the Field Museum changed quite dramatically, a point noted by Stanley Field, who remarked to Hoffman nine months after the hall opened: "Some change what?"[30] The early sketches offered no explicit hierarchy of racial types shaped by the spatial experience of the hall or the placement of the sculptures. The first versions interspersed informational charts and displays among the racial figures in a single hall. Even when subsequent designs grouped the technical displays at the end of the hall, the placement of the commanding "Unity" sculpture would have effectively pulled visitors toward those special exhibits. This early vision created the sense that technical information on races, their relationships and evolution, were as important as Hoffman's figures, and that physical anthropology might have more to say about race than merely reinforcing

[30] Stanley Field to Malvina Hoffman, March 2, 1934, Box 16, Field Stanley, President of Chicago Natural History Museum, Correspondence (c. 1932–1935), 850042-1, MHC.

ethnographic stereotypes. Instead, from 1927 to 1933, Europeans had been moved from a relatively egalitarian placement as three races among many to a central and culminating position, indicated by both the architectural form of the space and the placement of the monumental "Unity of Mankind" statue. As ultimately designed and promoted, the exhibition reinforced "common sense" hierarchical notions about race, validating it by exhibiting Hoffman's sculptures in a science museum while simultaneously de-emphasizing the importance of anthropological expertise and method in delineating it.

Why Bronze? Reifying a Racial Hierarchy

Apart from the design of the museum hall itself and the placement of the racial figures and other informational displays within it, the most critical decision that shaped the eventual outcome of the exhibition was the selection of an artist – Malvina Hoffman – and the casting of all the racial figures in bronze, rather than in a wax-and-plaster composite as the original contract stipulated. The Field Museum went to great lengths to impress upon the public the dual nature of Malvina Hoffman's sculptures as authentic, accurate scientific objects and as dramatic expressions of humanity, as products of both scientific rigor and artistic genius. Prior to the exhibition opening, a press release described the project as "a unique plan of combining art with science without sacrificing either the beauty of the former or the exactness of the latter," and reassured potential visitors that "while the strictest scientific exactitude in regard to details and measurements govern the creation of these sculptured figures, the highest degree of artistic expression is being permitted with the restriction that no license shall be taken with facts." The result, they claimed, would be "a series of works of art which are at the same time perfect examples of human racial types."[31] Curator Berthold Laufer compared the exhibit to a convention of live racial representatives who had been frozen in bronze for the benefit of science and education, "to facilitate study of their characteristic features and preserve them permanently."[32]

Bronze was an unusual, but not unprecedented, medium in anthropological exhibits. Henry Field was familiar with both Aleš Hrdlička's use of bronze face casts for the Panama-California Exposition in 1915 and

[31] Field Museum of Natural History, press release, May 31, 1933, Box 16, Correspondence, MHC.
[32] Berthold Laufer, "Hall of the Races of Mankind (Chauncey Keep Memorial Hall) Opens June 6," *Field Museum News* (June 1933), vol. 4, no. 6.

Herbert Ward's sculptures of Africans exhibited at the National Museum of Natural History in Washington, DC.[33] Inspired by Ward's bronzes and imbued with the traditional typological view, Field outlined an exhibit that would present the public with a room full of the world's races, artistically embodied for maximum impact.[34] Curator Berthold Laufer promised museumgoers the exhibit would offer "the most perfect representatives of all living races . . . transformed from life into bronze."[35] But there are a number of obvious drawbacks to the use of bronze as the primary medium for an exhibition on race and physical anthropology. Unpacking the decision process and rationales sheds light on bigger issues at work in the *Races of Mankind*.

Like so much else in the construction of the *Races of Mankind*, the voices and perspectives on the question of what medium to employ to best capture the "reality" of race were multiple and conflicting. Accounts vary. But it is clear that questions of willpower, institutional power, and money were at least as critical as anthropological concerns. Hoffman contended that she went ahead and cast two of her first five figures in bronze and surprised museum President Stanley Field with them when he came to Paris to see her work. She claimed he was immediately taken with them, saw the advantages of bronze, and promptly consulted with Marshall Field to raise the necessary additional funds.[36] Henry Field, on the other hand, claimed that the decision to go with bronze was made some time after Hoffman met with Arthur Keith in England. He claimed that he and Keith weighed Hoffman's objections and talked over the problem of the sculptural medium "time and time again," finally being "won over to the bronzes, with a few stone heads to break the monotony" (Hoffman's precise characterization of the exhibit in her 1936 account of the project, *Heads and Tales*). Field then supposedly returned to Chicago and consulted with Laufer, who agreed, particularly after Hoffman suggested using patinas to suggest skin color, and recommended to Stanley

33 Mary Jo Arnoldi, "Herbert Ward's Ethnographic Sculptures of Africans" in *Exhibiting Dilemmas, Issues of Representation at the Smithsonian*, ed. Amy Henderson and Adrienne L. Kaeppler (Washington, DC: Smithsonian Institution Press, 1997).

34 Henry Field intended to bring the Field Museum accolades and high attendance with an ambitious and accessible set of exhibits. Recalling his plans for the two halls, he wrote, "The visitor should be able to walk around this hall in half an hour and obtain an impression of the principal racial groups of the world's two billion people." Field, *Track of Man*, p. 192.

35 Laufer, "Projected Hall."

36 Hoffman, *Heads and Tales*, pp. 11–12.

FIGURE 4.5. Stanley Field and Malvina Hoffman in the garden of her Paris studio, with the "Senegalese Drummer" and "Sara Girl (Daboa)," (and Kiki, Hoffman's Siamese cat). Courtesy of The Field Museum, #GN80305.

Field that the contract be amended.[37] However, archival records suggest a more protracted and incremental process in which Henry Field played a minor role and Stanley Field an important one (Figure 4.5). Initially, only a few of the figures, in addition to the "Unity of Mankind," were to be done in bronze. An amended contract stipulating that all the life-size figures be cast in bronze was not drawn up until November 1930, nine months after Hoffman was hired. Stanley Field then cabled Henry Field, who was traveling in Europe, that this decision had been made.[38] The first bronze sculptures did not arrive at the museum until December 1930. These two – the Blackfoot Indian and Nordic – were installed in Stanley

---

[37] Field, *Track of Man*, p. 197.
[38] Stanley Field to Henry Field, Western Union Cablegram, November 12, 1930, Races of Mankind Correspondence, Henry Field, vol. 12, 1920–1950, FM-HF. The cable reads:

> Your cable not sufficiently explicit what part of hall do suggested changes refer to and do they affect miss hoffmans work for your information all full length figures are to be in bronze with duplicate cast in plaster for study collection busts as originally planned masks to be full heads have changed location keep hall and pre historic hall to first two halls on north side ground floor larger and better in every way miss hoffman advises you will see her london eighteenth do nothing to change her plans.

Field's office until the exhibit opened. Stanley Field wrote Hoffman that they had arrived, telling her that he liked them "immensely in bronze" and that "when Dr. Laufer saw them he was most enthusiastic and all I can say is that if it had not been previously decided to do these in bronze, there would have been no trouble getting his agreement to it, after he saw those two figures."[39] Nonetheless, a report dated June 1931, more than a year after the project was under way, suggests that museum trustees still needed to be convinced of the advantages of bronze. Hoffman wrote that she would ship over from Paris the first bronzes along with two colored plaster heads, "so that the Trustees may have before their eyes both versions of the same subject."[40]

From the outset, Hoffman was involved in the arrangement and appearance of the *Races of Mankind*: she convinced Field Museum staff that she alone should be responsible for sculpting all the figures, rather than hiring four artists as Henry Field had originally planned; she insisted on collaborating in the design of the museum hall to hold her sculptures, which entailed unprecedented construction, unusual wall and floor materials, and new lighting styles[41]; throughout the modeling process Hoffman had great latitude in selecting individuals to pose as racial types; and perhaps, most crucially, she prodded and agitated for an entire gallery of bronze sculptures.[42] Although Hoffman's original contract called for composite wax-and-plaster figures, she had no desire to create such things and expressed her dismay at the prospect at the outset. Her artistic sensibility (and perhaps her professional pride) seems to have been offended by the practice.[43] She found the plaster-and-wax figures ugly and lifeless as well as inaccurate, and knew they were prone to decay without proper maintenance. She blamed poor attendance at anthropological exhibits on the dreary character of such displays, a sentiment that some

[39] Stanley Field to Malvina Hoffman, December 22, 1930, Box 3, Correspondence F–G, Field, Stanley (Field Museum), 850042-51, MHC.

[40] "Consultations with Anthropologists," June 1931, Box 16, Hall of Man, MHC.

[41] Correspondence documents the detailed and frequent discussion on all aspects of the gallery: floors, ceilings, walls, alcoves size and construction, bases, number and color of benches, lighting, and positioning of figures on bases. See, for example, Stanley Field to Malvina Hoffman, Jan. 30, 1933; Feb. 15, 1933; April 28, 1933; May 2, 1933; May 16, 1933; May 18, 1933; Box 16, Hall of Man, MHC.

[42] Field, *Track of Man*, pp. 193, 197, 209; Hoffman, *Heads and Tales*, pp. 4, 9–12.

[43] Ironically, Hoffman objected to the hyperrealism of wax-and-plaster figures desired by Henry Field and Berthold Laufer. According to Field, she objected to "an international Madame Tussaud's waxworks with nearly nude figures," the very epithet with which her bronze figures were later dismissed by some critics. Linda Nochlin, "Malvina Hoffman: A Life in Sculpture," *Arts Magazine*, vol. 59 (November 1984), pp. 106–110.

anthropologists shared. Berthold Laufer eventually concurred with her comparison, characterizing Hoffman's bronze figures as "dramatically conceived and intense with life and motion . . . far removed from the ordinary plaster busts of racial types."[44] Hoffman continued to remark on the poverty of plaster or wax figures and extol the virtues of bronze long after the decision had been made in favor of bronze. The summer after she accepted the Field Museum commission, Hoffman visited the Museé Trocadero in Paris and saw painted plaster figures, "some with real hair glued visibly to the faces and some with modeled hair." She was typically appalled by such "monotonous dummies,"[45] noting, "most casts are badly finished, quite ghastly and repelling. The clothes have been collecting dust for 30 years so you can guess the result. – We have a great chance to do something new and better in this line!"[46]

Bronze was not, however, without problems. It was much more expensive than any other material commonly used in natural history museum displays, and, although it doesn't decay the way plaster-and-wax figures do, it does require special expertise to maintain over time, expertise that natural history museums do not typically possess. More significantly, bronze was hardly the ideal medium to convey racial characters through sculpture. Although the *Races of Mankind* bronzes may have been more compelling, lively, and life-like in their active poses and carefully detailed features than inanimate plaster figures ever were, they remain patently artificial. Strikingly, bronze failed to convey many physical features crucial for racial comparisons, including skin color. To partially remedy that deficiency, Hoffman had the figures coated in chemical patinas to suggest variation in skin tone: the native Americans are reddish, African figures are nearly black, and East Indian and Asian figures are shades of tan.[47] Field and Laufer also attempted to ameliorate the failure of bronze sculptures to convey physical features crucial to racial comparisons, particularly colors and textures, with some of the displays in the technical exhibits. In the end, the museum exchanged relative accuracy and authenticity for dramatic impact and the unique qualities of bronze sculpture.

Ultimately, museum staff accepted the use of bronze to model race for a variety of reasons, not least of which was the impact a gallery of

[44] Laufer, "Projected Hall."
[45] Malvina Hoffman to Stanley Field, March 22, 1932, Box 3, Correspondence F–G, MHC.
[46] Malvina Hoffman to Stanley Field, July 3, 1930, Box 3, Correspondence F–G, MHC.
[47] Hoffman, *Heads and Tales,* p. 12.

101 life-size sculptures would have on the public. Beyond a spectacle that would "make the world take notice,"[48] the literal and metaphorical permanence of bronze clearly appealed to anthropologists, museum officials, and Hoffman. Echoing the preservationist rationale anthropologists repeatedly offered to justify their practice, Hoffman argued that her sculptures would preserve people on the verge of extinction. "Before the end of this century," she wrote, "many of the primitive races which are now represented in bronze in this hall and modeled from living subjects will have disappeared into the dim records of history."[49] And from Hoffman's point of view, the use of a fine art medium positioned her as a prominent portraitist tackling an enormous commission to create a series of serious artworks for a natural history museum, rather than an artist for hire creating a "wax-works" gallery in a museological composite medium. As an artist, Hoffman wanted to leave behind dramatic bronze portraits, not a series of wax-and-plaster dummies destined to a dusty, crumbling future.

Trouble with Typology

An essential tension in constructing typologies is the balance between generalized archetypes and specific instantiations. The tension in typologies between unity and diversity, generality and specificity, was embodied in the "Unity of Mankind" statue, the exhibit's "central and dominating motif."[50] Representing humanity as "a well-defined, fundamentally uniform species, which has spread all over the surface of the earth,"[51] the "Unity" figures were intended to represent all members of their "stock" and yet also be minutely accurate in their particular features, just as each of the racial types in the exhibit were supposed to be both precise copies of live individuals and representatives of broader natural kinds that encompassed a range of variation. Whereas the life-size figures lining the exhibition hall, such as the "Kashmiri Man," (Figure 4.1) emphasized the localization of typical racial characteristics in single individuals, and by implication the existence of racial categories in nature, the heroic-size "Unity" figures emphasized archetypal racial generalization as idealized bodies, cobbled together from typologically "perfect" parts

---

48   Stanley Field to Malvina Hoffman, February 22, 1930, Box 3, Correspondence F–G, MHC.
49   Hoffman, *Heads and Tales*, p. 11.
50   Field, *Track of Man*, p. 192.
51   Laufer, "Projected Hall."

of various individuals to properly represent the entirety of their racial division. The legs of the white figure, for example, were based on those of Davis Cup champion Malcolm Whitman; the torso was modeled on another man.[52] As ideal figures, each embodied "the highest qualities of his race" and was "worthy of minute study," according to Laufer.[53] In contrast, each racial type, as a member of a hierarchy, a rung on a ladder, was identified with labels indicating the specific type (e.g., "Ainu Man" or "Jicarilla Apache"), the country of origin, and the main racial stocks to which it belonged. Whereas the black "Unity" figure represented all "Negro" races, the white figure all "Whites," and the yellow figure all "Mongoloid" races, the individual racial types were often categorized as mixtures of the primary stocks: the Ainu of Japan were characterized as a "White-Mongoloid Mixture"; the Hawaiians and Samoans were described as a "White-Mongoloid-Negro Mixture."[54] The "Unity of Mankind" sculpture was intended to emphasize mankind's common humanity, a message of unity in diversity, a unifying focal point to an exhibition otherwise preoccupied with differences. But that message was muffled by the relentless logic of a racial typology and representational strategy that offered the Field Museum no method for highlighting the qualities that joined human beings of every stripe in the single class *Homo sapiens.*

The acknowledgment in exhibit labels that the races depicted represented mixed "stocks" reveals the conceptual confusion underlying the scientific study of race. By the late nineteenth century, anthropologists despaired of ever achieving consensus on racial classifications of clearly delineated types. As data on supposedly racial traits and racial groups rapidly accumulated, it became increasingly difficult to define types with confidence, divine their origins, or elucidate their relationships. Racial science increasingly became a search for the abstraction of pure races among the messy heterogeneity of living (and even prehistoric) people. The omission in the exhibit of all such conundrums was consistent with the Field Museum's depiction of science as a powerful tool for organizing the world based on reason and quantified observation, and of racial types as unproblematic natural categories. How mixed types fit into the exhibit's three-legged racial scheme was left unclear. The exhibit sidestepped questions

[52] Field, *Track of Man*, p. 198.
[53] Laufer, "Projected Hall" and "Hall of the Races of Mankind."
[54] Henry Field, *The Races of Mankind*, Popular Series, Anthropology Leaflet 30 (Chicago: Field Museum of Natural History, 1933).

about the evolution of races – such as how sufficient isolation was maintained to perpetuate racial types, where these types arose and when, or how they might have changed over time. The *Races of Mankind* was a snapshot of contemporary world racial types, arranged according to continents and countries in which they were found in 1933. By removing the dimension of time, the Field Museum could leave unaddressed questions of evolution and all the attendant conceptual, definitional, and evidentiary problems. As a collection of contemporary types, the exhibit instead emphasized the nature of racial essences and the living relations between the world's peoples, framed in terms of primitivism and "civilization."

The Field Museum's romantic typology reflected cultural and linguistic lineages as much as apparent physical variations. The particular conflation of culture and biology in the *Races of Mankind* exhibit reflected the hierarchical racial typology embraced by Henry Field and the anthropologists he turned to for advice. Human types were defined by a concept of race that linked skills, character traits, and level of "civilization" to bodily forms, arrayed in a range from the most primitive to the most civilized. The sculptures themselves implied this, both in the way they were arranged in the hall, with "avenues of primitives" leading to the central space featuring the "Unity of Mankind" and European types, and also in the way nearly all the figures, including heads and busts, were portrayed clothed and accompanied by objects such as drums or weapons, and posed so as "to appear in lively action befitting the behavior of [their] particular group."[55] For example, a Tamil man from India, described by Hoffman as climbing a palm tree to collect toddy for wine, is wrapped in textured cloth, climbing with the aid of ropes around his feet and chest, and carries a basket slung around his waist (Figure 4.6). The sculptures' active poses and cultural trappings encouraged viewers to believe that they were encountering actual individuals caught in an authentic moment of life, a series of voyeuristic snapshots. No doubt the elements of material culture aided visitors in readily identifying the racial types on display, particularly given the stereotypical nature of the figures (a Hawaiian surfing, an African warrior with a spear, an Australian boy holding a boomerang, etc.). Poses, such as tribal dances, hunting stances, and praying postures, were presented as racially characteristic, implying a biological basis for culturally specific behaviors. Although anthropologists at the Field Museum were not convinced Hoffman's Senegalese drummer was strictly typical of West Africans in his facial features, his place in the *Races of Mankind* was justified by "his musicianship,

---

[55] Laufer, "Projected Hall."

FIGURE 4.6. "Tamil Climber," *Races of Mankind* (1933). Courtesy of The Field Museum, #MH43_A.

especially his complete absorption in his rhythms, [which was] typical of Negro players in many parts of the continent"[56] (see Figure 4.5).

For her part, Hoffman claimed that she "watched the natives in their daily life" and then selected for her figure's pose "the moment at which I felt each one represented something characteristic of his race, and of no

[56] "Senegalese Drummer in Bronze by Malvina Hoffman," *Field Museum News*, September 1936, p. 4.

FIGURE 4.7. Malaysian cockfighters, including Hoffman's Javanese boy and a "girl" from Bali, *Races of Mankind* (1933). Courtesy of The Field Museum, #MH88.

other."[57] Indeed, so crucial were the cultural accoutrements and behaviors in identifying race, they had to be improvised when particular models were insufficiently accessorized or animated. For example, Hoffman expediently acquired an Indonesian model from the French Colonial Exposition. Whisked from the service entrance of a Balinese restaurant to her studio in Paris, the boy was put in a batiked loincloth and directed to strike a prearranged pose[58] (Figure 4.7). Other models, such as the

[57] Hoffman, *Heads and Tales*, p. 12.
[58] Ibid., pp. 172, 174.

FIGURE 4.8. Samoan man, *Races of Mankind* (1933). Courtesy of The Field Museum, #MH73.

Javanese boy, were posed with artifacts and clothing that did not reflect the reality of the state in which Hoffman found them, but instead served to portray them as primitives. A young Samoan man, for example, was posed holding an ancestral war knife and presented as a warrior, as if he was still wielding it in 1931, when in fact he was a member of a Mormon colony in Hawaii[59] (Figure 4.8).

The types themselves demonstrate the blurred nature of the supposedly purely somatic categories offered in the *Races of Mankind*. Most of the racial names were in fact tribal or regional ethnic names, such as Shilluk, Mangbetu, Somali, Vedda, Tamil, Navaho, Dyak, and so forth. Many of the types also incorporated information of a linguistic or purely socio-logical nature: the Toda and Tamil types were listed as members of the "Dravidian group," a language group; two of the Chinese figures were listed as a "type of scholar," another is described as a "jinriksha coolie."

[59] Ibid. For a discussion of race and the Paris Colonial Exposition, see Jennifer Ann Boittin, *Colonial Metropolis: The Urban Grounds of Anti-Imperialism and Feminism in Interwar Paris* (Lincoln: University of Nebraska Press, 2010). Another interesting discussion of empire, colonialism, museums and expositions is Matthew G. Stanard, *Selling the Congo: A History of European Pro-Empire Propaganda and the Making of Belgian Imperialism* (Lincoln: University of Nebraska Press, 2011).

FIGURE 4.9. Malvina Hoffman's "Nordic" man, *Races of Mankind*, 1933 (later relabeled "American Man, Brooklyn, New York" after reinstallation in the 1970s). Courtesy of The Field Museum, #MH25.

Some Europeans also were labeled with linguistic or regional epithets – Anglo-Saxon, Basque, Georgian, Lapp – reflecting continuing anxiety about ethnic distinctions among whites. The exceptions to this sort of labeling were, not surprisingly, exclusively among the European types, some of which were designated only as "Mediterranean," "Alpine," or "Nordic." Indeed, it is the exceptions to the culturally and linguistically configured bodies – and the fact that those exceptions were Europeans – that makes the racial typing of culture, behavior, and language under the guise of bodies and biology in the other figures so striking. Chief among these Europeans was the full-length Nordic (Figure 4.9), representing, according to the museum, Northern Europe and the United States. In the racial geography portrayed at the Field Museum, Nordic was the only race associated with a place called the United States, blatantly effacing the patent diversity encountered all over the country, not least outside the museum's front door. Other groups that shared that territory were either grouped among the American figures, a designation restricted to members of Native American tribes from North, South, and Central America, or were not represented as living in North America at all. American visitors of Asian, African, or Polynesian descent would have looked in vain among the types associated with their ancestral homes for figures representing the United States. But what is truly remarkable about the Nordic figure, what set it starkly apart from the rest of the exhibit and lays bare the racial hierarchy offered to museumgoers, is its contrast to every other full-size figure in the exhibit. He is stripped of any clothing or cultural trappings, implying a transcendence of culture, class, even of race. He recalls no moment of life, but rather classical statuary. The code we read here is not one of race or ethnology so much as an artistic trope for the height of civilization, much like Michelangelo's *David* or Auguste Rodin's *Age of Bronze*. In contrast to the Tamil, who was clearly a worker, and an emaciated one at that, or the Chinese "jinriksha coolie," posed in the act of pulling his rickshaw, "Nordic" represented an idealized masculine whiteness as a classless, raceless human pinnacle. Had he, like the others, been caught in a telling moment of life, he would have looked more like Tony Sansone, the Brooklyn bodybuilder Hoffman modeled for her figure.[60]

Constructing and Contesting the Races of Mankind

Conceiving of a physical anthropology exhibit on the scale of the *Races of Mankind* is one thing. Actually mounting it is something else. Field Museum promotions of the hall would have had visitors believe that each

[60] Arthur B. Caldwell to Malvina Hoffman, n.d., Box 12, Hoffman, Malvina, Correspondence, Fan Letters–Lectures and Books (f.2), 850042-214, MHC; Malvina Hoffman to Henry Field, April 6, 1930, Races of Mankind Correspondence, Henry Field, vol. 12, 1920–1950, FM-HF. Hoffman's bronze figure wasn't a sufficiently transparent statement of racial supremacy for some people. Upon seeing the bronze for the first time, Marshall Field complained to Stanley Field that he much preferred the plaster version he'd seen in Hoffman's New York studio and wondered if she might be able to do it in a different medium, perhaps white marble. Stanley Field to Malvina Hoffman, April 17, 1931, Box 3, Hoffman, Malvina, Correspondence, Field, Stanley (Field Museum), 850042-51, MHC. Museum officials later thought better of labeling this figure "Nordic," presumably after such racial categories fell out of favor in the post–WWII era, and at some point re-labeled it "American, from Brooklyn New York."

of Hoffman's racial figures represented the equivalent of a biological type specimen, carefully selected, measured, and vetted to assure that each was utterly typical in all the requisite characters that defined its race. In fact, the selection of particular individuals and even certain subtypes was left to Hoffman and local experts – and the vagaries of pragmatism, social and professional contacts, and chance. The reality was a much more contingent, compromised affair than museumgoers realized, one that was made no simpler by the lack of consensus in the field.

As Hoffman got underway with her commission, spending the first months studying physical anthropology, consulting with European anthropologists, collecting references and photographs, and working up bronze figures in her campaign to convert museum officials to an exhibition in bronze, Henry Field and Berthold Laufer also collected data and consulted with other anthropologists about how to proceed. Earnest Hooton not only disagreed with the assumptions about racial definition and representation underlying the *Races of Mankind* hall, but he, along with others, was also extremely skeptical about Hoffman's ability to select individuals who typified particular races. Hooton hoped that a physical anthropologist could be hired to precede Hoffman to the locales where she was to study local races, to compile adequate data and help her select "really characteristic types." In Hooton's opinion,

The addition of such a member of the expedition will enormously enhance the scientific value of Miss Hoffman's work. Indeed, such a feature may well prove to be the most worthwhile part of the project, and should contribute not only to the ultimate interest of the exhibition which you plan but also to the solution of the problem of racial classification.[61]

He noted that anthropologists Harry Shapiro and Alfred Tozzer, Hooton's senior archaeologist colleague at Harvard, both agreed with his recommendations. Asked by Henry Field for advice in planning an efficient itinerary for locating the principal racial types in India, Hooton deferred to Roland Dixon, another senior colleague regarded as "undoubtedly one of the most erudite ethnographers of all time."[62]

---

[61] Earnest Hooton to Henry Field, September 4, 1931, Box 3, Correspondence F–G, MHC.

[62] A. M. Tozzer and A. L. Kroeber, "Roland Burrage Dixon," *American Anthropologist*, vol. 47 (1945), pp. 104–118. Dixon died December 19, 1934, at 59. According to Tozzer, "Almost literally he knew everything that had been written on the primitive peoples of Asia, of Oceania, and of North and South America."

Hooton concurred with Dixon's assessment that the proposed list of racial types was overly reliant on "purely linguistic" criteria, and bluntly told Field what he thought of an expedition that could not include "a competent physical anthropologist":

I shall fear that the results will be far from satisfying from a racial point of view, if the selection is left to any artist. Were a really serious attempt made, with a physical anthropologist to select the models, one could make out a fairly detailed program of where to go for certain types. As it is, the labor involved in making such a program is not worth it.[63]

Hoffman herself reported this criticism, noting that "certain museum authorities" told her

in no uncertain terms...that the idea of an artist...going around the world to reproduce the living races of man in sculpture was an absurdity – that it took a lifetime to measure up and really know even one single race, so what earthly use was there in attempting such a wild goose chase? The result would be unscientific or inartistic, or both.[64]

The Field Museum seriously considered sending an anthropologist with Hoffman, but in the end decided against it, in part because the man thought best qualified was unavailable and because it would have been too costly. Furthermore, not everyone inveighed against Hoffman's anthropological skills or judgment. Arthur Keith dismissed the necessity of an extensive program of measurement,[65] although he agreed that there ought to be photos and measurements taken of individuals modeled. As for Hoffman herself, Keith had great confidence in her instincts:

Miss H. has the eye of the artist as well as a flair for seeing & rendering the outward racial character of every man & woman she works on. A cold-blooded anthropologist is likely as not to pick a subject her artistic soul will rebel against. The primary selection I would leave to her as long as she makes certain the person is of the required race.[66]

---

[63] Ronald Dixon to Earnest Hooton, January 17, 1932, in Earnest Hooton to Henry Field, January 19, 1932, Races of Mankind, volume 13, FM-HF.

[64] "'Man as I Found Him, Malvina Hoffman Lecture – Orchestra Hall – Nov. 6, 1939, Chicago," Clippings Folder, Box 30A, MHC.

[65] Stanley Field to Malvina Hoffman, September 9, 1931, Box 3, Correspondence F–G, MHC.

[66] Arthur Keith to Stanley Field, April 13, 1931, Field, Stanley (Field Museum); Box 3, Correspondence F–G, 850042-51, MHC.

When questions about the appropriate races and individuals to select in India arose, Laufer echoed Keith's confidence in Hoffman, arguing that she was

the leader of the expedition and should not be hampered or confused by instructions and directions coming from this quarter. We could naturally make only vague suggestions to Miss Hoffman while she was here, and indicate in general what we wanted.... India has such a bewildering number of races that scholars differ widely as to the importance of the many hundred different types. Doctor Guha is an excellent anthropologist, trained by Doctor Hooton of Harvard, and we believe that Miss Hoffman can thoroughly depend on his judgment.[67]

Later, Laufer indicated his willingness not only to have Hoffman select individuals who typified the races that they instructed her to find, but that she might actually select "some excellent types that will attract her that are not included in the present program.... The main objective for Miss Hoffman is to find the right types; to establish the scientific names for them when on exhibition is our task."[68] Hoffman's own popular account of her travels and work on the Field Museum commission, *Heads and Tales,* as well as her voluminous correspondence and personal journals, give us some insight into how, in fact, she viewed her subjects, and the sort of racial insights she possessed. Despite Hoffman's consistent avowals that all people were essentially the same, in her letters and autobiography she persistently characterized the native people she encountered in stereotypical, frequently derogatory terms. Her descriptions of her travels and the people she encountered reveal her privileged location in a white upper class, often imperial, world in which, through her connections to other members of her class and race, including colonial officials, merchants, and scientists, she was afforded access to a wide variety of people around the world. Although Hoffman regarded her contact with foreign cultures and people as an adventure, and delighted in recounting her experiences, as a

---

[67] Stanley Field to Malvina Hoffman (citing Laufer), December 15, 1931, Stanley Field Folder, Box 3, Correspondence F–G, MHC. Biraja Sankar Guha (1894–1961) was one of the leading anthropologists in India in the interwar period. Trained at the University of Calcutta and at Harvard University for his doctorate, he was a founding Director of the Anthropological Survey of India from 1945 to 1954. In 1931, when Hoffman met him, Guha worked as an anthropologist for the Zoological Survey of British India, the precursor to the Anthropological Survey. Biraja Sankar Guha, Ranjit Kumar Bhattacharya, Jayanta Sarkar, *Anthropology of B. S. Guha: A Centenary Tribute,* Anthropological Survey of India, Ministry of Human Resource Development, Dept. of Culture, Govt. of India, 1996.

[68] Stanley Field to Malvina Hoffman (citing Laufer), December 24, 1931, Stanley Field Folder, Box 3, Correspondence F–G, MHC.

rule she moved in Westernized, upper-class circles, staying in hotels or the homes of her anthropological or governmental hosts, traveling in the best accommodations (although these were often much less comfortable than those in Europe or America, a point she never tired of making), and constantly aided by various sorts of servants. Hoffman's instinct for classing people she encountered as primitive or civilized was unerringly consistent with the "scientific" anthropological classification portrayed in the *Races of Mankind*. Although Laufer and his successors at the Field Museum flinched at Hoffman's broad generalizations and colorful descriptions, the basic sentiments were not so different from those of Arthur Keith, Henry Field, or even Laufer himself. In her account of the places she traveled, the people she met, and the models she acquired, Hoffman did find much to praise. She was especially taken with indigenous arts – dance, music, textiles, painting, poetry – and there were many people she met whom she admired (although these were almost exclusively her social peers). But, perhaps following the Field Museum's lead in its emphasis on "primitive" races, Hoffman's correspondence and her autobiography are preoccupied with an enumeration of all the varied ways less "civilized" people revealed their primitiveness.

In Shanghai, Hoffman modeled Hu Shih, a prominent philosopher and writer, as the Japanese were invading, and remarked not on the frightening circumstances and uncertain future of the people, but on the benefits of martial law, which "made little difference to our daily program. In fact, it added many incidents, all of which helped us to study and understand the Chinese temperament and character."[69] In India, as millions of people demonstrated for independence under Mohandas Gandhi's leadership, Hoffman's response was an ugly caricature of the protestors and relief at being able to escape to genteel comforts:

The swarming millions seem to preserve an endless supply of mental energy and saliva when it comes to expressing rather inconsistent programs for reform and freedom of thought. It was, therefore, quite a relief . . . to find the "old homestead" filled with quiet, white-robbed youths and venerable scholars, who extended us a most gracious hospitality amidst clouds of incense and collections of ancient paintings and sculpture.[70]

She seemed oblivious to the genuine hardships of the servants and workers who aided her and often served as her "victims." When confronted by their difficulties she tended to either remark on her own good fortune or

[69] Hoffman, *Heads and Tales*, p. 220.
[70] Ibid., p. 307.

treat their misfortune as yet another example of their indomitable spirit. Confronted with local people who approached her as an equal, Hoffman was nonplussed. She and her husband were stunned when her Japanese Ainu guide, Yae Batchelor, the adopted daughter of British missionary John Batchelor, spoke fluent English. "Actually seeing and talking with so gentle-voiced a member of the aboriginal race filled us with dismay," she wrote.[71]

Africa, where Hoffman traveled in 1926, three years before the Field Museum offered her the *Races of Mankind* commission, particularly entranced her. Hoffman studied the local people and later used some of the research material she gathered to create museum figures, work Arthur Keith subsequently called her "highest and happiest flights."[72] Paris in the 1920s, where Hoffman had one of her studios, was famously infatuated with all things African and African American.[73] Artists from Aaron Douglas to Pablo Picasso were inspired by African arts, African-American dance revues such as Josephine Baker's Revue Negre drew huge crowds, and "scientific" promotional tours such as Georges-Marie Haardt's Citroën Expedition captivated Europeans with the written, photographic, and filmic accounts of their adventures in "darkest Africa."[74] Karen Dalton and Henry Louis Gates have argued that the French found in the music and dance hedonistic and voyeuristic pleasure, and looked to "primitive" art for ways of remaking their self-image and defining

[71] Ibid., p. 205. *Annual Report of the Director to the Board of Trustees, 1931* (Chicago: Field Museum of Natural History, 1932), p. 70.

[72] Field, *The Races of Mankind*, 1942, p. 13.

[73] Karen C. C. Dalton and Henry Louis Gates, Jr., "Josephine Baker and Paul Colin: African American Dance Seen through Parisian Eyes," *Critical Inquiry*, Summer 1998, vol. 24, no. 4. Ann Laura Stoler's work on the intimacy and erotics of empire are also relevant here. See *Race and the Education of Desire: Foucault's History of Sexuality and the Colonial Order of Things* (Durham, NC: Duke University Press, 1995), and *Carnal Knowledge and Imperial Power: Race and the Intimate in Colonial Rule* (Berkeley: University of California Press, 2002).

[74] Jennifer Boittin, *Colonial Metropolis: The Urban Grounds of Feminism and Anti-Imperialism in Interwar Paris* (Lincoln, NE: The University of Nebraska Press, 2010); Jeannette Jones, *In Search of Brightest Africa: Reimagining the Dark Continent in American Culture, 1884–1936* (Athens, GA: University of Georgia Press, 2010); Patrick Brantlinger, *Dark Vanishings: Discourse on the Extinction of Primitive Races, 1800–1930* (Ithaca: Cornell University Press, 2003); Marianne Torgovnick, *Gone Primitive: Savage Intellects, Modern Lives* (Chicago: University of Chicago Press, 1990); Richard J. Powell and David A. Bailey, eds., *Rhapsodies in Black: The Art of the Harlem Renaissance* (London and Berkeley: University of California Press, 1997); Georges-Marie Haardt and Louis Audouin-Dubreuil, *The Black Journey: Across Central Africa with the Citroën Expedition* (New York: Cosmopolitan Book Corporation, 1927).

themselves in the modern world. Certainly, Hoffman seems to have caught this African fever, as well. In *Heads and Tales*, she rhapsodized about Africa, recalling a performance she attended at the Colonial Exposition in 1931,

> The tom-toms began to beat their barbaric rhythms and African singers joined in a sort of chanting wail. In single file the ebony dancers came in . . . the music in the clouds ceased and the musicians on the stage began their wild pounding and fanatical drumming. The black warriors jumped into the glaring light and each group outdid their predecessors in variety and agility of savage dancing. Spears flashed in the moonlight, shouts of wild ecstasy rent the air, and every one present forgot completely that they were in the heart of Paris . . . we had heard the pulse of darkest Africa, transplanted, but authentic, a direct life-current from the jungle.[75]

In central Africa, Hoffman stayed on the reservation of an English game hunter, where she was afforded access to the pygmies of the Ituri Forest. Unlike the staged African spectacle at the French Exposition, where Hoffman perceived the dancers as thrilling, evocative, eroticized primitives, the Ituri people near the reservation in Africa provoked her to articulate a different kind of primitivism. She compared them unfavorably to monkeys, whose "power of observation and natural ingenuity had far surpassed that of the pygmies." She observed, with some surprise, "They are not as deformed as one might suppose," and wondered if their condition might be similar to that of Shetland ponies, which "never grow up into real ponies."[76] Describing a meal of "dead elephant, killed by one of their diminutive tribal heroes," she told readers they "treat themselves to a field-day of overeating until their little tummies swell up, and after a good deal of scratching and chattering the happy families fall asleep" (Figure 4.10).

Hoffman's perception of primitiveness was also reflected in her aesthetics. Hoffman's treatment of her African subjects demonstrates how she, and the Field Museum staff who authorized her work, imposed their own culturally contingent definition of beauty on the models and then read it back out of the sculptures as a creation of nature. Hoffman was perfectly open in applying her aesthetic prescriptions to her subjects. Indeed, she regarded it as her job. Of the Tasmanians, who were not included in the exhibit of living races because they were extinct, Hoffman wrote, "It was a great relief to me, for they were ugly enough to make celibacy an easy

---

[75] Hoffman, *Heads and Tales*, pp. 173–174.
[76] Ibid., p. 149.

FIGURE 4.10. Ituri African family, *Races of Mankind* (1933). Courtesy of The Field Museum, #MN17.

task, and sculpture an impossible one."[77] When she could not avoid modeling an "ugly" race altogether, such as the Bushmen of Africa, whom the Field Museum insisted be included in the exhibition because they typified primitiveness, she sculpted them in forms that both minimized characters she found ugly and at the same time evoked the requisite primitivism

[77] Ibid., p. 13.

through a highly detailed, heightened naturalism. In her search for Bushman models, Hoffman visited the Musée de L'Histoire Naturelle in Paris to see the remains of the "Hottentot Venus" and was so dismayed by the encounter she "came very near to abandoning the project."[78] The woman immortalized as the "Hottentot Venus" was Saartjie Baartman, a Khoi woman from southern Africa brought to Europe in 1810 as a curiosity. Europeans considered her a convincing and grotesque example of the inferiority of native peoples, particularly curious because of her protruding buttocks, a common physical feature among the Khoi.[79] She was paraded about Europe and much was written about her, under the derisive name Venus. After her death at twenty-five in 1815, she was autopsied and dissected, her genitalia preserved and displayed in the Musée de l'Homme until the 1970s. Henri de Blainville and Georges Cuvier published widely read studies of her genitalia and buttocks, comparing her to an orangutan. Baartman, as the Venus, came to represent for nineteenth-century Europe the fundamental difference from and inferiority to civilized Europeans.[80] Hoffman's response to viewing the cast of this poor woman, this

---

[78] Ibid., p. 155.

[79] The technical term for this feature is "steatopygia."

[80] There has been quite a bit of popular and scholarly work produced on the history and significance of the "Hottentot Venus." Among the best work remains an early essay by Sander L. Gilman, "Black Bodies, White Bodies: Toward an Iconography of Female Sexuality in Late Nineteenth-Century Art, Medicine, and Literature," *Critical Inquiry*, vol. 12 (Autumn 1985), pp. 212–223. More recently, another excellent treatment is Sadiah Qureshi, "Displaying Sara Baartman, The 'Hottentot Venus,'" *History of Science*, vol. 42 (2004), pp. 233–257. See also, Londa Schiebinger, *Gender in the Making of Modern Science* (Boston, MA: Beacon Press, 1993); Anne Fausto-Sterling, "Gender, Race, and Nation: The Comparative Anatomy of 'Hottentot' Women in Europe, 1815–1817," in Jennifer Terry and Jacqueline Urla, eds., *Deviant Bodies: Critical Perspectives on Difference in Science and Popular Culture* (Bloomington, IA: Indiana University Press, 1995), pp. 19–48; Partha Mitter, "The Hottentot Venus and Western Man: Reflections on the Construction of Beauty in the West," in E. Hallam and B. Street, eds., *Cultural Encounters: Representing "Otherness"* (London: Routledge, 2000), pp. 35–50; Clifton Crais and Pamela Scully, *Sara Baartman and the Hottentot Venus: A Ghost Story and a Biography* (Princeton: Princeton University Press, 2008).

After protest by the South African government, including pressure from President Nelson Mandela, France returned her remains in 2002. They were buried on Women's Day at Hankey in the Gamtoos River Valley in the Eastern Cape, thought to be the area of her birth. The grave is designated as a National Heritage site, and Sara Baartman has become a new kind of contested symbol, now of colonialism and racism. See Qureshi, "Displaying Sara Baartman," pp. 233, 250–251. For press accounts of the repatriation, see Chris McGreal, "Remains of 'Hottentot Venus' Finally Returned to Her Homeland," *The Guardian*, Wednesday, Feb. 20, 2002, sec. G2, p. 6; Lucille Davis, "Sarah Baartman, at Rest at Last," SouthAfrica.info, retrieved from: http://www.southafrica.info/about/history/saartjie.htm#.Ue2ytWTuU5s (accessed July 22, 2013).

FIGURE 4.11. Bust of Nobosodru, Hoffman's "Mangbetu," *Races of Mankind* (1933). Courtesy of The Field Museum, #MH96.

"monsterpiece of female ugliness," as she called her, was to insist "that I be given artistic freedom to select at least the best possible representative of a race and not the ugliest, even if every anthropologist in the world preferred the latter."[81] The resulting sculpture of a Bushman family from the Kalahari Desert in southern Africa is an example of what art historian Linda Nochlin has referred to as the "relentless naturalism" of Hoffman's racial sculptures. Nochlin's derogation of Hoffman's "appalling concentration on surface detail, resulting in a kind of waxworks realism in bronze,"[82] unwittingly reveals the technique by which Hoffman complied with museum demands while remaining faithful to her own standards of beauty.

The bust of a Mangbetu woman, whom Berthold Laufer hailed as a "Negro" type of beauty, reveals in a different manner how aesthetics conveyed and undergirded the assumptions about race and civilization that informed the Field Museum's exhibition[83] (Figure 4.11). The Mangbetu

[81] Hoffman, *Heads and Tales*, p. 155.
[82] Linda Nochlin, "Malvina Hoffman: A Life in Sculpture," *Arts Magazine* (Nov. 1984), p. 106.
[83] Laufer, "Hall of the Races of Mankind," p. 3.

figure depicted Nobosodru, a young woman from Central Africa, photographed by George Sprecht during the Citroën Expedition, who became
an icon of Africanness when her likeness graced an advertising poster
for a popular film documenting the expedition.[84] The appropriation of
Nobosodru's likeness served a variety of imperial, scientific, and cultural
purposes, its meaning and effect heavily dependent on the broader purposes and context. Greg Foster-Rice has argued that traditional artistic
representational styles, and antimodern aesthetics, were allied in the twentieth century to essentialized, hierarchical racial categories that promoted
the kind of fixed racial types and notions of superiority seen in classificatory schemes such as Aleš Hrdlička's third-generation "Old Americans"
of European ancestry, and also at the Field Museum. In contrast, Foster-
Rice argues, a more modern aesthetic was associated with Boasian cultural relativism and a rejection of primitivism, as seen in the modernism
of the New Negro movement's representations of blackness.[85] In the case
of Nobosodru, she seems to have been appropriated for disparate ideological purposes. Initially, her likeness was reproduced as a product of
Haardt's expedition, in "ethnographic" photographs taken on the expedition, circulated publicly (and privately to allies such as Hoffman), as

[84] For a first-hand account of the expedition by its leader Georges Marie Haardt, see
The Black Journey. Hoffman also mentions Haardt in *Heads and Tales,* although not,
interestingly enough, in her discussion of Africa and African types, despite having drawn
at least two of her figures directly from his expedition photographs – Nobosodru for her
Mangbetu bust, and Daboa for her life-size dancing Sara figure, pp. 303–304.

[85] Aleš Hrdlička's "Old Americans" referred to whites, generally of Western European,
especially British, ancestry, who could trace their American descent back at least three
generations, who formed a superior class of Americans. Gregory Foster-Rice, "The
Visuality of Race: 'The Old Americans,' 'The New Negro' and American Art, c. 1925"
(PhD diss., Northwestern University, 2003). The relationship between primitivism, modernism, and anthropology is a rich topic that has produced a number of scholarly volumes, including: James Clifford's influential, *The Predicament of Culture: Twentieth-
Century Ethnography, Literature, and Art* (Boston: Harvard University Press, 1988);
Marianna Torgovnick, *Gone Primitive: Savage Intellects, Modern Lives* (Chicago: Uni-
versity of Chicago Press, 1991); and Elazar Barkan and Ronald Bush, eds. *Prehistories
of the Future: The Primitivist Project and the Culture of Modernism* (Stanford: Stanford
University Press, 1995). On the idea of Old Americans and Hrdlička's racial science, see
also Aleš Hrdlička, *The Old Americans* (Baltimore: Williams & Wilkins, 1925); Gunnar
Myrdal, *An American Dilemma, the Negro Problem and Modern Democracy,* vol. 1
(New Brunswick: Transaction Publishers, 1996), p. 138; Michael L. Blakey, "Skull Doctors: Intrinsic Social and Political Bias in the History of American Physical Anthropology,
with special Reference to the Work of Ales Hrdlicka," *Critique of Anthropology,* vol. 7,
no. 2 (1987), pp. 7–35; Larry T. Reynolds and Leonard Lieberman, eds., *Race and Other
Misadventures: Essays in Honor of Ashley Montagu in His Ninetieth Year* (Rowman &
Littlefield, 1996), pp. 74–76. A similar contrast to the one Foster-Rice discusses can be
seen between the Field Museum's *Races of Mankind* hall and the WWII-era and postwar
exhibitions discussed later in this book.

well as in the documentary film *La Croisière Noire,* which depicted the
expedition for eager audiences all over France. Alice Conklin argues, in
A Mission to Civilize, that the kind of technological mastery over nature
represented by the Haardt expedition photographs and film was a critical
component of French imperialism in Africa.[86] Conversely, in the United
States in 1927, Nobosodru's visage was used by modernist painter Aaron
Douglas for the cover of the Urban League journal *Opportunity,* one of
the leading publications of the New Negro movement.[87] There, the use
of her distinctive African image buttressed efforts by Harlem Renaissance
writers and artists to promote African Americans' pride in their heritage
and culture, and engage them in the broader struggle for civil rights
and against discrimination. For the Field Museum, Hoffman depicted
Nobosodru as one of three African women with elaborate hairstyles,
sculpted in a simplified, classical form, shorn of distracting and person-
alizing detail in favor of formal lines and elegance. These African women
became ethnographic, racialized objects for display in a science exhibit,
and also, implicitly, subjects of imperial power. Nobosodru remained
an "exotic," alongside the other African women in the exhibition, an
alluring Hottentot Venus for Depression-era Americans. To the extent
Hoffman's sculpture of Nobosodru might have functioned as a marker
of African-American pride in their heritage, it would have been an effort
made by visitors or organizations outside the bounds of the hall itself and
the materials describing and promoting it. Inside the hall and its racial
logic, the elegance and beauty of the Mangbetu figure crafted by Hoffman
for the museum depicted not "Negro" beauty, but "tribal" adornment
interpreted through "Western" aesthetic tropes and definitions of beauty.
If museum staff had intended to represent indigenous forms of beauty,
rather than selected practices that conformed to Western aesthetics, they
also would have displayed busts of a so-called Ubangi woman with a

[86] Alison Murray Levine, "Film and Colonial Memory: *La Croisière Noire,* 1924–2004,"
in *Memory, Empire and Postcolonialism: Legacies of French Colonialism,* ed. Alec G.
Hargreaves (Lanham, MD: Lexington Books, 2005), pp. 81–97; Fabian Sabates, *La
Croisiere Noire* (Paris: Eric Baschet, 1980); Linda Kim, "Malvina Hoffman's *Races of
Mankind* and the Materiality of Race in Early Twentieth-Century Sculpture and Photography" (PhD diss., University of California, Berkeley, 2006); Alice Conklin, *A Mission
to Civilize: The Republican Idea of Empire in France and West Africa, 1895–1930* (Stanford: Stanford University Press, 1997). The George Sprecht photo of Nobosodru, from
the Haardt Expedition, and subsequent replicas by painter Aaron Douglas and Malvina
Hoffman are briefly discussed in Richard J. Powell, "Re/Birth of a Nation," in Powell
and Bailey, eds., *Rhapsodies in Black,* pp. 14–33.

[87] *Opportunity: A Journal of Negro Life,* New York: National Urban League, May 1927.

lip disc and a Burmese Padaung woman wearing neck rings, rather than exhibiting them as examples of physical deformations.[88] The selection and depiction of subjects for exhibition reflected the aesthetic philosophies of Field Museum staff and Malvina Hoffman, not those of the people they represented to the American public.

Once Hoffman got started with her figures, Laufer was less sanguine about her ability to register typical figures unproblematically. His initial criticisms focused on the symbolic figures for the "Unity of Mankind," in response to sketches and small models Hoffman produced of the central figure. Hoffman complained to Henry Field that she had received a "stiff" letter from Laufer, warning her not to let "ethnography" creep in.[89] As no record of the sketches or models Laufer was reacting to appears to exist, one can only wonder what Hoffman had proposed, given the state of the sculptures that received Laufer's approval. Indeed, concern with Hoffman's methods of selection and the accuracy of her depictions plagued the exhibit down to the last figure.[90] Adolph Schultz required Hoffman to alter the Australian man's toes and feet, to lengthen the

---

[88] "Plan for Technical Division of Hall B, January 11, 1933," Malvina Hoffman, History of the Hall of Races of Mankind 1930–1934, FM-ROM.

"Ubangi" is often a misnomer, spread by the Ringling Brothers Circus, among others, that was applied to a variety of people from central Africa who wore lip discs. Properly speaking, Ubangi refers to a river and a region. See Joe Nickell, *Secrets of the Sideshows* (Lexington, KY: University of Kentucky Press, 2005).

Hoffman notes in *Heads and Tales* that she later encountered the African woman she modeled at the Ringling Brothers Circus. In one short passage, Hoffman managed, in her usual style, to combine reference to her own elevated social status, the colonial subjugation of her subjects, and self-congratulation for her humanitarianism:

The year after I made her portrait, I happened to go to the circus in Mr. Ringling's box in New York City. As the procession passed by, I recognized my discs girl with three or four of her village companions. We exchanged greetings and waving of hands, much to the surprise of the other guests in the box, who regarded me as "not a very nice person." (p. 150)

Similarly, Hoffman procured Burmese "giraffe-necked" models from the circus to model in her New York studio, so that she might create the portrait "without the usual strain and fatigue of modeling while traveling." Mimicking their speech with her own version of their broken English, Hoffman called them "curious . . . creatures . . . as simple as children," and recounted with amusement a trip to St. Patrick's Cathedral undertaken because, as Catholics, they wanted to see the Virgin Mary and attend services (pp. 293–294).

[89] Malvina Hoffman to Henry Field, April 6, 1930, Races of Mankind Correspondence, Henry Field, vol. 12, 1920–1950, FM-HF.

[90] Ironically, the very last figure Hoffman completed was the Bedouin figure, Henry Field's area of expertise. After suggesting Hoffman simply copy a photograph of King Faisal, Field came up with photos and measurements of Atiyeh ed D'khill, one of the workers

FIGURE 4.12. Malvina Hoffman's "elusive Alpine" bust, *Races of Mankind* (1933). Courtesy of The Field Museum, #MH26.

Shilluk Warrior's arms, and to change the Andaman Islander's bow and navel. Hoffman complained that she encountered "endless contradictory opinions" about the Kalahari Bushmen and found it difficult to "satisfy all the experts," particularly regarding the appropriate size of her female figure's buttocks. She resolved the controversy by relying on the authority of anthropologists at the Cape Town Museum in South Africa, who sanctioned her figure as an "average medium example."[91] She ran into the same problem with "the facts" about "Alpine" Europeans. "The number of experts consulted on the subject of what constituted a pure Alpine type resulted in confusion and contradictions all along the line. Each anthropologist seemed to have his own pet idea about the elusive Alpine," she wrote[92] (Figure 4.12). The Zulu, Aztec, Australian, Kashmiri, Egyptian, Hawaiian, Igorot, and Burmese types all provoked controversy over the accuracy and/or representativeness of Hoffman's figures.

from his expedition to Kish. Henry Field to Malvina Hoffman, July 29, 1930, Races of Mankind Correspondence, Henry Field, vol. 12, 1920–1950, FM-HF.
[91] *Heads and Tales*, p. 155; "Consultations with Anthropologists," June 1931, Box 16, Hall of Man, MHC.
[92] *Heads and Tales*, p. 165.

Nonetheless, many of Hoffman's sculptures did meet with anthropologists' favor, and this was sought and conveyed to museum staff, sometimes rather pointedly by Hoffman herself. The respected Indian anthropologist Biraja Sankar Guha had nothing but praise for Hoffman's work: "In all these cases [races modeled in India], through influential local men true representative types of men and women were all made from living persons representative of the different races of India and from my knowledge and personal acquaintance with most of them I can testify that Miss Hoffman's portraits are genuine and exact representations of these peoples."[93] While in Hawaii, the first stop on her "world" tour, Hoffman asked American Museum physical anthropologist Harry Shapiro and his assistant William Lessa to evaluate a Hawaiian head: "They seemed very satisfied that it was a typical subject, and showed racial characteristics."[94] Stanley Field himself conveyed to Hoffman Laufer's and Director Stephen Simms's pleasure with the Hamite and Somali heads she had sent them, noting that they were "exceptionally fine pieces of work" and they "are very enthusiastic about them." He added that Wilfred Hambly, an African ethnologist at the Field Museum, "recognized them and at once gave them his whole hearted approval."[95]

Hoffman's social position meant she was able to obtain models through the auspices of her hosts, either modeling social peers who were acquaintances where she traveled, or coercing those in less powerful social positions, such as servants, to pose for her, expediencies that caused Field and Laufer concern about the scientific accuracy of the exhibit. The French "Mediterranean" type was her colleague and friend Eugene Rudier, owner of the foundry where most of her bronzes were cast.[96]

---

[93] Transcription of letter to Field Museum, from B.S. Guha, Indian Museum, Calcutta, July 4, 1934, Box 3, Correspondence F–G, MHC.

[94] Letter to Field (Stanley?) from Malvina Hoffman, Oct 17, 1931, Box 3, Correspondence F–G, MHC.

[95] Stanley Field to Malvina Hoffman, May 11, 1931, Malvina Hoffman Collection, Box 3, Correspondence F–G, MHC.

[96] Some years later, N. C. Nelson, an anthropologist at the American Museum of Natural History, critiqued a world map illustrated with photographs of Hoffman's figures, and, understandably not recognizing Rudier, asked, "Who would pick Anatole France as a typical Frenchman?" (He also mistook Hoffman's Burmese figure for a woman pictured in National Geographic. She was in fact employed by Ringling Brothers – Field Museum Director Stephen Simms even went to see her when the circus came to town.) Stanley Field to Malvina Hoffman, June 30, 1931, Box 3, Hoffman, Malvina, Correspondence, Field, Stanley (Field Museum), 850042-51, MHC; Malvina Hoffman to Stanley Field, March 17, 1933, Box 16, Miscellaneous Chicago Natural History Museum, Correspondence (c. 1932–1935), 850042-3, MHC; N. C. Nelson to E. J. Schmidt, c/o Hammond & Co.,

One of the Chinese types was a bust of Hu Shih, a prominent philosopher and social critic who had studied with John Dewey at Columbia University and had been a proponent of Western democracy and thought in the turbulent years before the 1949 communist revolution.[97] Hoffman met him through the social connections of her anthropological host Philip Fugh, a friend of Berthold Laufer.[98] Arthur Keith's self-proclaimed Nordic bust was installed near the exhibit's other specimen of Nordic manliness, the full-length bodybuilder from Brooklyn. In 1930 and 1931, Hoffman used her connections to obtain permission from foreign ministers and consuls to select models from the "village" exhibitions at the Colonial Exposition in Paris, and in this way Hoffman got a number of African and Indonesian models.[99] Of the two Chinese "scholars," one was a "charming young gigolo" procured through the Paris embassy; the Manchu figure was paleoanthropologist Davidson Black's "janitor and tea boy" at the Peking Union Medical College, where Hoffman was provided studio space.[100] Georges Marie Haardt, a celebrated adventurer and leader of the Citroën Expedition, offered Hoffman his advice and loaned her his store of African photographs, including images of the women used to model Daboa, the girl from the Sara tribe, and Nobosodru, the Mangbetu woman.[101]

As Hoffman neared the end of her commission, a flurry of the figures created problems. Not long before the *Races of Mankind* was to

Feb. 7, 1944, Box 19, Map of Mankind, Correspondence, Business Transactions, etc. (c. 1942–1954), 850042-37, MHC; Stanley Field to Malvina Hoffman, Aug. 8, 1933, Box 16, Field Stanley, President of Chicago Natural History Museum, Correspondence (c. 1932–1935), 850042-1, MHC, Malvina Hoffman, History of the Hall of the Races of Mankind, FM-ROM.

97 Jonathan D. Spence, *The Gate of Heavenly Peace: The Chinese and Their Revolution 1895–1980* (New York: Penguin Books, 1981), p. 206.

98 Hoffman, *Heads and Tales*, p. 232.

99 Ibid., p. 172.

100 Malvina Hoffman to Stanley Field, June 14, 1933, Box 16, Hall of Man, Misc Correspondence (c. 1933–1964), 850042-6, MHC. Black is best known for his discovery in 1926, along with Chinese scientists, of *Sinanthropus pekinensis*, more commonly known as Peking Man. Much of the fossil evidence infamously disappeared during World War II, either as they were shipped out of China or when the Japanese ransacked Peking Union Medical College. The mysterious fate of the Peking Man fossils was the subject of Harry Shapiro's most popular post–WWII book. Harry L. Shapiro, *Peking Man* (New York: Simon and Schuster, 1974).

101 Malvina Hoffman to Henry Field, February 23 and 25, 1930, Races of Mankind Correspondence, Henry Field, vol. 12 1920–1950, FM-HF; Henry Field to Stanley Field, August 8, 1930, Box 3, Correspondence F–G, Field, Henry (Field Museum), MHC.

open, Hoffman attempted to submit for exhibition a bronze head, entitled *Elephant Hunter*, that Laufer suspected she had done prior to receiving the Field Museum commission. Hoffman had traveled to Africa in the mid-1920s and had subsequently modeled a number of heads based on subjects from that trip. She asserted that the *Elephant Hunter* was done expressly for the museum, but when she failed to produce any records specifying the figure's tribe and location in Africa, Laufer ruled it out, insisting that Hoffman adhere to the contracted types.[102] Perhaps worse, Hoffman then submitted an "Armenian Jew," much to Laufer's consternation. Laufer was becoming very concerned that the exhibit would appear imbalanced. He complained that there were plenty of Africans, but not nearly enough American Indians, and no figures from the Philippines or Korea. Stanley Field, acting as go-between, informed Hoffman that Field and Laufer had reviewed the list of contracted figures and its amendments, and determined that an Armenian had initially been listed, but no Jew. Moreover, they had already substituted a Lapp for the originally stipulated Scandinavian (the Lapp not being considered Scandinavian, but, having been done by Hoffman, was grudgingly accepted), and added the Berber, although it was not on the list, as a substitute for the unacceptable *Elephant Hunter* (which had been temporarily displayed while Hoffman completed all the figures). They refused, however, to omit the Bedouin in favor of the Berber, as Hoffman suggested, because it would "create an embarrassing situation": the Bedouin belonged among the Asian types, and would leave a hole in the exhibit if it was not completed, while the Berber was African, a section that was already full. Field admonished Hoffman that there could be no further substitutions – the list of types had been put together after considerable thought and discussion. She had to conform to the needs of the anthropology department to "properly represent the races."[103]

Although Field Museum anthropologists, at least initially, had confidence in Malvina Hoffman's ability to select representative models, they nonetheless took pains to see that accurate anthropometric data was also

[102] Stanley Field to Malvina Hoffman, June 20, 1933 and Malvina Hoffman to Stanley Field, June 27, 1933, Box 16, Hall of Man Project, Misc. Correspondence (c. 1933–1964), 850042-6, MHC; Stanley Field to Malvina Hoffman, Aug. 8, 1933, Box 16, Hall of Man, Field Stanley, President of Chicago Natural History Museum, Correspondence (c. 1932–1935), 850042-1, MHC.

[103] Stanley Field to Malvina Hoffman, Sept. 1, Nov. 20, Dec. 4, Dec. 20, 1933, Box 16, Hall of Man, Field Stanley, President of Chicago Natural History Museum Correspondence (c. 1932–1935), 850042-1, MHC.

collected. Having decided against a traveling anthropologist, Hoffman was given a crash course in physical anthropology, visiting anthropologists and museums throughout Europe, being instructed in the proper use of instruments and how to take measurements, and reading articles and books on racial types. Her husband, Sam Grimson, was sent along as expedition photographer and filmmaker, a solution to both the high expense associated with paying the salary and travel expenses of an anthropologist or a professional photographer and a means for Hoffman to bring her husband along on the eight-month trip and provide him an occupation.[104] Gretchen Greene, an acquaintance of Henry Field, who had traveled extensively and had an interest in anthropology, was hired to relieve Hoffman of logistical details and travel ahead, arranging with local anthropologists to select individuals and provide data. It was hoped that she would also help Hoffman and her husband collect measurements, descriptions, and photographs not only of the individuals modeled but also additional members of each race for the Field Museum collection. Shortly before embarking on the trip to Asia, museum President Stanley Field, a banker by profession, reminded Greene of

the necessity for absolute accuracy in taking measurements and you should be sure that you have properly prepared yourself for this part of the work before you leave London. A large percentage of the physical anthropologists of the world are very fussy and technical gentlemen, and they are pretty certain to say that, if these measurements are taken by an amateur and not by somebody who they perhaps have personally trained, the measurements are unreliable.[105]

Henry Field gave Hoffman a set of forms for anthropometric information, color charts for eyes and skin, and instructions for taking hair samples and anthropometric photos. With the aid of anthropologists, government

[104] Malvina Hoffman to Stephen Simms, March 14, 1931, Box 16, Hall of Man, Simms, S. C., Director of Field Museum, Correspondence 1930–1935, 850042-2, MHC. Hoffman discussed the possibility of bringing along a professional photographer with Museum Director Stephen Simms, even getting leave from Helen Frick to hire "her" photographer. But in the same letter, she noted that Frick's photographer would expect an excessively high salary ($6,000 for ten months), and suggested that perhaps she and Sam could purchase a good camera and learn to use it, and possibly hire local photographers as they traveled. Sam Grimson had been a violinist prior to their marriage, but that career ended following an injury sustained during World War I. Undoubtedly, the opportunity to work with Hoffman while she was abroad would have been welcome to them both. See also Hoffman, *Heads and Tales*.

[105] Stanley Field to Gretchen Green, Sept. 11, 1931, Box 3, Correspondence F–G, Field, Stanley (Field Museum), 850042-51, MHC.

officials, doctors, police, and social contacts, Hoffman and Sam Grimson accumulated thousands of photographs of their subjects. But their efforts at data collection were intermittent, inconsistent, and incomplete.

Initially, Hoffman seems to have been eager to learn as much about physical anthropology as she could before she set off on her trip to find models in India, the Pacific, and Far East. She visited museums and universities across Europe, including institutions in Hamburg, Prague, Berlin, Vienna, and Munich to consult with anthropologists and view collections, and consulted repeatedly with Field Museum staff and their advisors, especially Arthur Keith.[106] While traveling in the Far East, her letters to Stanley Field suggested an enthusiasm for the anthropological side of her work. She wrote of the many photographs her husband was taking of her subjects and other representative individuals and of the measurements that they or local anthropologists took, at least in the beginning.[107] Later Hoffman balked at taking anthropological measurements and assessing proper skin tone, eye color, and hair texture and color. She wrote to Stanley Field:

The box containing samples of hair, colours and eyes (glass!) arrived safely from Germany. They look extremely scientific, accurate, and anthropological in every sense but between you and me and Dr. Laufer and God I'm not worrying about the use of these much as there is trouble enough in our camp without trying to match celluloid platinum blond curls to the shaven heads of my victims.[108]

In a letter to Henry Field, Hoffman resisted making anthropological observations and collecting data, complaining that it was hard enough to get people to pose and disrobe without poking them with instruments. She finally wrote the Curator that she simply could not fill out their forms, which took too much time and were too complex.

Very often we were not able to take measurements of the subjects. It was the best we could do to get photographs, and in some cases just persuade them to pose. It is very awkward to obtain these measurements quickly with compass, because people have a natural aversion to being touched and handled, and I never insisted on my models being measured until they became resentful or suspicious of me,

[106] See her report on her European tour of institutions in Malvina Hoffman to Stanley Field, May 13, 1931, Box 3, Correspondence F–G, MHC.
[107] See, for example, letters to Stanley Field from Malvina Hoffman, Nov. 20, 1931, March 14, 1932, March 7, 1932, March 22, 1932, Box 3, Correspondence F–G, MHC.
[108] Malvina Hoffman to Stanley Field, Dec. 17, 1931, Malvina Hoffman Collection, Box 3, Correspondence F–G, MHC.

otherwise I could never have obtained a natural and friendly cooperation for my modeling problems.[109]

However, Hoffman did continue to take measurements for her own use in constructing full-size figures later in her studio, which she used along with notes, sketches, and photographs, and sometimes live models as stand-ins. Hoffman noted that her measurements were not always the same as those taken by anthropologists, describing how, for example, making her subject laugh would alter the width of the nostrils, the span of the mouth, or the shape of the eyes.[110] They managed to get the most basic anthropometric data on most of their modeling subjects – information such as age, weight, and stature as well as descriptive information such as hair color and texture, eye color, and skin color – and accumulated thousands of photographs, many of which were useful for ethnological purposes if not for physical anthropology. But nothing like statistically complete records were taken, and Hooton's aspirations for data that might resolve problems of racial classification went unfulfilled.

Despite museum rhetoric, many of the individuals depicted in the exhibit were hardly the carefully, scientifically selected racial types museumgoers were led to expect. In many cases, Hoffman relied on acquaintances, friends, and easily accessible people for her models, a practice that created some consternation at the museum but ultimately did not preclude inclusion of the statues in the exhibit. Although Malvina Hoffman was far from the objective recorder of carefully selected racial types the Field Museum presumably sought, and her florid descriptions were at odds with the authoritative, dispassionate scientific image the museum sought to project, her sculptures were accepted by the curators and eventually hailed as accurate, insightful depictions of humanity.[111]

Conclusion: An Exhibit to Make the World Take Notice

The *Races of Mankind* was a famous and enormously popular exhibition seen by millions of World's Fair goers and generations of school children. It was created at the nexus of global colonial, commercial, and

---

[109] Malvina Hoffman to Berthold Laufer, May 13, 1933, Folder: Malvina Hoffman Collection, Miscellaneous, Including Material from Museum Archives, Malvina Hoffman, FM-ROM.

[110] "'Man as I Found Him, MH Lecture – Orchestra Hall – Nov. 6, 1939. Chicago," Clippings Folder, Box 30A, MHC.

[111] See the Conclusion in this book for a discussion of the reception of Hoffman's sculptures at the Field Museum in the 1990s.

anthropological networks, mobilizing the resources and expertise of philanthropists and scientists around the world. As a typological pastiche, the exhibition patched together old and new in an uneasy amalgamation that failed to fully resolve the persistent tensions in the conceptions of race promoted among physical anthropologists, museum officers and benefactors, and the Museum's strong-willed artist. The substantially different interpretations of Berthold Laufer and Arthur Keith about the content and message that visitors should have perceived highlights the tensions in physical anthropology between a historically and culturally contextualized vision of racial studies, grounded in a belief in human unity, and an older, more divisive racialist typology preoccupied with hierarchy, competition, and miscegenation. That Laufer and Keith could view the same exhibition in such contradictory terms is a testament to the lack of consensus in the field, the lack of overt arguments in the exhibit, and the nature of an exhibit centered not on textual explications of race and method but on a visual representation of race. Henry Field's "Story of Man" provided a narrative linking the museum's two physical anthropology halls, but that narrative, along with its key evolutionary, originary, and hierarchal preoccupations, remained a subtext to be read through the physical space, arrangement of sculptures, and juxtapositions of displays in the *Races of Mankind*. The tensions within physical anthropology were also evident in the execution of the plans for the hall: in the compilation of a list of figures; in the haphazard way Hoffman's models frequently were chosen and approved; in the intermittent, often post hoc, efforts to pursue an anthropometric program that would make Hoffman's sculptures scientifically sound; and in the dichotomous arrangement of the exhibition in which the series of racial figures had to be augmented by a technical section that informed visitors about the meaning of race beyond what they could grasp merely with their eyes and intuitions. Most fundamentally, the exhibition makes apparent how profoundly interwoven were the supposedly separate pillars of anthropological practice and theory: race and culture. In the *Races of Mankind*, an exposition on race was also a narrative of culture and "civilization," the nature of society seen through the lens of racialized bodies.

Despite the lack of a singular racial message, the *Races of Mankind* garnered enormous attention, attendance, and praise. In the months leading up to and following the opening, the museum made sure the exhibit was widely publicized. In addition to appearing on the radio himself to tout the exhibition, Henry Field arranged for local, national, and international newspapers and magazines to run editorials, news items, and full-length illustrated features. The list of publications included the *New York Times* and *New York Times Magazine*, *Fortune*, the *Illustrated London News*, *Science*, *Scientific American*, *Polity*, *TIME*, *The Literary Digest*, *National Geographic*, *Asia*, the *New York Herald Tribune*, the *Chicago Herald Examiner*, and the *Chicago American*. It seems to have worked. Boosted by the major attraction of the Century of Progress World's Fair next door, museum attendance set records that were not met again until the 1960s.[112] In 1933 alone, the museum had 3,269,390 visitors, more than 1 million of whom had visited the *Races of Mankind* within its first six months (this compares to 1.8 million visitors to the entire museum for the whole of 1932). Museumgoers were sufficiently taken with the exhibit to purchase postcards and informational leaflets, which went through three printings.[113] Visitors could purchase sets of postcards, offering thirty-five different views of the sculptures. In 1933, the Museum printed 16,000 and sold more than 500 sets in the first 6 months. A "Map of Mankind" featuring photographs of Malvina Hoffman's sculptures, published by C. S. Hammond & Co., was produced in the 1940s and circulated in the schools, libraries, homes, and offices of America for more than twenty years.

It was not only the fair-going museum public who admired the *Races of Mankind*. The list of anthropological and other scientific collaborators runs into the dozens. Just listing those who played the most significant part in formulating exhibition plans or who offered critical advice would include Earnest Hooton, Arthur Keith, L. H. Dudley Buxton, Alfred Haddon, Roland Dixon, A. R. Radcliffe-Brown, Aleš Hrdlička, Fay Cooper-Cole, William K. Gregory, Wilton Krogman, and T. Wingate Todd. Shortly after the exhibit opened, Melville Herskovits, a student of Franz Boas most widely known for his groundbreaking study of African Americans and the African diaspora, wrote Henry Field, "You are to be congratulated on the installations in the hall where Miss Hoffman's

---

[112] The museum garnered its highest attendance since 1933–1934 in 1964, with 1,511,495 patrons. *Annual Report of the Director to the Board of Trustees, 1964* (Chicago: Chicago Museum of Natural History, 1965).

[113] Field Museum of Natural History, *Annual Report of the Director to the Board of Trustees*, 1933, p. 88. In 1934, the museum sold more than 934 sets (some 27,500 postcards) grouped by regions of the world. Postcards were also produced for the *Stone Age* hall. You can occasionally still find these postcards at flea markets, used book stores, and antique shops. The museum sold 3,100 leaflets for both the *Stone Age* and *Races of Mankind* halls in 1934 (p. 230). In 1935, the Museum sold 800 of the *Races of Mankind* leaflet alone. *Annual Report of the Director to the Board of Trustees, 1935* (Chicago: Field Museum of Natural History, 1936), p. 370.

statues are placed. I shall find it highly useful in connection with the course I give on the Races of Man."[114]

In Chapter 5, I turn to Boston, New York, and the Pacific Islands, and to Harry Shapiro, who, following a Boasian model, sought to develop a form of physical anthropology that might untangle the mysteries of the human form and its varied environments amid the web of history and culture.

[114] Melville J. Herskovits to Henry Field, June 15, 1933, Malvina Hoffman, FM-ROM; Melville J. Herskovitz, *The American Negro* (New York: Columbia University Press, 1928), and *The Myth of the Negro Past* (New York: Harper, 1941).

5

Harry Shapiro's Boasian Racial Science

Introduction

Some anthropologists in the interwar period employed Boasian principles of racial science to refute deterministic studies used to promote discrimination. Among these was Harry Shapiro, curator of physical anthropology at the American Museum of Natural History in New York. Shapiro's Polynesian research program and interwar popular anthropology provide a clear vantage point to examine one set of practices within the range of physical anthropology that addressed racial questions in the 1920s and 1930s. His physical anthropology not only offers a contrast with the assumptions and methods on display at the Field Museum of Natural History in Chicago in the *Races of Mankind* exhibition, but also demonstrates continuities with a postwar framework for understanding race that is usually viewed as wholly disjunct from prewar racial science.

Shapiro saw the tools of the physical anthropologist – their calipers, charts of eye and nose shape, skin color, and hair form – as a critical piece in an overall strategy that combined consideration of culture and physical environment with the study of bodies. Historically, at least among its critics, physical anthropology has typically been viewed as a cold, clinical, objectifying practice. Anthropologists such as Melville Herskovits, Montague Cobb, Franz Boas, and Harry Shapiro reveal a more complex picture. Physical anthropologists, like sociologists, psychologists, and eugenicists, were concerned about the social, political, and economic dimensions of race in the United States and around the world. But their motivations for studying human variation, their judgments and interpretations, varied more widely than is commonly understood. All physical anthropologists (and many other biological and social scientists) in the interwar years believed that human variation could be sensibly sorted and comprehended, at least in part, by quantifying, comparing, and cataloguing variations in a wide variety of bodily characteristics. All employed theory, methods, and instruments that by their nature distanced the practitioner from his or her subjects and objectified them. Some, such as Charles Davenport, Arthur Keith, and Earnest Hooton, did indeed view the people they studied and classified as scientific objects, sets of data points in a project that would describe and thereby justify racialized social and political hierarchies. Others such as Boas, Shapiro, and Berthold Laufer were disturbed by the social and political ramifications of the way human variation was commonly construed in essentializing hierarchies of ability, constitution, and character. For these practitioners, physical variation offered no more rationale for essentializing or dismissing their subjects than did cultural variation. Just as Boas and others who increasingly pursued a historicist ethnology resisted viewing the people they studied as "primitives" and instead strove to interpret their lifeways as reasonable adaptations to existing conditions, so, too, physical anthropologists such as Shapiro and Boas resisted the notion that dark skin, "frizzly" hair, a broad nose, or a round skull signaled some kind of essential, inescapable inferiority that precluded genuine affinity with the subjects of their research. Like Boas before him, Shapiro did not reject biological notions of race, nor anthropometric methods, in favor of a cultural understanding. Rather, both men insisted that only by combining biological, cultural, environmental, and linguistic understandings of people would persistent questions about human differences be answered. In this Boasian racial science, we can see the roots of the cultural construct embraced by social scientists after WWII, as well as the rudiments of a science of bodies abandoned with the repudiation of racial science as scientific racism.

Foundations: Statistics, Culture, Environment, Genetics

Harry Lionel Shapiro got his doctorate in physical anthropology at Harvard in 1926, under Earnest Albert Hooton, at the age of twenty-four, becoming the first graduate of the first doctoral program in physical anthropology in the United States. He was hired to replace Louis Sullivan as assistant curator of physical anthropology at the American Museum

FIGURE 5.1. Harry Lionel Shapiro, in French Polynesia, 1929. "H. L. S. between a couple of natives on the island of Napuka in the South Seas." Courtesy of American Museum of Natural History Library, #shapiro1p.

of Natural History that same year.[1] He remained there until his death in 1990, eventually serving as chair of the department of anthropology for twenty-eight years, from 1942 to 1970. Shapiro also served as an adjunct faculty member in anthropology at Columbia University for thirty years, from 1943 to 1973[2] (Figure 5.1). Although rarely remembered today outside of professional anthropological circles, Shapiro was widely known and admired among his colleagues throughout his very long career, as well as among New Yorkers, particularly in the post–WWII period, via his prodigious popular anthropology in print, radio, television, and museum exhibitions. For scholars, Shapiro offers a window into the practice of physical anthropology and racial science over a period of more than sixty years. His interwar work, which applied a Boasian approach to questions of race mixing, heredity, and environment and offered conclusions that undermined hereditarian views, made an influential contribution to some of the central debates in racial science and genetics in that period. Moreover, his professional life spanned a period of great change in physical anthropology and racial science in the United States, changes that were reflected in Shapiro's own career and the ways he commented on and responded to those changes.

Harry Shapiro was among the first of a new generation of degreed physical anthropologists, a group who formed an institutional and disciplinary foundation for an increasingly professionalized field, and one from which Shapiro himself ultimately departed. Shapiro clearly admired Hooton as a teacher, vivid lecturer, and dedicated supporter of his students in their lives and careers.[3] As Hooton's "firstborn scientific son,"

---

[1] Shapiro was hired after a period of negotiation between Hooton and himself, and Clark Wissler, chair of the Anthropology Department. Harry L. Shapiro Papers, Special Collections, Library, American Museum of Natural History, New York (hereafter AMNH-HLS); Frank Spencer, "Harry Lionel Shapiro, March 19, 1902–January 7, 1990," *Biographical Memoirs*, vol. 70 (1996), National Academy of Sciences, pp. 369–387.

Hooton wrote an obituary for Shapiro's predecessor, Louis R. Sullivan, which appeared in *American Anthropologist* and created some consternation around the museum for its criticism of the low pay museum anthropologists received and the implication that this may have contributed to Sullivan's ill health and untimely death in 1925 at thirty-three. According to Hooton, "There is not one research anthropologist attached to a museum staff nor a single teaching anthropologist who, from his salary, can purchase for himself and his family the nourishing food and social relaxations enjoyed by a first-rate mechanic. The only anthropologists who can pursue their professions in comfort are those who have inherited or married money." E. A. Hooton, "Louis Robert Sullivan," *American Anthropologist*, New Series, vol. 27, no. 2 (Apr. 1925), pp. 357–358.

[2] Shapiro collection finding aid, AMNH-HLS.

[3] Shapiro maintained not only a close personal relationship with Hooton and his family, but also a professional one. He continued to consult Hooton about his work and used

Shapiro seems to have been influenced by his teacher in a number of areas: his belief in the importance of reaching a popular audience, the application of statistics to physical anthropology, the importance of the problem of racial mixtures, and, especially in later years, a biological perspective on the nature of humans. Shapiro credited Hooton with being the first American physical anthropologist to use the kind of statistical methods developed by Ronald A. Fisher in biometrics, despite the less than enthusiastic embrace of such methods by some in the profession (Shapiro noted that "one of his contemporaries" had pointedly warned him to "avoid the dangers and pitfalls" of this sort of analysis[4]). Shapiro credited Hooton with sparking his interest in problems of racial mixing, and the issue of hybrid vigor, in Polynesia. Hooton regarded the study of race mixture as "one problem of physical anthropology which so intimately concerns the present and future welfare of modern nations that its study requires no apologia" because it raised "the question of the inheritance of physical characteristics in the hybrid offspring, as well as their fertility, vitality, intelligence, and capacity for civilization."[5]

Hooton regarded the various anthropological subdisciplines as ultimately inextricably linked:

The ultimate purpose of the anthropologist is to provide a complete interpretation of the human animal in his environmental setting. The development of physical anthropological fields has resulted in progressive encroachments of each specialty upon the preserves of the others, in order to secure significant explanatory data. Thus the archaeologist is driven further and further into interpretation of his cultural data in connection with skeletal material; the physical anthropologist finds himself perforce delving deeper and deeper into the collection and correlation of

Hooton's Harvard Anthropometric Laboratory to perform statistical calculations on his data from Hawaii and Pitcairn Island. Earnest A. Hooton, "Development and Correlation of Research in Physical Anthropology at Harvard University," *American Philosophical Society, Philadelphia. Proceedings*, vol. LXXV (Philadelphia, 1935), p. 505. Many years later, Shapiro wrote a memorial entitled "In Memoriam cum Amore." "Earnest A. Hooton, 1887–1954 in Memoriam cum amore," *American Journal of Physical Anthropology*, 56 (1981), pp. 431–434.

4 This might have been Aleš Hrdlička, who was notorious for his distrust of statistical methods. Hrdlička regarded physical anthropology as an outgrowth of anatomy, and regarded medical studies as the best sort of training for the physical anthropologist, which was his own background and one typical in the nineteenth and early twentieth centuries prior to the advent of graduate programs in physical anthropology. Frank Spencer, *Aleš Hrdlička M.D., 1869–1943: A Chronicle of the Life and Work of an American Physical Anthropologist* (PhD diss., 2 volumes, University of Michigan, 1979).

5 Earnest A. Hooton, "Development and Correlation of Research in Physical Anthropology at Harvard University," (Read Apr. 19, 1935), *American Philosophical Society, Philadelphia, Proceedings*, vol. LXXV, Philadelphia, 1935, pp. 504–505.

sociological data or of archaeological facts; the ethnologist advances steadily into the fields of his colleagues for the same reason.... Thus the archaeologist insists upon working up the skeletal material which he has exhumed; the ethnologist prefers to correlate his own anthropometric data with his cultural findings, and the physical anthropologist raids in all directions and utilizes his miscellaneous booty.[6]

Indeed, Hooton concluded his 1935 review of recent work in physical anthropology by arguing that such work could not be done well in a vacuum because "man's organism and his behavior cannot be disassociated from each other." He continued, "I envisage a social science of the immediate future in which great problems will be studied not only by all varieties of anthropologists working in unison, but also, with them, sociologists, economists, historians, legal experts, psychologists, physicians, geographers, zoologists, botanists, bacteriologists, and even students of government and education."[7] Thus, Shapiro had been prepared for his work in physical anthropology by a mentor who regarded the problem of human differences as a question that required a wide context, not merely a morphological one.

Ultimately, the disciplinary proclivities for statistical analysis within a broader historical, cultural, and social scientific context that Shapiro first learned from Hooton provided a foundation for work modeled directly on the methods and convictions of another key figure in the development of American anthropology: Franz Boas. When Shapiro self-consciously spoke of work that influenced the types of questions he pursued and the methods and assumptions he brought to those questions, the work was that of Franz Boas. Shapiro adopted Boas's orientation toward a historicized, culture-bound view of race, what Shapiro called "the dynamics of human types," that was in many ways congruent with Hooton's views, but which differed in some basic assumptions about anthropology's object of study. This was a crucial shift in orientation for Shapiro. Hooton's description of anthropology's purpose as the study of "the human animal in his environmental setting" is telling. Hooton saw anthropology's object as a primarily biological one – the human animal or organism. Thus, anthropology was akin to zoology or evolutionary biology in its interest in the organism, its environment, and the way organisms change over time and space through interactions among themselves and with their environment, particularly the changes that had hereditary implications.

[6] Ibid., p. 511.
[7] Ibid., p. 515.

And for Hooton, the hereditary ramifications included human behavior and capabilities. Given a particular set of people to study, Hooton would emphasize the biological, hereditary aspects. Boas, on the other hand, viewed humans as primarily products of a particular historical trajectory, cultural creatures who also happen to be biological ones who possess heritable characteristics, some of which can be modified through the influence of various sorts of environments. Boas, and Shapiro, particularly in his work in the 1930s, stressed the cultural environmental influences in studies of human heredity and appearance. An important consequence of this difference in perspective was evident in Hooton's enthusiasm for a biometrical approach to questions of human difference and a tendency to focus on the fixity of hereditary traits instead of their plasticity. In contrast, Boas promoted rigorous statistical and anthropometric approaches as adjuncts to morphological descriptions, in order to prevent erroneous conclusions about the hereditary nature of human differences, whether in bodies or behavior. In general, Boas was also much more cautious than Hooton in drawing conclusions about causation, correlation, and the genetic nature of various traits.

Shapiro was also trained in genetics, as well as statistical analysis, by two of the leading geneticists in the 1920s, William Castle and Edward East. After graduating in 1923 with his bachelor's degree and a growing interest in anthropology, and spurred by coursework with Hooton, Shapiro spent another year at Harvard at the Bussey Institute for Applied Biology, where he studied with Castle and East in preparation for his first Polynesian studies.[8] Harvard's Bussey Institute was one of the primary centers of genetic research and theory in the first decades of the twentieth century, along with Columbia University and Johns Hopkins.[9] Both Castle and East were critical participants in the development of population genetics, and their work contributed to the eventual synthesis of mathematical population genetics and organismal Darwinian evolutionary biology, a development known as the Modern Synthesis that would have profound effects on the course of physical anthropology and racial science in the mid-twentieth century.[10]

[8] Spencer, "Harry Lionel Shapiro," p. 371.

[9] William B. Provine, "William Ernest Castle, Edward Murray East, and the Bussey Institution of Harvard," *Sewall Wright and Evolutionary Biology* (Chicago: University of Chicago Press, 1986), pp. 34–43.

[10] See Chapter 2 in this book for a discussion of genetics and biometrics in the context of Franz Boas's work, and Chapter 7 in this book for the effects of the Modern Synthesis and evolutionary science on physical anthropology.

In 1923, when Harry Shapiro studied with Castle and East, they held somewhat divergent views on the effects of race mixing. East was well known for his work on Mendelian inheritance, particularly work sorting out inheritance of characteristics related to multiple factors (or genes, in modern parlance). He was an expert on inbreeding and crossbreeding, co-author of the widely read *Inbreeding and Outbreeding: Their Genetic and Sociological Significance*, and a pioneer in researching and developing hybrid corn.[11] Work on world agricultural planning during World War I spurred an interest in the social ramifications of genetics, and in succeeding years he was outspoken on issues of overpopulation and eugenics, publishing three popular monographs in the interwar period on heredity and society.[12] East (who was politically liberal and an advocate of civil rights) opposed racial mixing on genetic grounds, in particular the mixing of blacks and whites because it would "lower" the white race. He believed the set of characteristics that defined racial groups represented the result of generations of adaptive evolution, a "harmoniously integrated genotype," that would be disrupted by interbreeding. The result, he believed, would be various deleterious structural and mental "disharmonies" in the offspring.[13]

In the 1920s, Castle also advocated the social separation of blacks and whites in the United States, arguing that whites' higher intelligence should not be diluted by race mixing, and thought that as a cultural matter, "wide racial crosses among men [humans] seem on the whole undesirable."[14] But Castle, who also had accepted eugenicist proposals to segregate and sterilize the "feeble-minded," disputed the view that race crossing had necessarily deleterious results. He argued that there was no scientific evidence of the so-called structural "disharmonies," and that differences in social status or ability were just as likely the result of the

---

[11] Edward M. East and Donald F. Jones, *Inbreeding and Outbreeding: Their Genetic and Sociological Significance* (Philadelphia: Lippincott, 1919).

[12] William B. Provine, "Genetics and Race," *American Zoology*, vol. 26 (1986), pp. 857–887. East's books were *Mankind at the Crossroads* (New York: Scribners, 1923); *Heredity and Human Affairs* (New York: Scribners, 1927); and *Biology in Human Affairs* (New York: McGraw-Hill, 1931).

[13] Provine, "Genetics and Race," pp. 869–870. For a good brief overview of East's work, see John P. Jackson, Jr. and Nadine M. Weidman, eds., *Race, Racism, and Science: Social Impact and Interaction*, Science and Society Series (New Brunswick: Rutgers University Press, 2006), pp. 153–154.

[14] W. E. Castle, "Genetic and Eugenics" (Cambridge, MA: Harvard University Press, 1930), p. 233, as cited in Provine, "Genetics and Race," p. 871. Provine notes that this language remained unchanged from the first, 1916, edition.

environment.[15] Castle, who was renowned for his breeding experiments with small mammals such as rats and rabbits, remained critical of eugenic and genetic studies that suggested interbreeding invariably led to deleterious results, coming out publicly in 1930 against Charles Davenport's conclusions on race crossing in Jamaica, despite his close relationship with his former zoology instructor. In a scathing review published in *Science*, Castle criticized not only the faulty logic, exaggerated conclusions based on slim evidence, and shoddy statistics, but also the sociopolitical presumptions that so obviously drove the study: "We like to think of the Negro as an inferior. We like to think of Negro-white crosses as a degradation of the white race. We look for evidence in support of the idea and try to persuade ourselves that we have found it even when the resemblance is very slight."[16] Castle warned that Davenport's "sweeping" pronouncements about the inferiority of racial hybrids "will be with us as the bogey men of pure-race enthusiasts for the next hundred years."[17] Harry Shapiro responded to Davenport's work with his own study of race mixing on Norfolk and Pitcairn Islands, first in a strictly academic venue in 1929, and then through *Heritage of the Bounty*, published in 1936, for both lay and professional readers.[18]

At Harvard, Shapiro learned from two of America's most distinguished geneticists not only the basics of Mendelian genetics – dominance, segregation, and the inheritance of discrete, discontinuous characters – but was schooled in the earliest population genetics and steeped in questions that were currently vexing the study of heredity, problems such as race mixing and hybrid vigor, what to make of continuous variation and "blending" inheritance, how heritable traits change, and the respective roles of selection and environment in evolution. Coming to the field in the mid-1920s, Shapiro's work contributed to a marked shift in scientific opinion about the consequences of race mixing, away from a pervasive belief in the early decades of the century that such mixing was invariably problematic, toward a broadly embraced assessment in the interwar period that no

[15] Jackson and Weidman, *Race, Racism, and Science*, pp. 153–154.

[16] W. E. Castle, "Race Mixture and Physical Disharmonies," *Science*, New Series, vol. 71, no. 1850 (Jun. 13, 1930), pp. 603–606; Charles B. Davenport and Morris Steggerda, *Race Crossing in Jamaica* (Washington, DC: Carnegie Institution of Washington, 1929).

[17] Ibid., pp. 605–606.

[18] Harry L. Shapiro, "Descendants of the Mutineers of the Bounty," *Memoirs*, Bernice P. Bishop Museum, vol. 11, no. 1 (1929), pp. 3–106; Harry L. Shapiro, *The Heritage of the Bounty. The Story of Pitcairn through Six Generations* (New York: Simon and Schuster, 1936). Shapiro also published a version of his Norfolk study in *Science*, "Results of Inbreeding on Norfolk Island," in "Science News," *Science*, New Series, vol. 65, no. 1693 (Jun. 10, 1927).

evidence of such disharmonies could be found, and ultimately, by the 1940s, to the view that mixes between even widely different races were harmless.[19]

Shapiro was hopeful that the kind of anthropology, genetics, and scientific stance he admired in Hooton, Boas, East, and Castle would be reflected in the practice of physical anthropology and racial science more generally. At the Anthropology Club in Philadelphia in 1935, Shapiro spoke optimistically about efforts within physical anthropology to examine and improve methods of study, about "new currents of investigation" that were "finding their way into the stagnant pools of inherited beliefs... washing away the obscuring scum of a century." Where, in 1933, Henry Field and the Field Museum mounted an exhibition on the subject of physical anthropology defined by a collection of apparently unambiguously classified racial types, Shapiro, speaking two years after the *Races of Mankind* opened, noted that the study of race was being profoundly affected by the application of "fresh viewpoints" and "improved methodologies," but deliberately declined to discuss classificatory systems such as that on display at the Field Museum: "The subject, is of course, exceedingly complex and I do not intend to go into a disquisition on the variety of classifications that have been applied to mankind. Like the heads of the Hydra monster such classifications appear to have a limitless capacity of regeneration."[20] Henry Field had turned with confidence to turn-of-the-century classifications by Joseph Deniker and William Ripley to inform their enumeration of what they regarded as the principal racial types, and continued to base their scheme upon a nineteenth-century methodology that combined assumptions about human morphology with ethnic, regional, or national identities, and language groups to create racial types. Shapiro credited Boas's work with throwing into question the assumptions about the unchanging nature of the morphological features upon which all such categorizing had been based. "Boas's Changes in the Bodily Form of the Descendants of Immigrants was like a bomb dropped upon a genteel afternoon meeting of a sewing circle," Shapiro remarked. The study, published in 1911, "completely... demolished the house of cards built by preceding generations."[21] Boas demonstrated that supposedly fixed characters could change with modified environment, and

---

[19] Provine, "Genetics and Race," p. 857.

[20] "Part of Paper Delivered at Anthropology Club, Philadelphia, 1935," Box 4, Anthropology, AMNH-HLS, p. 1.

[21] Ibid., p. 2; Harry L. Shapiro, *Migration and Environment. A Study of the Physical Characteristics of the Japanese Immigrants to Hawaii and the Effects of Environment on their Descendants* (New York: Oxford University Press, 1939).

that statistically significant changes can occur in the space of a single generation. Shapiro was inspired by Boas's study to undertake an extensive
project of his own, attempting to confirm, and possibly extend, Boas's
thesis that "geographical changes of habitat with their associated environmental alterations may exert an immediate and profound effect on
whole populations."[22] That study was *Migration and Environment*, an
examination of Japanese immigrants to Hawaii, their children, and their
relatives back in Japan.

In Shapiro's version of physical anthropology's history, there were
three key twentieth-century figures: Hooton, Aleš Hrdlička, and Boas,
but only Boas received credit for shaping the content and approach of
the discipline. Shapiro credited Hooton with developing the professional
training of physical anthropologists through his doctoral program at Harvard, and Aleš Hrdlička with his founding role in institutionalizing the
discipline with the creation of a journal and professional association,
as well as pushing for the establishment of sound technical standards
and a wider scope of inquiry. Looking back in 1958 on the development of physical anthropology, Shapiro credited Boas with introducing
and promoting a genetic understanding of "physical types," based on
anthropometric data. Shapiro translated Boas's work and theory into
the modern population genetics paradigm, claiming that his work on the
descendants of immigrants "shattered" an "older concept of a relative fixity of physical type," forcing the discipline to "think in terms of plasticity
and of phenotypical response to environmental conditions."[23] In terms
of the relationship of cultural anthropology to physical anthropology,
he noted, "Culture is becoming recognized as an increasingly significant
component in the biological history of man."[24]

---

[22] Ibid., p. 3.

[23] Harry L. Shapiro, "Symposium on the History of Anthropology. The History and Development of Physical Anthropology," *American Anthropologist*, vol. 61 (1959), pp. 375–
376. Shapiro noted in the same paper that Boas also made noteworthy contributions
in problems of growth, and in "investigations in the same area which he encouraged
students and colleagues to undertake." It's not clear if this statement is meant to include
studies such as that of Margaret Mead's *Coming of Age in Samoa*, p. 377.

[24] Ibid., pp. 373–374. In the course of his career, Shapiro would adopt a more biological
point of view, but retain the Boasian predilection for a culturally contextualized perspective. Shapiro's shift mirrors the shift in his discipline, which in the postwar period
began to call itself biological anthropology, rather than physical anthropology, and
to concern itself with questions of human physiology and population, in addition to
the long-standing interest in human origins and evolution. Shapiro's interests turned
in these directions, as well, so much so that he was responsible for creating an exhibit
at the American Museum of Natural History in 1960 called *Biology of Man*. Shapiro's

Throughout his long career, Shapiro consistently viewed human variation as an important problem in biological science. But like Berthold
Laufer, who insisted that race and racial science were confined to the
nature of human bodies, the "physical traits acquired by heredity,"
Shapiro did not think the delineation of human morphology and biology would provide ready answers to human social problems.[25] Harry
Shapiro consistently viewed his study of race as an attempt to sort out
the biological and genetic complexities of human variation, and to situate
that variety and its complex, changing nature within a broader cultural
and historical context. But in contrast to eugenicists such as Charles Davenport, biometricians such as his teacher Earnest Hooton, and geneticists
such as Edward East, Shapiro did not believe the variations in human
form caused or even marked variations in behavior, beliefs, or the capacity for "civilization." Like Boas or Laufer, Shapiro was inclined to see the
issue of causality in more complex terms. They were inclined to see human
physical, mental, and cultural variation as a complex web of traits and
practices, all of which varied in response to changing conditions, including new physical environments and contact with new peoples, practices,
and ideas. By the time Shapiro embarked on his own research, he had
embraced a Boasian view that rejected simple hereditarian causality for a
view of human beings as complex organisms evolving in a web of history,
culture, and environment.

The American Museum of Natural History Goes to Polynesia

In the interwar period, Harry Shapiro pursued his researches in Polynesia,
first on Pitcairn Island, and then in Hawaii. The choice of Polynesia was
a convergence of personal proclivities, institutional location, and disciplinary history. If his hiring at the American Museum of Natural history
could not be described as simply fortuitous, as Earnest Hooton and Clark
Wissler exchanged extended correspondence about placing Shapiro following his dissertation work on Norfolk Island, it was nonetheless true

---

shift also seems to reflect changes in the scientific understanding of heredity, especially a
clearer understanding of genetics and its relationship to evolution and natural selection,
giving rise to the theory of population genetics. Another key development in the field,
and in the world, was the organized rejection of scientific studies of race and racism,
embodied in the UNESCO Statements on Race, which Shapiro participated in drafting.
See Chapter 7 in this book for a fuller discussion of Shapiro and this history.

[25] Henry Field, *The Races of Mankind*, Popular series, Anthropology Leaflet 30, Fourth
Edition (Chicago: Field Museum of Natural History, 1942), p. 6.

that Shapiro's private desire to delve into the complexities of race in Polynesia fitted neatly with the American Museum program that Wissler had
supervised since 1920 and with a broader anthropological program to
survey Oceania that had accelerated in the 1920s.[26]

Although anthropological interest in Polynesia was usually framed in
terms of the discipline's chief legitimating projects – salvaging endangered
human and material resources for the study of man, exploring the variations and commonalities of humanity, and fulfilling a positivist, humanist, "universal knowledge" project[27] – it was also true that Polynesia was
much more than just another area on the globe for anthropologists to
appropriate for their purposes. It was, to begin with, an imperial space.
By the 1910s and 1920s, the Philippines were firmly under U.S. control
after centuries of Spanish rule, despite indigenous revolutionary movements and declarations of independence, and Hawaii had been annexed
as a U.S. territory. British and other European anthropologists still oper-
ated around the world in an imperial context, and frequently, although
generally ineffectually, argued for the importance of anthropologists and
anthropological training to colonial rule.[28] In broader historical terms,
it was hard for anthropologists of any nationality to escape centuries of

[26] The terms "Polynesia," "Melanesia," and "Micronesia" – the three areas comprising the
      Pacific Islands – date from the eighteenth and nineteenth centuries. Polynesia is a collec-
      tive name for numerous small islands in the East-Central Pacific, bounded by Hawaii at
      the apex, New Zealand at the base, and Easter Island on the east. Polynesia includes Pit-
      cairn Island, Tahiti, the Marquesas, the Tuamotu Archipelago, and the Society Islands
      (all formerly French Polynesia), as well as Tonga, Samoa, the Cook Islands, and the
      Austral Islands. Melanesia is a racialized term modeled after Polynesia (from the Greek
      words *melas*, "black," and *nesos*, "island") to refer to an area in the Western Pacific
      inhabited by dark-skinned peoples, including New Guinea, Fiji, New Caledonia, Vanu-
      atu (formerly the New Hebrides), the Solomon Islands, and the Bismarcks. Micronesia,
      also modeled after Polynesia, refers to a region of small islands and atolls in the North-
      western Pacific, including the Caroline, Marshall, Mariana, and Gilbert Islands. The
      terms "Pacific Islands" and "Oceania" are generally used interchangeably, and tech-
      nically exclude not only Australia and Japan, but also the Indonesian and Philippine
      archipelagoes. However, casual usage, as well as many of the scientific texts consulted
      for this book, fail to make these distinctions when discussing the Pacific or Oceania, and
      frequently mention the Philippines and sometimes Indonesia.
[27] I borrow the formulation "universal knowledge" as Mary Poovey uses it in *A History of
      the Modern Fact: Problems of Knowledge in the Sciences of Wealth and Society* (Chicago:
      University of Chicago Press, 1998), and "The Limits of the Universal Knowledge Project:
      British India and the East Indiamen," *Critical Inquiry* 31 (Autumn 2004), pp. 183–202.
[28] Henrika Kuklick, "The British Tradition," in Henrika Kuklick, ed., *A New History of
      Anthropology* (Oxford: Blackwell Publishing, 2008), pp. 60–63. Much has been writ-
      ten on the colonial associations of anthropology. In addition to Kuklick's anthology,
      recent notable volumes include H. Glenn Penny, *Objects of Culture: Ethnology and*

imperial conquest (and many had little or no desire to), whether it was the legacies of European empire or the empire of science, to which the discipline of anthropology had only recently been admitted.[29] Histories of "discovery," friendship, and exploitation by four centuries of Europeans and Americans lurked behind anthropological research proposals.

Moreover, Polynesia is a potent, contested, complex referent, an idea and a place suffused with imagination, for both those who visit – or may never visit – and those who live there, lands where a panoply of Euro-American desires have been played out. Of desire there were many varieties: imperial, missionary, commercial, scientific, sexual, escapist; seeking to see, know, name, own, feel. Polynesia in its very formation was a region defined in the imagination of Europeans, envisioned as a tropical idyll home to savages, noble or dangerous depending on the account. The Pacific Islands were lands purely of the imagination for the thousands who read of Captain Cook's exploits or the *Mutiny on the Bounty*. In some measure, anthropological narratives were works of native imagination, too. Mythic, imaginative popular and ethnological accounts drew not only on the expectations of Euro-American visitors, but were grounded in information provided by local inhabitants based on their understanding of the anthropological project, their place in it, and the sort of story about themselves they wished to purvey to the outside world. So when American and British anthropologists proposed to "attack" the problem of Polynesia, they were entering a field already rife with history, desire, and imagination.

The American Museum of Natural History entered the Polynesian field following collective national and international efforts of anthropologists

Ethnographic Museums in Imperial Germany (Durham, NC: University of North Carolina Press, 2002); Andrew Zimmerman, *Anthropology and Anti-Humanism in Imperial Germany* (Chicago: University of Chicago Press, 2001); Peter Pels and Oscar Salemink, eds., *Colonial Subjects: Essays on the Practical History of Anthropology*, 1999. Less recent but also of note is George W. Stocking, Jr., ed., *Colonial Situations: Essays on the Contextualization of Ethnographic Knowledge*, History of Anthropology Series, vol. 7 (Madison: University of Wisconsin Press, 1991). For an excellent account of the imperial context of early British anthropology, see Henrika Kuklick, *The Savage Within: The Social History of British Anthropology, 1885–1945* (Cambridge: Cambridge University Press, 1991). For a recent account that explores how interactions between Filipinos and American colonial administrators shaped the construction of race in the Philippines and the United States, see Paul Kramer, *The Blood of Government: Race, Empire, the United States, and the Philippines* (Chapel Hill, NC: University of North Carolina Press, 2006).

29 Pels and Salemink, *Colonial Subjects*; Peter Pels, "The Anthropology of Colonialism: Culture, History and the Emergence of Western Governmentality," *Annual Review of Anthropology*, vol. 26 (1997), pp. 163–183.

in the 1910s and early 1920s to organize a program of ethnology, physical anthropology, archaeology, and linguistics in Oceania. An early and important salvo in this campaign was a set of reports commissioned by the Carnegie Institution of Washington to guide its distribution of Andrew Carnegie's $10 million bequest in U.S. Steel stock. The Carnegie Institution was a "non-university, non-museum, non-Rockefeller," nongovernmental source of funding for the natural and human sciences, especially physical anthropology, archaeology, and eugenics, including extensive support for Charles Davenport's eugenics laboratory at Cold Spring Harbor. Like the Rockefeller Foundation, National Research Council, Social Science Research Council, and other nongovernmental or quasi-governmental funders, the Carnegie Institution was changing the way anthropological projects were being justified and organized. These novel organizations were predicated upon a liberal, positivist "governmentality" that leveraged private philanthropic resources that had previously been solicited by or donated to museums and individual anthropologists on a project-by-project basis. Under these new foundations and institutes, a combination of political actors, business leaders, and leading scientists granted substantial, often ongoing, funding for broad-based projects on a competitive, or at least comparative, basis. The Carnegie Institution not only funded particular research agendas, but established departments and committees of "experts to direct special lines of research," as well as funding laboratories, observatories, and libraries to support that research. The Institution's mission was to "encourage in the broadest and most liberal manner, investigation, research and discovery, and the application of knowledge to the improvement of mankind."[30]

Published in 1914, the Carnegie-sponsored "Reports upon the Present Condition and Future Needs of the Science of Anthropology" assessed which anthropological problems deserved the most attention, and by extension, which problems the institution should support. Authored by William H. R. Rivers, Albert Earnest Jenks, and Sylvanus G. Morley, the

---

[30] Richard E. W. Adams, *Prehistoric Mesoamerica*, 1996, p. 7; Quetzil E. Castañeda, "The Carnegie Mission and Vision of Science: Institutional Contexts of Maya Archaeology and Espionage," *Histories of Anthropology Annual*, vol. 1 (2005), p. 31; George W. Stocking, Jr., "Ideas and Institutions in American Anthropology: Thoughts toward a History of the Interwar Years," *The Ethnographer's Magic, and Other Essays in the History of Anthropology* (Madison: University of Wisconsin Press, 1992), pp. 130, 133; George W. Stocking, Jr., "Philanthropoids and Vanishing Cultures: Rockefeller Funding and the End of the Museum Era in Anglo-American Anthropology," *The Ethnographer's Magic, and Other Essays in the History of Anthropology* (Madison: University of Wisconsin Press, 1992), pp. 208–209; Pels, "The Anthropology of Colonialism," pp. 175–176.

reports are characterized by a rhetoric of salvage anthropology, noting the "urgency" of undertaking "intensive" studies of "material" that was rapidly disappearing. "In many parts of the world the customs of savage and barbarous man are undergoing rapid and destructive change... the death of every old man brings with it the loss of knowledge never to be replaced," the report warned.[31] William Rivers, the "single most influential British Anthropologist" in this period, concluded that southern Africa and Oceania were the regions most in need of intensive study. He argued that they "combine an extreme degree of the urgency of their needs with very inadequate attempts to meet those needs" and of the two areas, Oceania "should have the preference." Rivers regarded it as an area that had "interesting and important examples of human culture" that were "on the verge of extinction," which were especially suited to intensive study and collection of "a large mass of valuable material... with relative ease."[32] He also thought that study of Oceanic culture would shed light on questions about the original populations in the Americas because they were closely related.

Albert Jenks, formerly chief of the Ethnological Survey for the Philippines and professor of anthropology at the University of Minnesota, also focused on areas in the Pacific, as well as indigenous peoples in the Americas.[33] He regarded the main problems in Oceania as the origin and spread of the Pacific Islanders and their culture.[34] Although Rivers stressed the importance of intensive ethnological study that produces an

[31] W. H. R. Rivers, "Report on Anthropological Research Outside America," in W. H. R. Rivers, A. E. Jenks, and S. G. Morley, "Reports upon the Present Condition and Future Needs of the Science of Anthropology" (Washington, DC: Carnegie Institution of Washington, 1914), p. 6. See also a review of the reports by A. C. Haddon, "The Urgent Need for Anthropological Investigation," *Nature*, no. 2329, vol. 93 (Jun. 18, 1914), pp. 407–408.

[32] Rivers, p. 28; George W. Stocking, Jr., "The Ethnographer's Magic: Fieldwork in British Anthropology from Tylor to Malinowski," *The Ethnographer's Magic and Other Essays in the History of Anthropology* (Madison: University of Wisconsin Press, 1992), p. 32. Stocking also noted that Alfred Haddon held Rivers in equally high regard for his "concrete method" of immersion among his subjects and their culture, saying in 1914 that he was "the greatest field investigator of primitive sociology there has ever been."

[33] Alexander F. Chamberlain, review of Albert Earnest Jenks, *The Bontoc Igorot*, Department of the Interior, Ethnological Survey Publications, vol. 1 (Manila: Bureau of Public Printing, 1905), *American Anthropologist*, New Series, vol. 7, no. 4 (Oct.–Dec., 1905), pp. 696–701.

[34] Albert Earnest Jenks, "Report on the Science of Anthropology in the Western Hemisphere and the Pacific Islands," in W. H. R. Rivers, A. E. Jenks, and S. G. Morley, "Reports upon the Present Condition and Future Needs of the Science of Anthropology" (Washington, DC: Carnegie Institution of Washington, 1914), p. 55. Jenks also

intimate understanding of local cultural practices and knowledge,[35] Jenks regarded questions of physical anthropology as among the most important modern anthropological questions. Jenks cited three problems for study in the Pacific Islands – "ethnic heredity, influence of environment on mankind, and human amalgamation" – and proposed a permanent laboratory to deal with these studies.[36] Alfred C. Haddon, an influential Cambridge University anthropologist and, most famously, leader of the revolutionary 1898 Torres Straits Expedition, reviewed the "Reports" for *Nature*, and concluded his piece by noting that they

point clearly to Oceania as being probably that part of the world which most urgently needs ethnographical investigation, and if the Carnegie Institution could see its way to organize a commission for the intensive study of as many portions of that area as possible, combined with an investigation of the more general problems of racial and cultural movements, it would confer an incalculable boon on all present and future students of the history of human culture.[37]

By the 1920s, the American Museum took up the Polynesian anthropological project. Scientific interest in Oceania accelerated in the late 1910s and into the 1920s, following the end of WWI, led primarily by Western and Pacific institutions in the United States, Hawaii, the Philippines, Australia, and Japan. The question of Oceanic explorations percolated through major scientific gatherings throughout the 1910s, including the British Association for the Advancement of Science, the National Academy of Sciences, and the American Association for the Advancement of Science.[38] In 1919, the United States National Research Council established a Committee on Pacific Investigations, including American

thought Polynesia was a site where the "genesis of speech in man" might be discovered because Polynesian languages were "of the most elemental character," a view that Haddon reports with some skepticism in his review.

35 Rivers, pp. 6–11. The "Report" contrasted intensive work by expert individuals with "cursory visitors," including anthropologists, who merely collect objects without developing a deep understanding of the objects' significance or the "processes of manufacture... which are even more important than the finished article." The Report criticized museum anthropologists for being concerned primarily with collecting objects to display, coming away from their brief exposures to cultures with a superficial understanding of them. For a discussion of Rivers' method of "intensive study" and its place in British social anthropology and anthropological practice more broadly, see Stocking, "The Ethnographer's Magic," p. 32.

36 Jenks, p. 55.

37 Earnest A. Hooton, "Urgent Need for Anthropological Investigation," p. 408.

38 "Proceedings of the First Pan-Pacific Scientific Conference," Bernice P. Bishop Museum, Special Publication no. 7, part 1, 1921 (Honolulu: Honolulu Star Bulletin, Ltd., 1921), p. iii.

Museum Curator of Anthropology Clark Wissler and chaired by geologist Herbert E. Gregory of Yale University. Among its initial activities was organization of the first Pan-Pacific Scientific Conference in Honolulu in 1920, arranged under the auspices of the Pan Pacific Union and with funding from the Territory of Hawaii. Anthropological delegates[39] joined biologists, botanists, entomologists, geographers, geologists, and seismologists to map an international cooperative research strategy for Oceania.[40] Gregory, newly appointed as acting director of the Bernice Pauahi Bishop Museum in Hawaii, chaired the conference. If anyone doubted the way the legacies and realities of colonialism echoed through the endeavor, or the way the participants regarded themselves as privileged investigators, one only need attend to Gregory's opening remarks, which positioned the proceedings as a continuation of European discovery. "The study of natural history," Gregory opined, encompassing within that rubric not only the flora and fauna of the Pacific Islands, but also their people, "has proceeded with few interruptions since the memorable observations of Captain Cook made the ocean generally known."[41] Echoing William Rivers's concern in 1914 that the "vast mass" of anthropological data on the area, much of it collected by amateurs such as colonial officials and missionaries, was "incomplete and even misleading," conferees concluded that a program of professional social science was necessary to adequately document and study "rapidly disappearing native cultures."[42] Anthropologists were interested in the perplexing problems of where Pacific Islands peoples had come from, when and how they got to the widely scattered islands, and how the various, apparently racially different, populations were related. A program of physical anthropology,

---

[39] In addition to American Museum anthropologists Clark Wissler and Louis Sullivan, these included Robert T. Aitken, W. T. Brigham, Kenneth Emory, Gerard Fowke, Federick Wood-Jones, Alfred Kroeber, John F. G. Stokes, and Alfred Tozzer. "Proceedings of the First Pan-Pacific Scientific Conference," pp. 22–26.

[40] See Eckhardt Fuchs, "The Politics of the Republic of Learning: International Scientific Congresses in Europe, the Pacific Rim, and Latin America," in Eckhardt Fuchs and Benedikt Stuchtey, eds., *Across Cultural Borders: Historiography in Global Perspective* (Lanham, MD: Rowman & Littlefield Publishers, Inc., 2002), pp. 218–221, for a consideration of the international politics behind the establishment of the various Pacific research organizations, including fears that the United States would take over the ostensibly international research project, and the eventual success in founding a genuinely international Pacific science association.

[41] "Proceedings of the First Pan-Pacific Scientific Conference," p. iii.

[42] Rivers, p. 7; "Proceedings," p. 104; Te Rangi Hīroa (Peter H. Buck), *An Introduction to Polynesian Anthropology*, Bernice P. Bishop Museum, Bulletin 187 (Honolulu: Bernice P. Bishop Museum, 1945; reprint New York: Kraus, 1971), p. 44.

ethnology, linguistics, archaeology, and history was mapped out, outlined as a series of questions and answers modeled on the "Notes and
Queries" published by the British Association for the Advancement of
Science, and printed up as pamphlets to guide investigators in the field.[43]
At the same time, Wissler and Gregory managed to convince Yale graduate and stock financier Bayard Dominick to donate $40,000 to finance
a set of graduate fellowships (including one for Harry Shapiro in 1923)
and a series of expeditions, with the funds channeled through Yale to
the Bishop Museum. The actual research plans for the expeditions were
worked out in consultation with Wissler, Roland Dixon, Alfred Kroeber,
and William Churchill.[44] This was followed in 1926 by a matching grant
of $50,000 from the Rockefeller Foundation following a favorable report
from Wissler, Edwin Embree, and Edwin Conklin.[45]

Wissler's plan for Polynesian researches at the American Museum was
initially realized through Assistant Curator of Physical Anthropology
Louis R. Sullivan. Sullivan not only attended the Pan-Pacific Conference along with Wissler, but beginning in early 1920, Sullivan embarked
on a study of race in Hawaii under the auspices of both the American
Museum and the Bishop Museum, whose collections and anthropologists were focused on the history of Polynesia, especially Hawaii. As a
Research Associate at the Bishop Museum, Sullivan gathered his own
anthropometric data on Hawaiians and analyzed data collected by field
workers on the Bayard Dominick Expeditions in Tonga, Somoa, and the
Marquesas.[46]

---

43 Hīroa (Buck), p. 44. The "Recommendations for Anthropological Research in Poly-
    nesia" from the Pan-Pacific Conference were initiated at the Honolulu conference and
    finished under the Division of Anthropology and Psychology of the National Research
    Council. The final recommendations were formulated by Roland Dixon, Gerard Fowke,
    Albert Jenks, A. L. Kroeber, Robert Lowie, W. H. R. Rivers, Louis Sullivan, Louis Ter-
    man, J. Allan Thomson, Thomas G. Thrum, Alfred Tozzer, Frederick Wood-Jones, and
    N. Yamasaki. "Proceedings," p. 103.
44 "Proceedings of the First Pan-Pacific Scientific Conference," p. 51.
45 Ibid., pp. 44, 47; Stocking, "Philanthropoids," pp. 187, 188. Yale University and the
    Bishop Museum had a notably close relationship. Gregory was initially given a three-
    year leave to act as a director of the museum (Stocking described this as part of a plan
    devised by Yale and the American Museum to "revitalize" the Bishop Museum), which
    was then extended until his retirement, an arrangement that was copied for his successor,
    Peter Buck (Te Rangi Hīroa), following his appointment as professor of anthropology
    in the graduate school and appointment as a director of the museum in 1936 (upon
    Gregory's retirement). Moreover, the museum and its equipment were made available
    for Yale graduate students, a Yale-Bishop Museum Fellowship was established, as was
    a Bishop Museum Visiting Professorship at Yale.
46 Hīroa (Buck), p. 45.

Like many physical anthropologists in the early twentieth century, Louis Sullivan came to the field via training in a variety of biological sciences including zoology, botany, and comparative anatomy, first at Bates College in Maine, where he was from, and then by graduate work at Brown University. During WWI, Sullivan served as a first lieutenant in the Surgeon General's Office, Section of Anthropology, where he worked on reports concerning "Defects Found in Drafted Men" and "Army Anthropology," and made an anthropometric survey of recruits at Camp Grant outside Rockford, Illinois. Sullivan had been hired as an assistant in physical anthropology at the American Museum in 1916, while earning his doctorate in that field at Columbia University. In 1920, as he was completing his PhD, Sullivan was promoted to assistant curator of physical anthropology, and sent to work at the Bishop Museum on problems of race in Hawaii and Polynesia more generally.

Sullivan earned his doctorate in physical anthropology during a period when the field of anthropology generally, and physical anthropology in particular, were just becoming fully professionalized. Although the study of human beings as zoological organisms dates back hundreds of years, and physical anthropology itself is conceptually and methodologically conventionally dated to the eighteenth-century classifications of Carl Linneaus and Johann Friedrich Blumenbach, as a modern scientific practice characterized by a unified methodology, standardized training, institutional stability, and professional organs such as journals, associations, and conferences, physical anthropology only became fully professionalized in the 1920s through the efforts of people such as Franz Boas, Earnest Hooton, and Aleš Hrdlička in the United States and Bronislaw Malinowski, W. H. R. Rivers, and A. R. Radcliffe-Brown in England. Columbia and Harvard Universities began producing notable numbers of doctorates in physical anthropology only in the 1920s. The American Association of Physical Anthropology and the *Journal of American Physical Anthropology* were each founded in 1919, largely through the efforts of Hrdlička at the United States National Museum.[47]

---

[47] For additional work on Hrdlička, see Juliet Burba, "Whence Came the American Indians?": American Anthropologists and the Origins Question, 1880–1935" (PhD diss., University of Minnesota, 2006); Gregory Foster-Rice, "The Visuality of Race: 'The Old Americans,' 'The New Negro,' and American Art, c. 1925" (PhD diss., 2 vols., Northwestern University, 2003); Frank Spencer, "Aleš Hrdlička, MD, 1869–1943: A Chronicle of the Life and Work of an American Physical Anthropologist" (PhD diss., University of Michigan, 1979); Michael L. Blakey, "Intrinsic Social and Political Bias in the History of American Physical Anthropology: With Special Reference to the Work of Aleš Hrdlička," *Critique of Anthropology*, Oct. 1987, vol. 7, pp. 7–35.

Like Hrdlička, many of Sullivan's teachers and colleagues were trained as physicians or anatomists, not as anthropologists. Sullivan himself, although trained in anthropology as a physical anthropologist, and with a prior background in the biological sciences, approached his anthropology from an organismal, comparative point of view. He viewed his anthropological work as directly analogous to that of zoologists and anatomists, his object being "one of the most complicated of any of the animals groups – man."[48] Not surprisingly, he cast a jaundiced eye over the quality and quantity of work falling under the heading anthropology and addressing itself to the nature of the human animal in Polynesia. Surveying the state of physical anthropology in Polynesia for the first Pan-Pacific Conference, Sullivan termed "somatological research" – the investigation of Polynesian bodies – a "virgin field," in which the little data that had been accumulated in a handful of "serious" studies was "entirely inadequate for conclusions as to the racial or inter-insular affinities of the Polynesians from a physical standpoint." But it was not simply that data was meager and scattered that aroused Sullivan's dismay. One might wonder, he remarked, if such a paucity of information would result in a concomitant "poverty of theories and speculations as to the relationships and migrations of the Polynesians?"

No, indeed; the history of no other similar group of mankind has probably been so much speculated upon and maltreated as that of the Polynesians. Voyagers, missionaries, legislators and students of all subjects who have traveled in Polynesia or have talked with some one who has traveled in Polynesia, or who have seen photographs of the natives, have been generous in their contributions to the theories and speculations in vogue; nor have they been backward in adding new theories of their own manufacture. In the literature it is easy to find authority for including or excluding the Polynesians in or from all the races of Mankind.[49]

Some authorities argued the original inhabitants came from the South Island of New Zealand, the Hawaiki of tradition. Others thought they hailed from Saba, Arabia, by way of India (and were a branch of the Indo-European family) or from the Americas. Or that they first crossed the Pacific to America and then returned to Polynesia. Or perhaps there were two migrations: one via Europe, Northern Asia, and the Philippines; a second by way of India.

---

[48] Louis R. Sullivan, "The Status of Physical Anthropology in Polynesia," *Proceedings of the First Pan-Pacific Scientific Congress*, Bernice P. Bishop Museum, Special Publication, no. 7, part 1, 1921, p. 69.

[49] Ibid., p. 63.

Sullivan regarded this question of origins as impossible for physical anthropology – or any single branch of anthropology – to answer. In a theme reiterated by other speakers at the Pan-Pacific Conference, Sullivan contended that such questions would require information and cooperation from a variety of disciplines including geology, botany, and zoology as well as the other forms of anthropology: ethnology, archaeology, and linguistics. Physical anthropology could, however, find answers to problems of racial type, racial mixing, and relations between types, Sullivan argued. As he enumerated them, the manageable problems were to: 1) accurately define and describe the Polynesian racial groups; 2) prove beyond a reasonable doubt the racial origin and affinities of the Polynesians; 3) designate "fairly accurately" to what branch of a given race they belong; 4) determine the degree of relationships between various Polynesian groups; and 5) point out the probable types that Polynesians had come in contact with and intermixed with during their migrations, as well as recently.[50] Despite the alleged tractability of these problems to the methods and insights of physical anthropology, according to Sullivan the discipline had succeeded "not at all." He lamented that the literature on racial questions was as confused and muddled as that on origins. A number of authorities claimed Polynesians should be classed as Mongoloid, two regarding them as akin to American Indians, another Malays.[51] Another set of authorities regarded them as Caucasians, one asserting that it was "admitted by all competent observers" that the Polynesians were members of an Oceanic branch of Caucasians and were "in no respect inferior to the average European either in complexion, physical beauty or nobility of expression."[52] Sullivan cited one scholar who enthused that "as far as physique and appearance go [Polynesians] certainly gave one an impression of being a superior race to ours."[53] Sullivan noted that the disjunction in systematics reflected a disciplinary divide, as well: the biologists classed Polynesians as Mongoloid; the ethnologists regarded them as Caucasians.[54]

[50] Ibid., p. 64.
[51] Ibid. Sullivan cites Geoffrey Saint-Hilaire, Thomas Henry Huxley, William Flower, Paul Topinard, Oscar Peschel, William Ellis, John D.[unmore] Lang, and W.[illiam] T.[homas] Pritchard. The quotation is from A.[ugustus] H.[enry] Keane, "The World's Peoples," (1908).
[52] Ibid., p. 65. Sullivan cites Jean Louis Quatrefages, F.[rankling] H.[enry] Giddings, S.[tephen] Percy Smith, John Macmillan Brown, and A.[ugustus] H.[enry] Keane.
[53] Ibid. Sullivan cites Lord George Campbell but does not give a publication or year.
[54] At the Pan-Pacific Conference, Sullivan noted that amateur or ill-trained investigators were prone to emphasizing atypical traits and drawing unfounded conclusions on that

Sullivan hoped to begin sorting out the complexities of race in Polynesia, a project he started by collecting data in Hawaii. Clark Wissler
had a long-standing interest in psychology and its relationship to race,
and at his urging Sullivan, in addition to collecting anthropometric data
on Hawaiians and "working up" the data collected by ethnologists on
Bishop Museum expeditions, also collaborated with Kathryn Murdock,
a psychologist, on a study of children of various races in Hawaiian public
schools. In his correspondence with Wissler and William K. Gregory, a
paleontologist at the American Museum and Columbia University, with
whom Sullivan studied human evolution, he repeatedly highlighted the
importance of the Pacific Islands, and Hawaii in particular, for the study
of race mixing. "All over the world people are watching with interest
the outcome of the natural experiment in race mixture that is going
on in these islands," Sullivan wrote.[55] His 1921 report on his work at
the Bishop Museum noted that Hawaii presented anthropologists with
a "unique opportunity for the study of race mixture and the effects of
environment on the development of the races of man."[56]

Having spent eighteen months collecting data, Sullivan returned to
New York, but shortly after was forced to move to the Southwest in an
attempt to improve his declining health. Between his poor health and the
inability of the American and Bishop Museums to provide sufficient funds
to hire an assistant to make the thousands of hand calculations required
to prepare his data for publication, Sullivan never did finish his Hawaiian
project. The dry climate of Arizona failed to forestall his decline, and in
1925, at the age of thirty-three, Sullivan died of tuberculosis. Sullivan's

basis. As an example, he cites early observers who placed undue significance on the
purported existence of "yellow hair" among the Tongans, concluding that Polynesians
were Caucasian or displayed significant Caucasian heredity. "Surely this is not the
average form and color of the hair for any Polynesian or Melanesian group," Sullivan
remarked. Sullivan continued this line of argument about method and "Mongoloid"
races in "The 'Blond' Eskimo – A Question of Method," *American Anthropologist*, New
Series, vol. 24, no. 2 (Apr.–Jun., 1922), pp. 225–228. Sullivan cited with approval the
repeated use by Franz Boas of the "transverse cranio-facial index" and the "transverse
cephalo-facial index" (measures of facial width) to distinguish between "mongoloid
stocks and the white races" (pp. 226–227). See Franz Boas, "The Half-Blood Indian,"
originally published in *Popular Science Monthly* (Oct. 1894), reprinted in Boas, *Race,
Language and Culture*, pp. 138–148.

55 Louis Sullivan to William K. Gregory, March 18, 1921. Anthropology Department
Papers, Special Collections, Library, American Museum of Natural History, New York
(hereafter AMNH-ANTH). See also, correspondence with Clark Wissler Nov. 3, 1920;
Nov. 23, 1922; Jul. 4, 1923.

56 Report from Louis Sullivan to William K. Gregory on his activities as research associate
in anthropology at the Bernice P. Bishop Museum, March 25, 1921, AMNH-ANTH.

early death left an opening in physical anthropology at the museum, which Harry Shapiro soon filled.[57]

Harry Shapiro's Boasian Project in Polynesia

Harry Shapiro's Polynesian studies, like Sullivan's, were done in collaboration with the Bishop Museum, but his projects applied to the questions of racial types and race mixing the Boasian concern with history, environmental context, and contestation of prevailing assumptions about viability and plasticity. Following Boas's lead, Shapiro conducted a number of studies in physical anthropology in Polynesia. He published numbers of technical papers and more popular articles, but his two largest projects attacked two pillars of typological racial theory: the presumption that racial mixes produced inferior and defective offspring; and the faith in stable racial characteristics and unchanging racial types. These studies represented two of the four problems in "anthropometric and anatomical research" deemed critical at the 1920 Pan-Pacific Conference, although with a Boasian flavor not anticipated by the conferees.[58] In *The Heritage of the Bounty*, published in 1936, Shapiro used the isolated population on Pitcairn Island, descendants of the mutineers from the *Bounty*, to undermine widely held assumptions about the results of miscegenation. *Migration and Environment*, published in 1939, was a substantial study of bodily changes among Japanese immigrants in Hawaii and their relatives, both offspring born in Hawaii and relatives back in Japan, along the lines of Boas's study "Changes in Bodily Form in Descendants of Immigrants."

---

[57] Shapiro and Wissler would later see to it that Sullivan's Hawaiian data was finally published. "Observations on Hawaiian Somatology, by Louis R. Sullivan, Prepared for Publication, by Clark Wissler," *Bernice P. Bishop Museum. Memoirs*, vol. 9 (1927), pp. 267–342.

[58] The other two problems were accurately defining and describing the "Hawaiian type or types," through a study of living Hawaiians as well as skeletal material "systematically collected," and a study of growth. Of the latter, the Conference "Recommendations" claimed:

The schools of Honolulu offer the opportunity of studying, side by side, large groups of Hawaiian, Japanese, Chinese, and Portuguese children, and various types of crosses all growing under the same environmental conditions and as nearly as possible of about the same social and economic status. The conditions for this experiment are as nearly ideal as we may hope to find anywhere.

They further noted that such a study would "be of direct economic interest to the Territory." "Proceedings," p. 115.

Race Mixing in Oceania

Harry Shapiro dedicated himself in the 1920s and 1930s to the study of racial mixing in Polynesia as a means of understanding the nature of races and the results of interbreeding, knowledge that he thought could be applied to other parts of the world, especially the United States, and as a means of untangling the complex racial history of Polynesia itself. The question of "hybrids" had taken on increasing significance in popular and anthropological speculations about human variety, in part because of the growth of evolutionary and genetic science in the late nineteenth and early twentieth centuries, and in part because of growing anxiety in the United States about immigration and an increasingly heterogeneous population. From anthropologists' point of view, there was a great deal of speculation and anecdotal observation but little in the way of systematic explorations of the subject. By the time Harry Shapiro began to develop his projects, he could look to only Boas's 1892 work on Native and white Americans and Eugen Fischer's 1913 study of Boer and indigenous African offspring in Southern Africa, the so-called Rehobother Bastards, both of which demonstrated evidence of hybrid vigor rather than degeneration. Shapiro's work on Pitcairn Islanders in the late 1920s joined a flurry of studies in that period, all of which questioned the deleterious effects of race mixing. These included three studies published in 1928 – Reginald Ruggles Gates's work on Amerindian crosses in Canada, Melville Herskovits's study of black-white crosses in the United States, and L. C. Dunn's analysis of data collected by Alfred Tozzer in Hawaii – as well as Ernst Rodenwaldt's 1927 study of mixed race "mestizos" on Kisar, in the Malay Archipelago.[59] Rodenwaldt declared, for example, "We are surely entitled to conclude that men in the past have been too hasty in ascribing to the consequences of consanguinity what were really the result of environmental influences."[60]

Physical anthropologists felt some urgency about the Polynesian project proposed at the 1920 Pan-Pacific conference and taken up by

---

[59] Boas, "The Half-Blood Indian"; Eugen Fischer, *Die Rehobother Bastards und das Bastardierungsproblem beim Menschen* (Jena: Gustav Fischer, 1913); Ernst Rodenwaldt, *Die Mestizen auf Kisar* (Batavia, Dutch East Indies, 1927); L. C. Dunn, "An Anthropological Study of Hawaiians of Pure and Mixed Blood," *Papers of the Peabody Museum*, vol. 11 (1928), pp. 90–211; Melville J. Herskovits, *American Negro: A Study in Racial Crossing* (New York: Knopf, 1928); Reginald Ruggles Gates, "A Pedigree Study of Amerindian Crosses in Canada," *Journal of the Royal Anthropological Institute*, vol. 58 (1928), pp. 511–532.

[60] Rodenwaldt, *Die Mestizen auf Kisar*. Cited in Anthony M. Ludovici, "Eugenics and Consanguineous Marriages," *The Eugenics Review*, vol. 25 (1933–1934), pp. 147–155 (p. 152).

Shapiro. A record of Polynesians' physical characters needed to be compiled "promptly," Shapiro argued. Far from a concern that extensive cultural contact and racial mixing would spur decline and widespread "disharmonies," Shapiro feared the very success of Polynesia's heterogeneous societies and thriving hybrid populations would soon so muddy racial genealogies and the mix of racial traits that all hope for mapping and tracking them would be lost. "[E]ven now in many islands the difficulties of finding unmixed natives are very great," he wrote. "Those familiar with Polynesian life anticipate a future when some foreign blood will flow in the veins of most Polynesians," particularly that of Chinese and Europeans, who were a constant presence. The pressures of a large, "prolific and vigorous hybrid group" threatened to "dissolve the remnants of the aboriginal population," Shapiro and his coauthor Peter Buck argued.[61] Shapiro, following Sullivan and Roland Dixon, regarded Polynesians as a heterogeneous population whose patterns of variations suggested several types distributed across the archipelago. But he nonetheless argued "there is a residium of common traits which generally

---

[61] Harry L. Shapiro and Peter H. Buck (Te Rangi Hīroa), "The Physical Characters of the Cook Islanders," *Bernice P. Bishop Museum – Memoir*, vol. 12, no. 1 (Honolulu: Bernice P. Bishop Museum, 1936).

Peter Buck, or Te Rangi Hīroa, as he was also known, is an interesting figure in Pacific anthropology. Maori on his mother's side and raised in part in their traditions, he became an ethnologist following six years' service in New Zealand's parliament and four years' service as a medical officer in the First World War, including service with the First Maori Contingent at Gallipoli in 1915, and service with the Pioneer Battalion on the Western Front, from 1916 to 1918. In his popular book on Polynesian history and ethnology, *Vikings of the Sunrise*, he briefly recounts how he navigated his dual identity as an indigenous scientist at Otago University, where he was the first Maori graduate: "I remember well when a fellow Maori student and I first entered the taboo precincts of the Medical School and saw at the top of the stairs a notice offering various prices for Maori skulls, pelves [pelvises], and complete skeletons. We read it with horror and almost abandoned our quest for western medical knowledge." But in the very next sentence, Hīroa marks his transformation, seeming to attribute it to the effects of scientific training. "After acquiring a knowledge of anatomy, however, I determined to contribute to the material on Polynesian somatology by measuring the heads of a number of my living countrymen," something he first accomplished on the troop ship carrying the Maori Pioneer Battalion back to New Zealand. Hīroa notes that after struggling to acquire a craniometer, he measured the heads of 424 "full-blooded" Maoris while a medical officer on the troop ship. Prior to joining the Bishop Museum as an ethnologist in 1926, Hīroa served as Director of Maori Hygiene in New Zealand. He went on to become director of the Bishop Museum in 1936, a post he retained until his death in 1951, as well as enjoying a joint appointment on the anthropology faculty at Yale University. Peter H. Buck (Ti Rangi Hīroa), *Vikings of the Sunrise* (New York: Frederick A. Stokes Company, 1938), p. 14; Hirao (Buck), *Introduction*, p. 46; M. P. K. Sorrenson, "Buck, Peter Henry 1877?–1951," *Dictionary of New Zealand Biography*, updated Jun. 22, 2007, retrieved from: http://www.dnzb.govt.nz/.

characterizes these people."[62] The American and Bishop Museums hoped that by thoroughly studying and comparing various island populations in Polynesia, the wide variability that characterized "Polynesians" could be understood, perhaps shedding light on where and how the original Polynesian population arose.

In one of his studies in this series, Shapiro compared the Ontong Javanese with a dozen groups of Polynesians, Melanesians, and Micronesians, with the hope that a study of the physical characteristics of the geographically and racially marginal Ontong Javanese might reveal something about the origins of Polynesians. Anthropologists regarded Melanesians and Micronesians as physically distinct from Polynesians. Shapiro viewed the Ontong Javanese, living at the edge of the Melanesian archipelagoes, with their lighter skin, straight or wavy hair, and narrower noses, as a striking contrast to Melanesians characterized by black skin and "frizzly" hair. "Obviously," Shapiro argued, the Ontong Javanese were "what they appeared to be, an intrusive element."[63] Ethnologists such as Raymond Firth and H. Ian Hogbin, seeing cultural similarities with Polynesians, regarded the Ontong Javanese and others living along the eastern edge of Melanesia as "Polynesian outliers," and had proposed various migrations to explain their odd distribution. Many regarded these isolated pockets of "Polynesians" along the edge of the Melanesian archipelagoes as "footprints which mark the route by which the migratory Polynesians reached their present home," linking the heart of Polynesia to the Asian mainland via Indonesia. But Shapiro was interested in the more basic question of whether or not these people were in fact Polynesians. If they were, they could "throw a strong beam

---

[62] Harry L. Shapiro, "The Disappearing Peoples of the South Seas," *Natural History* (New York: American Museum of Natural History, 1930), pp. 253–266; Louis R. Sullivan, "Race Types in Polynesia," *American Anthropologist*, New Series, vol. 26, no. 1 (Jan.–Mar., 1924), pp. 22–26; Roland B. Dixon, *The Racial History of Man* (New York: Charles Scribner's Sons, 1923); Roland Dixon, "A New Theory of Polynesian Origins," *Proceedings of the American Philosophical Society*, vol. 59, no. 4 (1920), pp. 261–267; Louis R. Sullivan, review of *The Racial History of Man* by Roland Dixon, *American Anthropologist*, New Series, vol. 25, no. 3 (Jul.–Sep. 1923), pp. 406–412. For criticism of Dixon's reconsideration of the distribution of racial traits and types, see Franz Boas, review of *The Racial History of Man* by Roland B. Dixon, *Science*, New Series, vol. 57, no. 1481 (May 18, 1923), pp. 587–590; Aleš Hrdlička, "*The Racial History of Man* by Roland B. Dixon," *The American Historical Review*, vol. 28, no. 4 (Jul. 1923), pp. 723–726.

[63] Harry L. Shapiro, "The Physical Characteristics of the Ontong Javanese: A Contribution to the Study of the Non-Melanesian Elements in Melanesia," *Anthropological Papers of the American Museum of Natural History* (New York: American Museum of Natural History, 1933), vol. 33, part 3, pp. 227–278.

of light on the functioning of Polynesian culture which has largely vanished from Polynesia itself," and more importantly, Shapiro argued, they might "reveal data on the physical type which has become seriously diluted in Polynesia."[64] Although conceding that ethnological and linguistic data did suggest a Polynesian resemblance, "on the racial side," Shapiro thought the connection was frequently asserted without much foundation. Shapiro argued that an intensive study was needed to avoid "hasty identification or a too glib generalization."[65] In the end, Shapiro concluded, based on his synthesis and analysis of data collected across the Pacific Islands by a range of investigators, that these "outliers" were neither an isolated group of Polynesians residing at the edge of the Melanesian islands nor the result of older Polynesian migrants being "contaminated" by local "Melanesian admixture." Instead, Shapiro concluded that the limited data – including photos of the Ontong Javanese – suggested kinship with the Caroline Islanders, a Micronesian group. Given the prevailing Oceanic categories – Polynesian, Melanesian, Micronesian – the Ontong Javanese remained a mystery. The ethnology, linguistics, and physical anthropology provided no common answer for the origins and evolution of the Ontong Javanese people.

The Chinese immigrants who threatened the predominance of "pure" Polynesians in Hawaii provided Shapiro with an initial avenue for pursuing the studies of racial mixture and genetics that he later pursued more fully in his study of Pitcairn Islanders and Japanese immigrants in Hawaii. Between 1929 and 1932, under the sponsorship of the Rockefeller Foundation and University of Hawaii, Shapiro pursued a study of Chinese immigrants, their children, and their relatives back in China, all of whom came from a limited area around Canton, along with Hawaiian and Chinese "crosses." The Chinese-Hawaiians were ideal, Shapiro argued, for analyzing "the genetic behavior of human traits" that along with "psychological, sociological, and physiological studies" would offer "a complete picture of a mixed group in its biological and social setting."[66] He noted that the Hawaiian-Chinese mixes had a number of advantages, from a geneticist's point of view. Most of the "crosses" were unmixed with the many other racial "strains" that had populated Hawaii in recent

---

[64] Ibid., p. 233.

[65] Ibid.

[66] H.[arry] L. Shapiro, "Race Mixture in Hawaii," *Natural History* (New York: American Museum of Natural History, 1931), pp. 31–48. For a fuller study of Shapiro's Chinese-Hawaiian project, see Christine Leah Manganaro, "Assimilating Hawaiʻi: Racial Science in a Colonial 'Laboratory,' 1919–1939" (PhD diss., University of Minnesota, 2012).

years. In addition, the Chinese had immigrated long enough ago to have produced two generations, so a "full range of Mendelian combinations exist and permit a more complete analysis." Conversely, because most of the crosses did not date back further than 1870, they were still within the memory of living people, unlike the "Caucasian-Hawaiian crosses," which were so old that the essential genealogical data was "obscured and frequently lost."[67] Although he never published the results, Shapiro's study of the Chinese and Chinese-Hawaiians was a direct precursor for his major interwar study of the Japanese-Hawaiian population, *Migration and Environment*.[68] Meanwhile, as data on the Chinese in Hawaii continued to accumulate and preliminary results emerged from Hooton's Harvard anthropometric lab, Shapiro was also busy with his long-awaited study of race mixing on Pitcairn Island.

Mutineers and Tahitians: An Experiment in Hybrid Vigor on Pitcairn Island

Shapiro claimed that his particular interest in the people who inhabited Pitcairn Island was sparked by a colorful lecture he heard as an undergraduate at Harvard University. His instructor, and later the director of his doctoral work, Earnest Hooton, regaled the class with the tale of the mutiny on the *Bounty* and the peopling of Pitcairn Island by a handful of mutineers and their Tahitian companions. In the preface to his book, Shapiro described the imaginative lure of Polynesia and the Pitcairn Islanders as Hooton lectured:

Then with all the magic of the movies the scene shifted, and we were in the warm, vibrant Pacific, on the Cytherean shores of Tahiti, amidst the turmoil of a mutiny on board the Bounty and stranded on a forgotten speck of land called Pitcairn, where human folly was succeeded by inhuman virtue. That unique narrative of an eighteenth century breadfruit expedition resolving itself into mutiny, court-martial, hanging, crime, murder, and, finally, a new population of mixed bloods made a glorious text.[69]

---

[67] Ibid., pp. 47–48.

[68] Shapiro never published his analysis of the Chinese-Hawaiian study. The data were collected by Harvard graduate students William Lessa and Frederick Hulse, and University of Hawaii graduate student Margaret Lam from 1929 to 1932, and analyzed at Earnest Hooton's physical anthropology laboratory. Together, they measured thousands of Chinese, Hawaiian, and mixed individuals and drew blood samples from hundreds of people in the Hawaiian Islands. In 1932, Lessa also visited villages in Kwangtung Province, China, to measure hundreds of relatives of the Chinese immigrants to Hawaii. AMNH-HLS.

[69] Ibid., p. xi–xii.

Shapiro spent some twelve years studying the hereditary legacy of the mutiny on the *Bounty*. His goal from the beginning was to study the isolated population on Pitcairn Island, to see for himself and assess anthropometrically the living legacy of that eighteenth-century confluence of British sailors and Tahitian society. Following his senior year at Harvard and his training in anthropology with Hooton and Alfred Tozzer and in genetics and statistics with William Castle and Edward East, Shapiro was awarded one of the Bishop Museum research fellowships funded by Bayard Dominick to support a trip to Pitcairn in 1923. But it was not to be; travel to many of the Pacific Islands was a chancy business in the 1920s. Pitcairn Island was particularly remote (part of its original attraction to the mutineers), visited only every few months by ships bound primarily for other destinations. Prevented by bad weather from landing, Shapiro settled for Norfolk Island, where in 1856, a large number Pitcairn Islanders had resettled. His dissertation work on the Norfolk Islanders only peaked his interest in the remaining, still isolated Pitcairn Islanders.[70] After repeated disappointments on subsequent trips to Polynesia, Shapiro finally reached the island in 1934. His book, *The Heritage of the Bounty, The Story of Pitcairn Island through Six Generations*, published in 1936, offers a narrative of his long-anticipated trip, and his conclusions about the cultural and physical condition of the Pitcairn Islanders (Figure 5.2).

Earnest Hooton and Harry Shapiro found the story of mutiny on the high seas, the struggle to settle the island, and the establishment of an interracial community a romantic and intriguing tale, but the story of the *Bounty*'s mission to Tahiti also exemplifies a broader colonial context and the imperial, socially embedded nature of scientific practice in Polynesia. It starts with a group of Englishmen, with interests in West Indian plantations, who petitioned King George III, a patron of geographical exploration, to send an expedition to Tahiti, already famous after having been "discovered" by both Wallis and Bougainville in the late 1760s, to bring back specimens of the breadfruit tree. The breadfruit tree bears abundant fruit year-round, needs little care, provides a large crop, and was a staple food in Polynesia. The petitioners intended to plant the trees in the West Indies as a cheap way to feed slaves. Their petition was supported by Joseph Banks, then-president of the Royal Society and a member of Captain James Cook's first voyage to Polynesia. Lieutenant William Bligh, selected as captain, had been one of Cook's officers on

---

[70] This work was published as "Descendants of the Mutineers of the Bounty," *Memoirs*, Bernice P. Bishop Museum, vol. 11, no. 1 (1929), pp. 3–106.

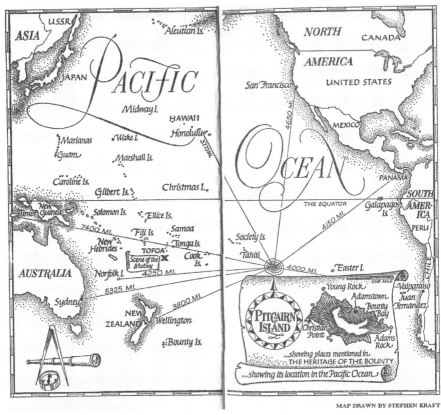

MAP DRAWN BY STEPHEN KRAFT

FIGURE 5.2. Stephen Kraft. Map of Polynesia and Pitcairn Island. *Source:* Harry Shapiro, *The Pitcairn Islanders* (1962). Courtesy of Harriet R. Shapiro and James E. Shapiro.

his "ill-fated" third voyage, which ended with Cook's death. Bligh and his crew spent six months in Tahiti collecting and growing more than 1,000 breadfruit trees to be carried back in the specially outfitted ship. Just after the *Bounty* set off from Tahiti in 1789, Fletcher Christian and twenty-four other crewmembers mutinied, setting Bligh and eighteen men adrift in a small boat with scanty provisions. Bligh and his men traveled 3,600 miles before reaching Timor.[71]

The mutineers returned to Tahiti, finally splitting into two groups: sixteen who stayed in Tahiti; and nine who took the *Bounty*, along with twelve Tahitian women and six Tahitian men, and sailed 300 miles south

[71] Ibid., pp. 37–53.

to Pitcairn Island in 1790, where, after unloading the ship of everything useful, it was run aground and burned. In the next few years, all the Tahitian men and eight of the nine mutineers murdered each other over land, women, and attempts by the mutineers to subjugate the Tahitian men. Under the leadership of the remaining mutineer, John Adams, order was established and the colony grew, eventually becoming remarkably religious, surprisingly harmonious, and notably egalitarian, including the equal division of property among all descendants, regardless of sex, and full suffrage for all adult islanders. The islanders spent their first eighteen years in total isolation, receiving their first visit in 1808 from a surprised Captain Mayhew Folger, who thought the island was uninhabited. In 1856, because of overcrowding, the colony moved to Norfolk Island, but a number of families returned to Pitcairn in the 1860s, forming the descendents of the population studied by Shapiro.[72]

Shapiro regarded the Pitcairn Islanders as an extraordinary biological experiment in human heredity and race mixture. A population descended from British mutineer fathers and Tahitian mothers (the Tahitian men having all been murdered before siring offspring), with carefully maintained genealogical histories, a generally harmonious and egalitarian society, inhabiting an isolated island where there had been almost no additional immigration, the Pitcairn Islanders offered Shapiro an unexamined breeding group ideal for research. Based on his earlier work on Norfolk Island, Shapiro anticipated his study would demonstrate that racial mixes did not produce inferior offspring (Figure 5.3).

Based on analysis of his measurements and observations, Shapiro concluded that the Pitcairn Islanders were taller than their parents; produced a healthy number of children; favored their English ancestors in some characteristics, their Tahitian ancestors in others; and showed a mixture in other cases. There were more blue-eyed and light-haired people in the mixed-race descendents than in a typical Tahitian population, but more with very dark hair than in an English population, for example. Shapiro concluded that there was no evidence, biological or cultural, to suggest the Pitcairn Islanders were inferior to either of their ancestral peoples, and some instances appeared to surpass them (Figure 5.3). These results reinforced his earlier conclusion, based on his field work at Norfolk Island, that the racial mixing among the English and Tahitians had produced offspring that showed improvements over one or both of their parental

---

[72] Ibid., William Bligh and Edward Christian, *The Bounty Mutiny* (New York: Penguin, 2001).

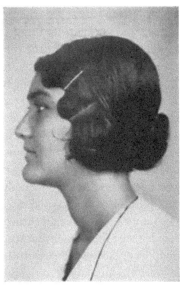

Burley Warren—The author's
host on Pitcairn

Representing Pitcairn's
Younger Generation

Parkins Christian, great-great-
grandson of Fletcher Christian

Mary Ann McCoy—The oldest
woman on Pitcairn

FIGURE 5.3. Pitcairn Islanders, ca. 1936. *Source:* Harry L. Shapiro, *Heritage of the Bounty: The Story of Pitcairn through Six Generations* (New York: Simon and Schuster, 1936). Courtesy of Harriet R. Shapiro and James E. Shapiro.

stocks in one or more of the observed traits, what geneticists referred to as "hybrid vigor" or heterosis. In his Norfolk study, Shapiro calculated that the descendents of the *Bounty* mutineers and their Tahitian wives produced an average of 11.4 children "per mating," with several women giving birth in their fifties, and that the men averaged six feet in height. Both of these indices significantly exceeded the averages for English and Polynesian groups.[73] Shapiro's conclusions bolstered claims made by William Castle, who also regarded racial mixes as genetically unproblematic, and Edward East, who had recently demonstrated the healthy quality of mixed strains of corn. Castle argued that "the widest possible human crosses are comparable with the crossing of geographical varieties of a wild species of animal or with the crossing of distinct breeds of domesticated animals."[74] Shapiro's conclusions about the Norfolk and Pitcairn Islanders confirmed Castle's contention that "offspring produced by crossing such races do not lack in vigor, size, or reproductive capacity."[75] In addition to the physical evidence of hybrid vigor, Shapiro found the Pitcairn Islanders had developed a vigorous culture, as well, accommodating, even inviting the wider world, while retaining elements of their disparate heritage. In his final analysis, he pronounced Pitcairn culture a productive and creative mixing of English and Tahitian cultures that produced its own innovations.

Although these hardly seem like extraordinarily interesting results from a twenty-first century point of view, Shapiro had reason to regard his study as extremely important. Unlike other instances of racially mixed populations, the Pitcairn Islanders were free from the social stigma of miscegenation, a stigma that other studies of race mixing not only did not recognize as a factor but perpetuated and accentuated. In his study, Shapiro was following Boas's lead. In 1892, Boas had studied children of white and Native American parents and found that the offspring were physically superior to either parent, being taller, and that "half-blood"

---

[73] Shapiro later claimed that his study of the Norfolk Islanders in 1923 was the first to show hybrid vigor among a mixed-race population, a phenomenon "then only recently identified by Professor [Edward] East in the miscegenation of separate strains of corn." "Human Adaptations to the Environment," 1980; "A Stroll Down Memory Lane," Typescript Box 44, Folder "Autobiographical," AMNH-HLS; "Results of Inbreeding on Norfolk Island," in "Science News," *Science*, New Series, vol. 65, no. 1693 (Jun. 10, 1927), p. x.

[74] William E. Castle, *Genetics and Eugenics*, p. 335, cited in Wilhelm W. Krauss, "Race Biological Impressions in Hawaii," *Mid-Pacific Magazine* (Oct.–Dec. 1935), pp. 301–303.

[75] Ibid.

women were more fertile than Native American women living in the same conditions, contradicting the commonly held view that "hybrid" races were less fertile.[76] In 1915, three years after publishing his seminal paper on the influence of the environment on heredity, Boas addressed the International Congress of Americanists, where he argued for the utter lack of evidence underlying widespread claims of racial inferiority among mixed races.

In setting out for Pitcairn Island, Shapiro had in mind not only Boas's work, but the hope of refuting the widely cited conclusions drawn by Charles Davenport and Morris Steggerda in their 1929 study *Race Crossing in Jamaica*.[77] Davenport is probably best known as the dean of American eugenics. He ran the Eugenics Record Office, later the Department of Genetics, at Cold Spring Harbor on Long Island. He was cofounder of the Galton Society, along with Madison Grant – author of *The Passing of the Great Race*, a lament for the fate of the Nordic race – and had been in charge of anthropology for the U.S. army during World War I.[78]

Race Crossing in Jamaica was a study intended to investigate the effects of racial mixing between blacks and whites. Three groups in Jamaica were compared: a group of descendants from West African natives imported as slaves by the Spanish over the course of three centuries, a group of "browns" thought to represent crossing between the original Africans and their descendents with white residents, and a group of white residents of varying backgrounds. These groups were measured and observed on

---

[76] Boas, "The Half-Blood Indian."

[77] Shapiro received a copy of *Race Crossing in Jamaica* from the Carnegie Institution in 1930 at his request and at the suggestion of Clark Wissler. W. M. Gilbert, administrative secretary at Carnegie Institution to Harry Shapiro, May 23, 1930. Anthropology Department, AMNH-HLS.

[78] For the history of American eugenics and Davenport's role, see, inter alia, Garland E. Allen, "The Misuse of Biological Hierarchies: The American Eugenics Movement, 1900–1940," *History and Philosophy of the Life Sciences* 5 (1984), pp. 105–128; Hamilton Cravens, *The Triumph of Evolution: American Scientists and the Heredity-Environment Controversy 1900–1941* (Philadelphia: University of Pennsylvania Press, 1978); Daniel Kevles, *In the Name of Eugenics. Genetics and the Uses of Human Heredity* (Cambridge, MA: Harvard University Press 1995); Kenneth Ludmerer, *Genetics and American Society* (Baltimore: Johns Hopkins University, 1972); Diane Paul, *Controlling Human Heredity: 1865 to the Present* (Atlantic Highlands, NJ: Humanities Press, 1995); John P. Jackson, Jr. and Nadine M. Weidman, eds., *Race, Racism, and Science: Social Impact and Interaction*, Science and Society Series (New Brunswick: Rutgers University Press, 2006); Jan Anthony Witkowski and John R. Inglis, eds., *Davenport's Dream: 21st Century Reflection on Heredity and Eugenics* (Cold Spring Harbor, NY: Cold Spring Harbor Laboratory Press, 2008).

scores of traits: anthropometric measures; observations of qualitative traits such as eye color and hair form; and study of children for developmental characteristics. In addition, Davenport had the subjects measured on a whole series of psychological tests, including the Army Alpha tests and tests of musical ability. He and Steggerda concluded the blacks and "browns" did considerably worse on these tests than whites (except for music, where blacks excelled). In the summary, Davenport noted that the browns were superior to both whites and blacks on seven mental tasks, and inferior on nine of them. Nonetheless, he concluded that the browns were clearly inferior. Although on average the "browns did not do so badly," there were apparently more individuals among them who were "muddled and wuzzle-headed." Although blacks "may have low intelligence," Davenport argued, they used what little they had more effectively than the browns, who "seem not to be able to utilize their native endowment," a "disharmony" resulting from their hybrid nature. Davenport explained away the nearly equal number of tasks on which the browns outperformed both whites and blacks by claiming that adult browns excelled in only one task. The seven measures on which browns were superior were all recorded among adolescents, and it was well known from experience in the United States, Davenport noted, that mulattoes, on average, failed to progress beyond that stage.[79] Although Davenport acknowledged that the samples for these variations were small, he did not question that they measured "disharmonies in the mental sphere." In an example of Davenport's analysis of results in the physical sphere, he concluded that the "exceptionally large proportion" of dark eyes among the white population could not have been a result of "Negro blood," but rather confirmation of the widely held belief that "blonds do not thrive in the tropics."[80] Davenport concluded that the browns were inferior to both their white and black parental groups, biologically and intellectually, and stressed that such crosses produced individuals with disharmonious traits.

Davenport and Steggerda ignored the work of Franz Boas in their discussion of studies of race mixing and hybrid vigor, although they did discuss two other studies that Boas regarded as evidence for hybrid vigor, but only in the context of the extent of variation among the groups compared. Davenport contended that low variability was a sign of racial

[79] Davenport and Steggerda, *Race Crossing in Jamaica*, pp. 471–472.
[80] Ibid., p. 254.

purity, discounting Melville Herskovits's recent evidence to the contrary. The two studies Davenport and Steggerda cited were Eugen Fischer's study in South Africa of crosses between Boers – men of Dutch descent – and indigenous women (the so-called Hottentots), who had been characterized by racial scientists as the "Rehobother Bastards," and a study of "Mestizos" in Kisar, in the Malay Archipelago, by Ernst Rodenwaldt.[81] Twenty-four years after its initial publication, Shapiro described *Race Crossing in Jamaica* as "one of the most outspoken attacks on miscegenation" in its field, and still felt it necessary to mount a detailed critique of Davenport's methods and claims.[82]

Shapiro criticized Davenport's study directly on a number of grounds. No attempt was made to establish any degree of relatedness between the three groups compared, and the groups themselves were problematic in terms of their representiveness of each class. There was some attempt to control for socioeconomic factors by excluding from the white pool merchants and people of the "governing class," but the isolation and ethnic background of many of the whites made it unlikely that they resembled those who contributed to the original crosses. The browns were not the products of recent mixing, and their actual heritage was unclear. The age composition of the three groups was not comparable either, with the browns having a significantly higher proportion of men between sixteen and twenty, a confounding factor because, Shapiro argued, men that age had not reached physical maturity.[83] Their "inferiority" that Davenport "discovered," Shapiro argued, "can easily be attributed to age alone." Shapiro also noted that the tests of intelligence and other mental traits might well have registered the effects of the environment, conditioning, and test-taking situation, as much as any innate qualities. Beyond these serious methodological problems, Shapiro found much of Davenport's analysis unwarranted, if not outright ridiculous. The supposed physical disharmonies that Davenport adduced as evidence of the browns' inferiority consisted primarily of their relatively shorter arms, a conclusion Shapiro described as "making a mountain out of a molehill." This was particularly "flimsy evidence," because the same proportions were equaled in a number of whites, and some of them showed an even more

[81] Fischer, *Die Rehobother Bastards*; Rodenwaldt, *Die Mestizen auf Kisar*; Herskovits, *American Negro*.

[82] Harry L. Shapiro, *Race Mixture* (Paris: UNESCO, 1953).

[83] Shapiro excluded two men from his study of Ontong Javanese for this reason, although data had been collected on them along with other subjects of the study. "Physical Characteristics of the Ontong Javanese," p. 238.

marked tendency along the same lines. "The ridiculousness of the effort is made apparent," Shapiro continued, when Davenport claimed that such a disharmony in the browns would make it hard for one so afflicted to pick up an object off the ground.[84]

Franz Boas was also a vocal critic of work on racial mixing that failed to consider the full social context in which human beings live and produce children, and in particular the effects of social stigmas on interracial couples and their offspring. As a contributor to a volume of essays in honor of Alfred Kroeber, Boas used the opportunity to discuss the relationship between physical and social anthropology, and the key relationship between bodies and their social environment. Boas discussed difficulties associated with the idea of a "type" in any but the largest and most distinct groups, "like Negroes, Mongoloids, Australians," and argued that in virtually all other attempts at classification, the results would be poor "without knowledge of the conditions that have made the type what it is." This was even truer, he continued, "[w]here it can be proved historically that a population is mixed, such as American Mulattoes, the half-blood Indians, or the half-castes of the Orient." Common claims that such mixed people had inferior physiques or were prone to disharmonies could not be established without knowledge of who the parents were. "Were they of normal value, or of inferior strains in the race to which they belong, and are the conditions under which the mixed population live equal to those of the two parental stocks?" Without such knowledge, Boas argued, "biological inferences have little value." Boas argued that Shapiro's work in Pitcairn Island, as well as his own study of Indian-white mixes, demonstrated that mixed populations "preserved full vigor," whereas studies such as that by Davenport and Steggerda made the erroneous assumption of "equality of all social groups," overlooking "group differences which can be evaluated only by those intimately familiar with the social life of the people."[85]

In contrast to Davenport and Steggerda, in *Heritage on the Bounty*, Shapiro stressed the importance of the influence of the social environment on both actual bodies and the outcome of human reproduction. In discussing ways that studies of race mixture can be compromised, Shapiro first noted that genealogies must be complete so that the investigator knows "precisely what elements have entered into the making of a given hybrid, when it occurred in terms of generations, what crossbreeding has

[84] Shapiro, *Race Mixture*, pp. 41–46.
[85] Franz Boas, "The Relations between Physical and Social Anthropology," pp. 172–175.

occurred since the original cross, and, finally, the matings in each generation." Also, the races crossed had to be sufficiently distinct to avoid "complex and blurred results by virtue of overlapping characters and genetic similarities." But beyond these criteria, the scientific investigator had to be aware of the effect of the environment on the subjects studied, environment in a culturally historicized sense. Thus, he warned, "In the crossings of races that show marked physical differences or in cases where deep prejudice exists the partners of such unions are often degenerate or socially inferior members of one or both groups." This is not always the case, "but," he argued,

the fact remains that the evils which are popularly assigned to the mingling of blood may with greater justice be attributed to the quality of the blood which produced the hybrid. It stacks the cards, to say the least, against the half-caste to attribute any defect in his heritage to the irregularity of his breeding.... The social, economic, and therefore, environmental background of the half-caste is almost always inferior to the best in the possession of one of his ancestral stocks, and frequently it is worse than the average to be found in both. The social stigmata attached to those unfortunates of mixed blood are often almost insupportable and may engender at their worst unfortunate physical and psychological consequences.[86]

In light of the sort of study produced by Davenport and the conclusions he drew, Shapiro's discussion of "degenerates" takes on the progressive cast its author intended. Shapiro contested a science of human differences that privileged quantified bodies without considering the impact of the social environment in which people lived, formed families and communities, and produced children. If, for Shapiro and Boas, race, language, and culture were not interchangeable markers because each had their own historical development that had to be conceptualized independently of the others, it was equally true that race, language, and culture could not be divorced from one another. Shapiro and Boas advocated not the utter divorce of biology from culture but a joining of those investigations, not to concoct spurious associations of skin color with moral fiber, but to understand the forces that shape the wide variety of human appearance and behavior that we see.

Changes in Hawaiian Immigrants: Confirming Boas

Harry Shapiro's second major interwar research project confronted another foundational problem in the science of race: how stable were

---

[86] Shapiro, *Heritage on the Bounty*, pp. 218–219.

the bodily characteristics upon which racial taxonomies were erected? All historical, archaeological, and paleontological evidence testified to the restlessness of human beings. People had migrated all over the world, moving into profoundly different environments, confronted with new peoples and cultures, throughout human history. Some of the most profound debates within anthropology hinged on how to explain the variation and change among cultures, languages, and bodies that resulted from all this motion. Most comparative anatomists and physical anthropologists had long assumed that whatever cultural or linguistic changes people went through, their bodies remained essentially intact. Indeed, this assumption drove part of the interest in racial mixing, because it seemed to explain the high degree and complexity of human variation without recourse to the complexities of evolution. By the interwar period, however, anthropologists such as Boas and Shapiro recognized that human variation had to be understood within an evolutionary framework that explained change as the result of interactions among organisms and their environments.

Thus, when Harry Shapiro was invited in 1930 by the University of Hawaii to continue studies of race and race mixing in Hawaii that had been begun by his predecessor, Louis Sullivan, he saw an opportunity to replicate and extend Franz Boas's seminal efforts to study the effects of environment and the fixity or plasticity of the human form. The results of Shapiro's initial study of Chinese immigrants, their children, and their relatives back in China were so promising that Herbert Gregory, director of the Bishop Museum, and University of Hawaii officials supported a more extensive study of Japanese immigrants to Hawaii and their relatives in Japan. The results of that study were published in 1939 as *Migration and Environment, A Study of the Physical Characteristics of the Japanese Immigrants to Hawaii and the Effects of Environment on Their Descendants.*

In *Migration and Environment,* Shapiro explicitly cited Boas's "pioneering" study of changes in immigrants' bodies as the conceptual and methodological precursor. Shapiro had made some initial efforts in French Canada to find a group of people with which he could study the effects of environment on race, with the intention of constructing a study that would confirm Boas's key study.[87] Although Shapiro was familiar

---

[87] H.[arry] L. Shapiro, "The French Population of Canada: A Study of the Possible Effects of Environment on the Stability of Human Types," *Natural History* (Jul.–Aug. 1932), vol. 32, no. 4, pp. 311–355.

with Aleš Hrdlička's study of "old Americans" and the way descendants of initial European immigrants to the United States had changed in appearance over the course of generations and acknowledged its influence on his work, he regarded Boas's work on immigrants and head form as a superior study. Shapiro asserted that until Boas's "classic investigation," no one had challenged physical anthropologists' "article of faith" that average racial characteristics were stable over long periods of time. Yet despite widespread discussion of Boas's "incontrovertible" conclusions, Shapiro argued that "the criticism of current practices implied in it failed to affect the traditional procedures of racial studies. Even the vistas of research which it opened remained neglected." Moreover, Shapiro lamented, recent and growing critiques of physical anthropology, rooted in dissatisfaction with traditional adherence to a doctrine of stable racial characters, had failed to actually impact methods of study. Shapiro argued that physical anthropologists divorced theory from practice, acknowledging in theory that environment played a significant role in shaping organisms, but failing to operationalize that theory in any form in their comparative studies of racial groups.[88] Shapiro's work continued Boas's efforts to change the theory and practice of physical anthropology by argument and example.

In *Migration and Environment*, Shapiro examined three contemporaneous populations to "discover what immediate effects a change of environment might produce in the physical habitus of a migrating population."[89] Shapiro switched to the Japanese in Hawaii because he thought they were likely to offer even better results than the Chinese. In Shapiro's view, there were two key criteria for any investigation of the "stability of physical type." The population had to have been settled in one area long enough to have adjusted completely, and they had to have "remained unmixed with other elements" and be "fairly homogenous in origin in the beginning."[90] The Japanese in Hawaii not only fit these criteria well, but because they had more recently immigrated and came from an even more restricted area in their home country, it was likely to be easier to find their relatives back home.[91] In order to assess

---

[88] Harry L. Shapiro, *Migration and Environment. A Study of the Physical Characteristics of the Japanese Immigrants to Hawaii and the Effects of Environment on Their Descendants* (New York: Oxford University Press, 1939), pp. 184–185.

[89] Ibid., p. 5.

[90] Shapiro, "The French Population of Canada," p. 347–348.

[91] Shapiro, "Quality in Human Populations," *Scientific Monthly*, vol. 45, issue 2 (Aug. 1937), p. 112.

Japanese immigrants to Hawaii, Shapiro had measurements taken not only on the immigrants themselves, but also their offspring – replicating Boas's study – and their relatives in the villages in Japan from which they came, a total of more than 2,500 individuals, a very large study sample for that era.[92]

The Japanese still living in Japan, whom Shapiro referred to as "sedentes," provided a control group for assessing changes among the immigrant and Hawaiian-born populations, and included as many relatives as possible, something that Boas had been unable to do in his seminal study. Shapiro coined the term "sedentes" to refer to the population that remained in the original geographical "home" and from which a migrant population in another location, in this case Hawaii, was derived. Shapiro described the sedentes as Japanese who "were born in Japan and have remained there all their lives, moving within a narrow geographical orbit and marrying into the local population.[93] Shapiro's addition of this third population represented a crucial change from the method applied by Boas in his study of changes in head form among children of immigrants, in which he studied only foreign-born immigrants to the United States and their American-born children.

It was critical to compare immigrants and their offspring not just with each other in their new environment, but also with the ancestral population in the original environment, to determine whether any differences among the three groups were the result of selection or environment. Without the ancestral comparison it would not be possible to say with

---

[92] The number of subjects included in Shapiro's study was very large by contemporary standards. His study of the Ontong Javanese, by comparison, relied on the work of other investigators and represented a total of 155 Ontong Javanese and between 5 and 424 subjects for the other groups. "Physical Characteristics of the Ontong Javanese," pp. 238, 251, 255, passim. Frederick Hulse, a young anthropologist, was sent by Shapiro to Hawaii and Japan to collect the data. For each subject, he made forty-three separate measurements, calculated twenty-one indices, and made forty-one qualitative observations. Hulse had done a study of pure and mixed races in Cuba in 1928, in addition to his work for Shapiro on the Chinese-Hawaiian study. Hooton claimed that Hulse's study revealed the necessity of studying the Spanish in its home country to appraise possible changes in physical status in descendants of Spanish immigrants. In 1957, Hulse published a study of hybrid vigor among the Swiss, "Exogamie et heterosis," *Archives Suisses d'Anthropologie General*, vol. 22, pp. 103–125; "Research in Physical Anthropology at Harvard," p. 506; Michael A. Little, "The Development of Ideas on Human Ecology and Adaptation," in Frank Spencer, ed., *A History of American Physical Anthropology, 1930–1980* (New York: Academic Press, 1982), pp. 405–434.

[93] "Sedente" is derived from the Latin *sedere*, "to sit." This term later came into the lexicon of "ecological" studies of human evolution and adaptation. "[T]he convenience" of his somewhat awkward term "over the tedium of a frequently repeated periphrasis," Shapiro noted, "is its own apology." *Migration and Environment*, p. 7.

confidence whether any changes among immigrants or their children represented novel differences spurred by a new environment, or whether such differences merely reflected the emigration of a distinctive subset, in the case of the immigrants, or of a reversion to the form of the grandparents, in the case of the children. Shapiro used contemporaneous relatives in Japan, rather than the grandparents and parents of the immigrant generations, in part for practical reasons – it would have been time and cost prohibitive to find them, there might not have been enough of them as many would no longer be living, and it would have introduced a confounding age factor. The most important reason for not comparing immigrants to their grandparents and parents, however, was methodological. Shapiro was cognizant that recent research demonstrated considerable changes in bodily development of successive generations of the same population, and wanted to avoid that confounding factor.[94] He reasoned that "the problem was fundamentally not a comparison of Hawaiian born grandchildren with Japanese born grandparents, but rather an investigation of the deviation, if any, of the Hawaiian born Japanese from what they might have been had they remained in Japan."[95] In other words, Shapiro wanted to keep the age of the immigrant and sedentes populations, their gene pool, and their respective environments in Hawaii and Japan as constant as possible, to see what changes might have been produced in populations of related individuals who lived in divergent environments.[96]

In titling his monograph *Migration and Environment*, Shapiro signaled the twin factors that he and Boas regarded as keys to understanding human diversity, both sociocultural and physical. By characterizing his study as an attempt to determine the physical consequences of migration, Shapiro linked his work to a much larger body of anthropological and historical work concerned with the cultural and linguistic consequences of migrations. The importance of migrations in human history was widely acknowledged, but in the study of human difference and development, Shapiro argued, it was frequently ignored as a source of change because most physical anthropologists proceeded methodologically and conceptually as though human physical characteristics were essentially unchanging over space and time, assumptions that facilitated the construction of typologies but which were at odds with the principles of human

---

[94] Ibid., pp. 8–9, 186.
[95] Ibid., p. 8.
[96] Boas, "Changes in the Bodily Form of Descendants of Immigrants."

evolution.[97] Although movement of humans across the globe in vast numbers was a fairly recent phenomenon of the last few centuries – millions of Europeans had migrated to the New World; more millions of Africans had been forcibly transported there to serve as slaves on plantations; Indians and Chinese spread over wide areas in Southeast Asia, the Pacific and later the Americas and Europe; and the Japanese had "reached into the far corners of the earth" – the migratory process was "coeval with mankind," Shapiro argued. "Man has always been migratory.... The present magnitude of the phenomenon serves to focus our attention on an ancient and significant fact."[98] Boas had also noted this migratory feature of human groups, arguing that the mixing seen in the United States was hardly unique. European nationalities were also "highly complex in origin," he argued, with multiple migrations and various mixes dating back to ancient times – Celts, Phoenicians, Greeks, "Teutonic tribes," "Moors," Jews, and Slavic peoples all migrated across Europe and mingled with existing populations. The fact that European populations were lately regarded as relatively pure was itself a historical artifact, Boas argued, a result of the development of hereditary land tenure that tethered people to particular locations for generations, until the development of cities, when internal migrations again commenced.[99] Like Boas, Shapiro understood race and human diversity as a variable feature subject to a variety of influences, situated in time and space, instead of static, ahistorical features of human existence. Migration formed the historical dimension of the racial process Shapiro hoped to elucidate.

From the physical anthropologist's point of view, population movements were critical in the study of race because of the environmental changes that resulted from migration and the effect of those changes upon the constitution of the migrating peoples. Migration, Shapiro argued, lacked "all social meaning" if it was divorced from its consequences. Evolutionary theory and the mechanism of natural selection suggested that environment affected all organisms, whether they migrated or not. Species were subject to the selective forces of whatever environment they found themselves in; there was no life absent an environment. For Shapiro and Boas, this applied equally to humans. But the effect of the environment was extremely difficult to assess in populations remaining in the same

---

[97] *Migration and Environment*, p. 184.
[98] Ibid., p. 4.
[99] Franz Boas, "Report of an Anthropometric Investigation of the Population of the United States," originally published in the *Journal of the American Statistical Association*, vol. 18 (Jun., 1922), pp. 181–209, reprinted in *Race, Language and Culture*, pp. 28–59.

location over time because both physical characteristics and the local environment were presumably linked in a complex web of change. The enormous complexity of the morphological, genetic, and environmental factors themselves and the processes of change only compounded this difficulty. Only by holding some of these factors constant could investigators hope to begin to see associated cause and effect. It was the combination of the migration of large groups with new physical and social environments that afforded Shapiro an opportunity to test the effect of environment on human variation and possibly determine what if any changes were the result of changing environments. The mixed populations on Hawaii, like the Pitcairn Islanders, were to Shapiro a marvelous biological experiment.

In designing his study in this way, Shapiro was trying to isolate environment as an independent variable. Boas had articulated the importance of considering environment when attempting to understand changes in bodily characteristics in 1915, following his description of the "incontrovertible" changes to head form in immigrants under changing conditions. "[W]e have to consider the investigation of the instability of the body under varying environmental conditions," Boas argued, "as one of the most fundamental subjects to be considered in an anthropometric study of our population."[100] Shapiro understood the human environment to be both crucial to a complete understanding of human variability and a methodologically difficult piece of the human puzzle. He was utterly aware of the complexities of the environment and its effects that he sought to study. He noted that "environment" was commonly used loosely to "cover the large number of variable factors that constitute the milieu in which an organism exists," which in its broadest sense includes everything other than the organism itself. For humans, this included not only features of the constantly changing natural setting,[101] but also changing spiritual, social, cultural, economic, medical, and intellectual "currents."[102] Some of these changes were slight and presumably had little effect on human diversity, while others were likely to be important, but "[w]hat precise effect differences in any or all of these variables may exert on the biology of man is practically unknown." Eventually, Shapiro hopefully suggested, biologists might be able to determine which environmental factors caused

---

[100] Ibid., pp. 58–59.

[101] Shapiro lists: changes in seasons; the succession of night and day; temperature; humidity; barometric pressure; atmospheric circulation; solar activity; food supply; water supply; mineral content of the soil; presence and concentration of bacteria; rainfall; as well as other features that ecologists use to divide the environment into geologic, geographic, climatic, and other zones. *Migration and Environment*, pp. 5–6.

[102] Ibid., pp. 4–5.

changes and precisely how, as well as how much such factors needed to vary in order to produce particular modifications. In the meantime, Shapiro could merely acknowledge the "vast number of variables" in the physical environment, and even more so in the social and cultural environment, that changed when his subjects moved from Japanese village life to Hawaii. Given the state of current knowledge and the resistance in physical anthropology to any evidence that supposedly stable racial characters were in fact changeable, Shapiro was content to rigorously demonstrate "whether or not bodily modifications do occur under environmental changes, whatever may be the specific environmental factor involved."[103]

In the end, Shapiro found "an astounding degree of significant divergence"[104] between the Japanese immigrants and their relatives in Japan: immigrants deviated from their relatives in roughly 70 percent of the measurements. Among the Japanese born in Hawaii, Shapiro found they varied significantly from their parents in 46.4 percent to 55.2 percent of characteristics measured, a finding he termed an "arresting modification"[105] for two such closely related groups. In addition, the Hawaiian-born population deviated even further from the population in Japan, the males differing from the sedentes in a remarkable 79.3 percent of the measurements and 90.5 percent of the indices (females differed in nearly 68 percent of the measurements and 80 percent of the indices.) This suggested that the Hawaiian-born population was not reverting to forms found among their Japanese relatives, but rather diverging from them even further than their parents, the immigrants. Shapiro's study had produced a robust confirmation of Boas's discovery that the children of immigrants could diverge rapidly and significantly from both their parents and their foreign relatives, and thus that key anthropometrically studied racial traits were not stable characters upon which to build static racial typologies.

In addition to these important findings, Shapiro also sounded a few notes of caution. He was loath to speculate beyond what he felt his data and analysis supported and was in general agnostic with respect to causes. Thus, he was at something of a loss as to what might have been the source of the differences he found between the immigrant Japanese in Hawaii and the sedentes in Japan. He noted that recruiters had not exercised any kind of physical selection among the men who immigrated

---

[103] Ibid., p. 6.
[104] Ibid., p. 187.
[105] Ibid., p. 190.

to Hawaii. Economic pressures shaped the character of immigrants, favoring those able to engage in manual labor, but this could not be a factor differentiating them from the home population because they were all related. He rejected selection based on psychological factors because it required positing an association of physical characteristics with innate psychological ones, which seemed implausible. "While innate psychological drives no doubt exist," he argued, "it is extremely doubtful that they are genetically linked with special physical variations within a mixed population." In the end, all he could conclude was that further investigation was necessary to figure out how such changes arose.

Shapiro concluded from his findings that "the assumption of stability in man's physical characters is no longer tenable without qualification." Moreover, he argued, "man emerges as a dynamic organism which under certain circumstances is capable of very substantial changes within a single generation." These were conclusions with profound consequences for the study of racial variation, modification, and classification. If this had happened with the Japanese migration to Hawaii, he reasoned, it also must have happened to other groups in other parts of the world in the course of frequent migrations and changes in environment. Further, if populations could be altered by changes in the environment, then identical populations reared in different conditions might diverge. In that case, merely comparing group averages would not reveal their true relationship; historical information about their connection would be required, as well. Like Boas, Shapiro thought human racial groups were conditioned by their history as well as their heredity. Equally significant, in Shapiro's view, were the implications of his findings for the study of human evolution. "If migratory groups represent selected strains of a population which may later undergo additional modification through environmental influences, then a mechanism is provided here to explain some of the variation which is encountered among related populations." "Selective migration" and "environmental modification" were two factors among many contributing to the production of "group variations" and the "complex process of differentiation and evolution." It is anachronistic to call Shapiro's interest in group variation and populations in an evolutionary context "population genetics," but both his discussion of the genetics of breeding and the relationship of environmental forces to variations among human populations reveal a concern with a contextualized, historicized view of human heredity one also finds in Boas.[106]

[106] Ibid., pp. 201–202.

Shapiro's study, like Boas's, also raised new questions for the study of human variation and evolution. Much more needed to be known about the factors in the environment that elicited physical changes, the kind of changes that were possible, and what degree of modification was possible. As it stood, Shapiro could only remark that "although it would be extremely desirable to be able to correlate modifications in bodily characteristics with specific environmental forces, in the nature of the data we are unable to make any such deductions." But Shapiro had definite ideas about what was not possible. "The available evidence," he argued,

suggests that a given type is characterized by only a limited plasticity, and that the patterns of change are fixed by the nature of its fundamental structure. Consequently we may hardly expect to find that any population may be altered in any direction, or that by some form of transmutation through the agency of environmental alchemy we may transform one stock into the semblance of another.[107]

He noted that there was no evidence to support such wholesale changes, and abundant evidence in nature to contradict it. Although the stasis in bodily characteristics that underlay the old typological view of races had to be abandoned in favor of a "dynamic human structure, plastic to environmental influences," he was not suggesting that there was no such thing as race. Making this perfectly clear, he ended his study by stating, "I emphatically do not believe that the Japanese will ever become identical with Hawaiians as a result of enjoying an identical environment and I do not expect to find that the Japanese in Hawaii will eventually lose all similarity to the stock from which they came." Human beings may be plastic, he argued, but there were biologically bounded limits to the extent of change, and it was these very limits that made a science of human variety potentially possible and profitable.[108]

Shapiro echoed concerns and conclusions voiced by Boas three years earlier in an essay devoted to the kinds of social knowledge required of the physical anthropologist who seeks to differentiate human races. There Boas argued:

The attempts of certain anthropologists to analyze on the basis of measurements and observations a population and to discover the constituent races is, at present at least, a hopeless task. Without the most detailed knowledge of the laws of heredity of each feature considered, as well as of the effects of environment, the task is like that of a mathematician who tries to solve without any further data a single equation with a large number of unknown quantities. If anything is to

[107] Ibid.
[108] Ibid., pp. 199, 202.

be done along these lines the historical composition of the population has to be known in detail.... Any attempt at a morphological classification of the races, excepting the very largest groups, like Negroes, Mongoloids, Australians, does not lead to satisfactory results without knowledge of the conditions that have made the type what it is.[109]

Like Shapiro, Boas's discussions of the puzzle of human diversity and the challenges of physical anthropology were cast in the basic trinity that guided his work: history, environment, heredity. Notable, too, is his emphasis on the population as the relevant unit of analysis, along with the laws of heredity, an insight that links Boas to the population genetics that took off theoretically and institutionally in the 1920s. We also see, here in Boas as in Shapiro, this populational, statistical, genetic perspective joined to a lingering belief in racial types and the persistence of group differences despite the complexities of heredity, environment, and migration. Indeed, as Boas formulated his critique, the existence of more narrowly defined racial types than the Negro, Mongoloid, or Australian that he enumerates are not dismissed as unfounded, but merely inaccessible given the current state of knowledge. Both Boas and Shapiro had faith in the sciences and faith that the questions to which they sought answers were appropriate and important questions to ask. Boas, in concluding a report on his study of changes in head form, noted that far from invalidating either the study of head form or the anthropometric method, it proved their great value, particularly the method, for distinguishing the effects of heredity from those of the environment.[110] The point of their work was not to eliminate the study of race, but to do a more rigorous job of elucidating the heterogeneity of human physiques they continued to see in terms of racial variation.

Exoticism and the Antihero: Writing Popular Anthropology

Like Henry Field, Franz Boas, and Ruth Benedict, Harry Shapiro felt a desire and need to offer the American public an account of his anthropological knowledge that would not only entertain them, but bear on pressing issues. In the course of his long career, Shapiro published dozens of popular articles, especially in the American Museum's *Natural History* magazine, gave public lectures, appeared on radio and television, published an account of his attempts to locate the mysteriously missing

---

[109] Franz Boas, "The Relations between Physical and Social Anthropology," pp. 172–173.
[110] Boas, "Changes in the Bodily Form of Descendants of Immigrants," p. 562.

Peking Man fossils, and mounted a major exhibition in the 1960s.[111] In the interwar period, his most widely noted and also most historically interesting venture in popular anthropology was his account of his work on race mixing on Pitcairn Island, *Heritage of the Bounty*.

From a historian's perspective, *Heritage of the Bounty* was unusual in two respects. To begin with, it actively contravened hoary stereotypes and segregationist dogma, offering the American public a view of race mixing as entirely unproblematic. In contrast to the Field Museum's *Races of Mankind* exhibition, which presented stereotypically primitive racial types, and a decade worth of eugenicist publications and pronouncements about the dangers of race crossing, Harry Shapiro offered an account of a racially mixed community, the product of a mythic British naval mutiny set in an exotic locale, as sympathetic, healthy, and admirable. With *Heritage of the Bounty*, Shapiro put before an eager reading public the same sorts of conclusions about miscegenation that Franz Boas had arrived at some forty years earlier, but that had never gained wide popular currency.

Heritage of the Bounty was also unusual in the tone Shapiro adopted. Not only did he openly admire the Pitcairn Islanders, his narrative paints many of them as three-dimensional individuals, beginning with his host, Burley Warren, introduced on page seven. Many other men and women are named within the text, in his appendices, and in photographs, including the front and side "type" photos that invariably accompanied texts purporting to offer scientific conclusions about race in this period. Perhaps even more unusual than Shapiro's treatment of his subjects was his self-presentation. Few scientists in 1936 were willing to present themselves to the public as fallible investigators, confounded by their assumptions and experiences, groping their way toward their conclusions. Shapiro did. Perhaps he learned more than genetics from William Castle, who, according to historian William Provine, was "honest and direct about his data, interpretations, misgivings, prejudices, and changes

---

[111] Shapiro also offered his services to local government. He assisted the New York City medical examiner in identifying human remains (both recent and historic), work that led him to develop techniques in what became forensic anthropology. He also provided his racial expertise, working with city orphanages to help them place children by identifying their race. Often he was asked to assess children who were of mixed race, or were suspected of being mixed race. Shapiro's papers at the American Museum of Natural History include a number of files on these "Foundlings." On the Peking Man fossils, see *Peking Man* (New York: Simon & Schuster, 1974). For further discussion of Shapiro's exhibition, *Biology of Man*, as well as Ruth Benedict's wartime efforts to discuss race with the American public, see Chapters 6 and 7 in this book.

of mind."[112] Although Shapiro is not known for repeatedly retracting theories too boldly stated, as was Castle,[113] instead following a more cautious Boasian model, Shapiro was indeed open about his misgivings, prejudices, and changes of mind in a way that set him apart from most other physical anthropologists. Shapiro saw himself as a scientist approaching his research problems with methodological and theoretical rigor and wanted his peers and a broader public to perceive him that way (which they did). But Shapiro was also clearly constantly aware, not only of the contemporary social and political context in which he conducted his work, but of his own subjectivity.

Since the earliest voyages of exploration, Polynesia has been a place of both science and imagination, of reason and desire. When early twentieth-century anthropologists such as W. H. R. Rivers and Albert Jenks in 1914 or the Pan-Pacific conferees in 1920 proposed to map Polynesians, they framed it as a place of absence, a lacuna in the body of anthropological knowledge previously filled only with unreliable myth and anecdote. Anthropologists would impose their scientific, rational rigor on the imagined landscape, co-opt and control a new, last terrain. Embarking on an anthropological iteration of the discovery narrative, they entered an ongoing dialectic between an older, imaginative map of Polynesia and the world and the newly "scientific" one they were constructing. This dialectic is plainly visible in the way Harry Shapiro seesaws between what he had anticipated in his imagination, what he finds on Pitcairn Island, and the place he seeks to construct as a rigorously described and explained anthropological landscape. Working in Polynesia forced Shapiro to constantly confront the persistent tension between reason and desire in the work of anthropology. The anthropologist in the "field" becomes deeply involved with the subjects of his study, and fashions himself and his life at the same time he is fashioning theirs; he and his subjects are continually and simultaneously being constructed along with the "work."

As the voluminous literature on Bronislaw Malinowski attests, the inescapably subjective nature of the anthropologist as participant observer, revealed so fully in Malinowski's journals, was rhetorically obscured in the professionally "objective" scientific accounts crafted for public consumption. Today, following a generation of epistemological, self-reflexive anguish and analysis, the practice and descriptions of

---

[112] Provine, *Sewall Wright*, p. 38.
[113] William B. Provine, "Geneticists and the Biology of Race Crossing," *Science*, New Series, vol. 182, no. 4114 (Nov. 23, 1973), pp. 790–796.

anthropological work more openly confront the relative position of the anthropologists and their informants in collaboratively crafting an ethnological account. Anthropologists openly display their own particularity and consider what effects their presence and pursuit of knowledge have on the people and place they seek to know.[114] Harry Shapiro's Depression-era account of his yearning for exotic Pitcairn Island, his fieldwork there, and his conclusions about its people betray a concern with his own and the Islanders' subjectivity that may help explain why his book was still being reissued in 1968.

From its very first pages, Shapiro positions his study of Pitcairn Islanders alongside an autobiographical narrative that unsettles his scientific persona and the quest to objectively gather data on a carefully defined scientific problem. His account opens with an "Epistle Dedicatory" to Earnest Hooton, and a tale of Shapiro's own imagination and desire for the "magic" of Polynesia. On his first trip to Polynesia in 1923, Shapiro intended to go to Pitcairn Island, but unable to find a ship that would stop there, was forced to settle for a secondary colony of descendants of the mutineers on Norfolk Island. On subsequent trips to Polynesia, there was never enough time to make it to Pitcairn. "After each of my successive trips to Polynesia, I sighed regretfully that I was still unable to voyage to that isolated rock in the middle of the south Pacific. Pitcairn seemed like an unrealizable dream." Then, in 1934, Templeton Crocker, an independently wealthy explorer who liked to sponsor expeditions across the globe, invited Shapiro to join him on his yacht, the *Zaca*, as an anthropologist representing the American Museum of Natural History. Asked where he wanted to go, Shapiro said Pitcairn Island.[115]

Shapiro's tale of desire and imagination continues in the first chapter, "My First Day on Pitcairn," which opens with his anxiety that the island and its people might not measure up to his expectations:

I was really terrified that I might be disappointed. The numerous books I had read, the many pictures I had created from this reading, had shaped in my mind

---

[114] This kind of self-reflexive practice has been significantly less common among biological anthropologists, the current disciplinary descendants of the interwar physical anthropologists, who remain much more comfortable with conventions of scientific practice and knowledge production. See Chapter 7 in this book for a discussion of the biological turn in postwar physical anthropology.

[115] Shapiro, *Heritage of the Bounty*, p. xiv. For another discussion of Shapiro's Pacific excursions aboard Templeton Crocker's *Zaca*, see Warwick Anderson, "Hybridity, Race, and Science: The Voyage of the *Zaca*, 1934–35," *Isis*, vol. 103, no. 2 (Jun. 2012), pp. 229–253.

a Pitcairn that I couldn't bear to have altered by the inexorable impact of reality. But there it was at last before me, ready for me, with unguessed impressions and experiences to be added to my store.[116] . . . My first impression was slightly disappointing. I had expected to see definite indications of the Tahitian contribution to their mixed Anglo-Polynesian origin. Instead, the men, en masse, were more like a group of Englishmen – dock workers – with ugly, knobby hands and feet, roughened and calloused by labor. They wore non-descript garments, the gifts of passing vessels or bartered from the crew of the New Zealand Shipping Company steamers. Battered officers' caps were clapped on mops of shaggy hair. Blue sea-jackets rubbed shoulders with ancient tweed coats. And hardly a pair of trousers matched its companion jacket.[117]

Shapiro betrayed an awareness of the grip this potentially distorting imagined Pitcairn might have on his perceptions and work, and throughout the book recounted his imaginings and desires along with his adjustment to what he found. He narrated the tensions between the pull of his imagined Polynesia and his professional and intellectual desire to proceed in a rigorous scientific manner. Immediately after detailing his initial negative impression of the Islanders' appearance, Shapiro shifted to the job at hand, cataloguing the Islanders' "strong muscular necks and hairy chests," their "rough, calloused, padded feet . . . where nature had so amply fulfilled her function." The "prominent" noses, heavy brows, ruddy skin, sloping foreheads, dark hair, and "almost universal loss of teeth" of these "English-looking, though varied" islanders were enumerated. Shapiro reminded the reader, as perhaps he reminded himself, that he "was not interested in their clothes," but "with their faces and their physical structure. . . . examining them with a professional eye, picking out traits reminiscent of their English and Tahitian ancestors."

In his journal, excerpts of which are reprinted in *Heritage of the Bounty*, Shapiro set down further impressions of the impoverished inhabitants, at first critical and disappointed, then self-critical:

There is perhaps too much of a suggestion of shanty white about these islanders – the not quite neatly built houses, the cast-off clothing, the necessarily makeshift furniture, the air of utilizing the junk shop – which makes them too close to our seamy side to be truly romantic. One has constantly to be whipping the imagination with scenes from the Bounty or with glamorous names like John Adams or story-book ones like Thursday October Christian to keep from forgetting that these are Pitcairn Islanders. And yet their kindness is very touching. And I have great affection for inarticulate Burley and Eleanor. And the clean brown paths carefully besomed each Friday, the terraced village, the cocoanut-bole-lined roads

---

[116] Shapiro, *Heritage of the Bounty*, p. 3.
[117] Ibid., p. 6.

do make a nice picture. The fault lies with civilization. We have taken away their fresh, crackling tapa and offer only discarded clothing in their stead, we have shown them the uses of tin, and destroyed the beauty of thatch, we have sent them our broken-down furniture and displaced their simple benches.[118]

Shapiro struggled to reconcile competing impressions, conflicting desires: the rigorous, objective observer of racial subjects and the eager voyager, imagination stoked with lessons and tales of the *Bounty*'s descendants in their village on a verdant isle; the middle-class, educated American repelled by the shabbiness of poverty and the anthropologist keenly aware of the promiscuous and invidious movement of Western wealth and power; the disappointed romantic robbed of his imaginary Pitcairn Islanders; and the man, who despite his intrusive professional objectives and disappointed dreams, found the islanders generous, welcoming, and forthcoming.

Shapiro's lament for the demise of a simple and lovely Pitcairn Island life untainted by Western detritus, and his indictment of civilization, suggests that Shapiro retained vestiges of a mildly evolutionary anthropology. Shapiro's vision drew on a comparison of civilization with a romanticized, aesthetic idealization of the simple, harmonious, healthful primitive life, rather than the more common opposition of civilization with a backward, degenerated, or stagnated primitiveness that characterized the standard evolutionary view. His criticisms of "civilization" demonstrate his discomfort with an evolutionary hierarchy that defined "civilized" societies as superior by definition. But Shapiro's sadness at the corruption of this supposed way of life did suggest that he imagined it might have remained in some "original" state were in not for the intrusion of the modern world. This sort of narrative leaves the impression that Shapiro was engaged in what had been one of anthropology's chief projects: preserving "vanishing" cultures.

The rhetoric and practice of this "salvage" anthropology have been roundly, and justifiably, critiqued in recent years in the context of anthropology's ontological and epistemological crisis. The salvage rhetoric commonly used by ethnographers to justify their studies was consonant with European and American imperialist notions of Manifest Destiny used to justify overrunning less technologically equipped people in a headlong rush to acquire territory and wealth, rationalized as the inevitable outcome of civilized sophistication and ingenuity overwhelming savage simplicity. Moreover, it was often grounded in the perception of native

---

[118] Ibid., p. 268.

culture as a static entity and the role of the anthropologist merely one of salvaging what could be retrieved before the culture inevitably vanished.[119]

The iconic example of this rationale for the study of "primitive" peoples can be found in one of the discipline's canonical texts (today, for quite different reasons than in Harry Shapiro's time): *Argonauts of the Western Pacific*, by Bronislaw Malinowski. In the foreword, Malinowski wrote:

> Ethnology is in the sadly ludicrous, not to say tragic, position, that at the very moment when it begins to put is workshop in order, to forge its proper tools, to start ready for work on its appointed task, the material of its study melts away with hopeless rapidity. Just now, when the methods and aims of scientific field ethnology have taken shape, when men fully trained for the work have begun to travel to savage countries and study their inhabitants – these die away under our very eyes.[120]

George Stocking has analyzed the way Malinowski rhetorically positioned himself and the practice of extended participant observation as a detached, empirical, objective scientific process by employing a variety of "artifices" or "devices," especially engaging narratives, dramatized events, use of the "ethnographic present," and writing in the first-person singular as the "heroic Ethnographer."[121] Like Malinowski, Shapiro was simultaneously creating a complex narrative of Pitcairn Islanders' life

---

[119] This is not to say that indigenous populations did not suffer devastating losses following contact with Europeans and Americans. Louis Sullivan, noted, for example, in his study of Marquesans, that their population fell from 100,000 in the late eighteenth century to scarcely 2,000 by 1930. Shapiro cited Captain Cook's estimate that 400,000 people inhabited Hawaii in 1779, whereas a mere 23,000 "pure" Hawaiians (and 51,000 "part-Hawaiian") lived there according to the 1930 U.S. Census. Even if the eighteenth-century estimates were overstated, the population decline was dramatic. Louis R. Sullivan, "Marquesan Somatology," *Bernice P. Bishop Museum – Memoir*, vol. 9, no. 2 (Honolulu: Bernice P. Bishop Museum, 1923); Shapiro, "Physical Characters of the Cook Islanders," p. 4.

[120] Bronislaw Malinowski, *Argonauts of the Western Pacific: An Account of Native Enterprise and Adventure in the Archipelagoes of Melanesian New Guinea*, with a preface by Sir James G. Fraser (New York: E. P. Dutton & Co., Inc.: 1961), p. Xv; A. R. Radcliffe-Brown, *The Andaman Islanders* (Glencoe, IL: Free Press, 1948), is another key text in the establishment of scientific participant observation in American anthropology. James Clifford sees the methodological innovations in these texts as precursors to "what may well be the tour de force of the new ethnography," *The Nuer*, by Edward E. Evans-Pritchard (1940). James Clifford, *The Predicament of Culture: Twentieth-Century Ethnography, Literature, and Art* (Cambridge, MA and London: Harvard University Press, 1988).

[121] George W. Stocking, "The Ethnographer's Magic," pp. 12–59. See also, by Stocking, "Maclay, Kubary, Malinowski: Archetypes from the Dreamtime of Anthropology," *The Ethnographer's Magic, and Other Essays in the History of Anthropology*,

and physical anthropological fieldwork,[122] featuring himself along with select local people. But unlike Malinowski, Shapiro offered a much more nuanced and self-contradictory picture of the science of anthropology and himself as a scientific observer, a more self-deprecating and honest portrayal of the complex, human process of knowledge production. Stocking has noted that Malinowski's account of his work in the Trobriand Islands was deliberately constructed as an authoritative demonstration and legitimation of an ethnographic method, the ethnographer as scientist, and resulting conclusions that could be drawn about "savages," and not as a candid account of his actual experiences among the islanders. Nonetheless, other scholars faced with the discrepancies between the experience and attitudes recorded by Malinowski in his diaries and his account in *Argonauts*[123] have regarded Malinowski's efforts with a more jaundiced eye, some going so far as to call him a "fraud." Shapiro's book, written some seventeen years later, displays a different set of anxieties, a different agenda. Confronted by the Pitcairn Islanders of the 1930s, Shapiro constructed an account that revealed his uncomfortable recognition of the unexamined, romanticized notions he had harbored, and his ambivalence about his "civilization" and his role and practice as the Anthropologist.

(Madison: University of Wisconsin Press, 1992), pp. 212–275 (see especially, pp. 239–275 for discussion of Malinowski's personal and professional development and his work). There is a large literature that has analyzed the form and content of Malinowski's contributions to anthropology. Major texts in this genre include: James Clifford, "On Ethnographic Self-Fashioning: Conrad and Malinowski," in *The Predicament of Culture*; James Clifford and George E. Marcus, eds., *Writing Culture: The Poetics and Politics of Ethnography* (Berkeley: University of California Press, 1986); R. F. Ellen, et al., *Malinowski between Two Worlds: The Polish Roots of an Anthropological Tradition* (Cambridge: Cambridge University Press, 1988). The term "ethnographic present" is taken, via Stocking, from J. W. Burton, "Shadows at Twilight: A Note on History and the Ethnographic Present," *Proceedings of the American Philosophical Society*, 1988, vol. 132, pp. 420–433. It refers to the habit of ethnologists to treat their subjects as though their cultures were unchanging, frozen in the past, usually a "primitive" past. For a meditation on the nature of anthropological writing from one of anthropology's most reflective figures, see Clifford Geertz, *Works and Lives: The Anthropologist as Author* (Stanford: Stanford University Press, 1990). For a consideration of another formative figure in ethnological fieldwork, see Joann Hoffman, "A. C. Haddon's Original Vision: An Ethnography of Resistance in a Colonial Archive" (PhD diss., Fielding Graduate University, 2008).

[122] Clifford, *Predicament of Culture*, p. 29. "*Argonauts* is a complex narrative simultaneously of Trobriand life and ethnographic fieldwork." Clifford goes on to argue that Malinowski's text is "archetypal of the generation of ethnographies that successfully established the scientific validity of participant observation."

[123] Stocking credited the publication in 1967 of Malinowski's diaries with helping precipitate the subsequent "crisis" in anthropology. Stocking, "The Ethnographer's Magic," p. 15.

Disappointed and moved by the displacements created through contact
with the West, aware of the social costs and changes the modern world
imposed, Shapiro was confronted with the social and political context in
which he, as an American anthropologist, was visiting, indeed intruding
upon, the people of Pitcairn Island.

Shapiro constructed a narrative of his fieldwork that self-consciously
undermined the persona, so carefully constructed by others, of anthropologists as detached observers, capable of unproblematically comprehending and judging their "primitive" subjects. Shapiro repeatedly offered
readers vignettes of his own failings, his own humiliations. His narrative
of his initial encounter with the Pitcairn Islanders related not only his real-
ization of how thoroughly he had romanticized them, but also how little
he appreciated the extent of their abilities. Shapiro's description of his trip
ashore reveals his fallibility, contrasting his anxiety, impotence, and clum-
siness with the islanders' self-possession, knowledge, strength, and skill.
The story begins with Shapiro leaving the *Zaca* with a cargo of instruments and cameras loaded into a native dorie for the short trip to shore,
through the notoriously treacherous Bounty Bay, strewn with boulders.
As the boat approached the shore, Shapiro recounted his growing anxiety, certain that the crew, who seemed "phlegmatic, even disorganized,"
were not the skilled boatmen they were reputed to be, but "foolhardy
idiots." As they got closer and closer, the surf pounding the rocks, the
islanders dug in their oars, and matching an incoming swell, turned the
boat to navigate among boulders into a nearly invisible, narrow channel
through which the boat was "thrown straight and true" onto the small,
pebbly beach. As men leap out of the boat, they offered to help Shapiro.
"Scorning to use the arms lifted up to bear me dry-shod to the beach,"
he "leaped for a point beyond the nibbling waves" and promptly fell flat
on his face. "No one laughed," Shapiro noted. "Instead, several rushed
up to assist me."[124] In *Heritage of the Bounty*, Shapiro strove to balance
the requirements of scientific observation, data collecting, judgment, and
classification with the complexity of his experience and his relationship
to the people he traveled thousands of miles to study, people whom he
respected and enjoyed. As James G. Frazer said of Bronislaw Malinowski,
he "sees man in the round, not in the flat."[125] Indeed, Shapiro developed
such a warm relationship with his hosts that he maintained a correspondence with them for years after he returned to New York.

[124] Ibid., pp. 7–9.
[125] Frazer, preface, in *Argonauts*, p. ix. A tribute perhaps bestowed upon Shapiro with
   more justice.

Heritage of the Bounty was a scientific response to persistent, widely commented on but little-studied problems in interwar physical anthropology – race mixing and inbreeding – and a public intervention in a racialized narrative of miscegenation and primitivism that deemed people such as the Pitcairn Islanders exotic degenerates. Shapiro was trying to offer authoritative answers to the kind of "impertinent and ill-advised questions" visitors from passing steamers would pose to the islanders, couched in a narrative of his own fallibility and humanity.[126] Perhaps Shapiro's hope was that in making himself and the Pitcairn Islanders seem human, rather than like the detached "observer" and "observed" in an anthropological text, American readers would see the islanders as they saw themselves and would embrace his conclusions. His final comment on his study combined all the elements of his approach – his own tentative yet authoritative presence, the sympathetic individuality of the islanders, the scientific racialization that motivated the work, and his conclusion that the descendants of British mutineers and Tahitian women showed no ill effects of that ancestry:

As I think back to the individuals I knew on Pitcairn, I am impressed by the relatively large number of men such as Parkins Christian, Fred Christian, Edgar Christian, Norris Young, and Arthur Herbert Young and of women such as Mary Ann McCoy, Ada Christian, Margaret Lucy Christian, and Harriet Warren, who possessed qualities of leadership or traits of personality that raised them above the level of their neighbors. All in all, therefore, I can only conclude that inbreeding, as far as my evidence goes, has not caused degeneration among the Pitcairn Islanders. To that extent, this confirms the results of experimental inbreeding, that it is the presence of latent defects which makes inbreeding a dangerous thing and not any mysterious punishment consequent to the process itself.[127]

Shapiro presented the people he found at Pitcairn not as cardboard "Natives," but as individuated people with a living culture, responding to the wider world. In his final analysis he saw the people and culture of Pitcairn Island as a productive and creative mixing of English and Tahitian bodies and cultures, combined with innovations of their own. In Shapiro's judgment, the unique sociological and biological "experiment" on Pitcairn Island was a success.

Conclusion

Harry Shapiro's anthropology of race was rooted in culture and environment, shadowed by desire and imagination. He approached the problem

---

[126] Shapiro, *Heritage of the Bounty*, p. 133.
[127] Ibid., p. 254.

of human variation as a question not only of race – of bodies and their measurable traits – but of culture. For Shapiro, following both Franz Boas and Earnest Hooton, patterns of human physical change and variation were a matter of race *and* culture, of bodies in their full environmental and historical context. In Shapiro's view, bodies and their cultural and physical environments were inextricably interwoven. Both he and Boas understood human morphological difference fundamentally as a result of the continual convergence of heredity and environment, an evolutionary process that resulted in variable populations, over both space and time. Indeed, both recognized that variability was in fact necessary if one were to understand humans as products of natural selection and evolution, just like other animals and plants. Shapiro's work was predicated on a belief that with scientific methods such as anthropometry, biometrics, and genetic theory he and others might eventually sort out the complex relationship between heredity and environment, and map the history and pattern of human variation. In hindsight, Shapiro's understanding of the "plasticity" he observed, the variability of groups of "family lines" joined in a population of common descent that Boas discussed, was circumscribed by a limited understanding of how human genetics and heredity worked. In this he and Boas were no different from their peers, and indeed, they were a good deal more informed and more concerned with the problems of genetics and race than many of their colleagues. But it was also true that Shapiro, like Boas, clearly believed in the reality of racial types, that humanity had evolved into recognizable, if shifting and changeable, groups. The question of mixed races that motivated Boas, Shapiro, and so many other physical anthropologists (recall Hooton's enthusiasm, as well as Davenport's) was rooted in a fundamentally typological notion: there could be no mixed races without races to mix, however ill-defined. From a historical and historiographic point of view, the persistence of both Boas and Shapiro in continuing to think of lines of heredity in a broadly typological manner locates them conceptually between romantic typologists such as William Ripley and Arthur Keith, who clung to a vision of static racial characteristics and types, and a late twentieth-century conception of difference informed by population genetics, rooted in a sense of human variation as a wide range of independently sorting traits, permanently shifting and mixing, redistributed across ever-changing populations. The recent understanding of race as purely a social construction would have struck Boas and Shapiro as incongruous, even incoherent.

That is not to say Shapiro was unconcerned or unaware of the profound consequences of race and his racial studies. Interwar physical anthropology, popular and professional, framed as an objective, dispassionate, scientific exploration, was in reality shot through with desire, fear, and imagination. Debates about race constantly invoked imagined pasts and futures, filled with fears of miscegenation and hopes for racial harmony, fears of imagined savagery and yearning for native simplicity, the simultaneous desire and loathing for exotic others, all in the context of pressing social, economic, and political concerns. Working in Polynesia, Shapiro confronted these tensions directly, both in his own encounters and in presenting his work to his peers and the public. He was acutely aware of the heavy weight of romanticism and primitivism that threatened to overwhelm his efforts to present his subjects accurately and counteract invidious assumptions and stereotypes. But he also shrewdly took advantage of his audiences' fascination with the Pacific Islands. He enticed his popular readers with tales of mutiny and verdant isles, his scientific readers with tantalizing insights into the heterogeneity of Hawaii and Polynesia. Shapiro, like Boas, was acutely aware of the history of racism and the unjust, even horrific, acts that had been committed in the name of racial superiority. Both men located in a science of human bodies a means for rejecting biological determinism and the kind of racial typology that reasoned from morphological distinctions to mental and physical inferiority, and onward to social and political policy.

By the Second World War, many anthropologists were becoming increasingly anxious about racism rooted in what they viewed as erroneous understandings of race and human variation. Anthropologists, including Harry Shapiro, made concerted efforts during and shortly after the war to communicate with the public about race via publications, lectures, and exhibitions, and to intervene internationally via collaborative efforts such as the UNESCO Statements on Race. In the following chapters, I turn to this transitional period and examine the way these anthropologists, confronted by war and divisive racial policies at home and abroad, attempted to grapple with how to present human physical and cultural diversity, and the science of human variation, for public consumption.

6

Rejecting Race, Embracing Man? Ruth Benedict's Race and Culture

Introduction

Contrary to a common narrative that highlights the "rejection of scientific racism" in work such as Ashley Montagu's popular book *Man's Most Dangerous Myth* and the 1950 UNESCO Statement on Race, early efforts to reject racism frequently retained essentialist and biologically determinist notions of race. The intermingling of bodies and culture, heredity and society that characterized the study of race in the 1920s and 1930s continued into the 1940s and beyond. Rather than a complete rejection of race and racial science during and shortly following the war, we find the mix of cultural and biological perspectives continues, sometimes in concert and sometimes in tension. Despite anthropologists' growing anxiety about racism, racial misconceptions, and what they regarded as misguided or malevolent misuse of their work, most were not interested in abandoning a biological, hereditary definition of race.

By the late 1930s, U.S. scientists expressed growing concern about the suppression of open discourse under the National Socialists in Germany, and even more anxiety about their abuse of supposed racial science to foment patriotism and abuse marginalized populations, especially Jews. Their concern about the latter reflected not only revulsion at the violent and willful distortion of decades of racial studies, but also a fear that what many regarded as legitimate studies of race and a proper understanding of human variation were jeopardized by the bastardized version being promoted by the Nazis. The growing threat posed by the Nazi regime, to their own people and those subject to their aggression, as well as to freedom and democracy more broadly, galvanized scientists in a variety of fields to speak out much more forcefully and in a more united fashion than they had in the face of earlier hereditarian assaults on racialized populations, including the eugenic promotion of immigration restriction in the 1920s and the use of racial studies to justify discrimination and segregation in the United States.

Yet despite nearly universal opposition to Nazism and Nazi racial doctrine, U.S. and international scientists found it difficult to reach perfect unity on the broader subject of race. Provoked into crafting more explicit, sometimes collective, statements on the nature of race, culture, and society, they debated the conclusions of racial science, the nature of race, its relation to culture and environment, the nature of racism, and how to combat it. Anthropologists were not simply attempting to intervene in society to promote social justice and combat discrimination; they were also attempting to reorient their discipline, grappling with how to discuss and study race (and heredity) in changing political, international, and scientific circumstances. Ethnologists such as Ruth Benedict sought to counter misapprehensions of race and society by promoting the pluralistic Boasian culture concept, one that stressed the relativistic quality of human cultures and rejected an older evolutionary hierarchy of fixed developmental stages, primitive and civilized.[1] At the same time, anthropologists strove to erect a conceptual separation between the study of human biology and morphology, and the study of society and culture. Anthropologists sought to establish firm distinctions between "race" and "culture," to uncouple them as a means to undermine racism. The anthropological culture concept was disseminated alongside a delimited, increasingly biologized concept of race, but not as a replacement for it. In the 1930s and 1940s, nearly all parties agreed upon the existence of groups of physically distinct human beings that could be characterized as races. During WWII and in the years following the war, anthropologists strove to define, rather than reject, race and racial science, and to directly combat racism.

The period during and shortly after WWII was an era when the discipline of anthropology was transforming itself, under pressures from within and without, from a somewhat loose confederation of ethnologists and physical anthropologists all interested in "the study of human beings

---

[1] George W. Stocking, Jr., "Ideas and Institutions in American Anthropology," *The Ethnographer's Magic and Other Essays in the History of Anthropology* (Madison: University of Wisconsin Press, 1992), p. 164; Daniel Rosenblatt, "An Anthropology Made Safe for Culture: Patterns of Practice and the Politics of Difference in Ruth Benedict," *American Anthropologist*, vol. 106, no. 3 (Sep. 2004), pp. 459–472.

as creatures of society" into increasingly estranged subdisciplines of cultural and biological anthropology, one modeled on the social sciences, the other on the natural sciences.[2] In the early and mid-twentieth century, despite their particular cultural or racial focus, many ethnologists and physical anthropologists felt comfortable with each other's specialties. It is hard to imagine a cultural anthropologist trained in recent decades who would write as forcefully about the nature of human physical variation as Ruth Benedict did in *Patterns of Culture* and *Race: Science and Politics*, or a biological anthropologist who would feel comfortable addressing the public on the history and qualities of human cultural development, as Harry Shapiro did throughout his career.

In this chapter, I examine works by Ruth Benedict, one of the most active and prominent anthropologists attempting to influence how Americans thought about race and culture. I examine not only her major monographs, *Patterns of Culture* and *Race: Science and Politics*, but also *The Races of Mankind*, a pamphlet by Benedict and her collaborator Gene Weltfish intended to combat racism and racial misinformation, and its reincarnation as an exhibit at the Cranbrook Institute of Science in Michigan, in a brief animated film, and as a short illustrated book.

Ruth Benedict and Race: *Patterns of Culture*

Ruth Benedict was a crucial figure in the wartime effort of anthropologists and other scientists to communicate with the American public on the subjects of culture, race, and racism. An anthropologist trained first by Elsie Clews Parsons and Alexander Goldenweiser at the New School for Social Research, and then at Columbia University under Franz Boas, where she earned her PhD in 1923, she became a leader in the Columbia Anthropology department, taking over the day-to-day administrative activities from an aging Boas in the 1930s.[3] Despite her stature in

---

[2] Ruth Benedict, *Patterns of Culture*, with a new preface by Margaret Mead (Boston: Houghton Mifflin Company, 1961), p. 1; George W. Stocking, Jr., "Paradigmatic Traditions in the History of Anthropology," *The Ethnographer's Magic and Other Essays in the History of Anthropology* (Madison: University of Wisconsin Press, 1992), pp. 342–361 (p. 346–347); Regna Darnell, *And Along Came Boas: Continuity and Revolution in Americanist Anthropology* (Amsterdam and Philadelphia: J. Benjamins, 1998); Regna Darnell, *Invisible Genealogies: A History of Americanist Anthropology*, Critical Studies in the History of Anthropology (Lincoln and London: University of Nebraska Press, 2001).

[3] Margaret M. Caffrey, *Ruth Benedict: Stranger in This Land* (Austin: University of Texas Press, 1989), pp. 92–115, 259–281; Judith Schachter Modell, *Ruth Benedict: Patterns*

the field and her dedicated service as an administrator, teacher, and graduate mentor (including occasionally funding student research from her own pocket), Columbia refused to grant Benedict the professional recognition she deserved, passing her over for department chair when it came open and refusing to grant her a permanent tenured position until shortly before her death at sixty-one in 1948. As an anthropologist, Benedict was regarded in her lifetime, as well as by posterity, as one of the most influential theorists of her generation. She is best known for her comparative "socio-psychological" approach to the study of cultures, most famously presented in two anthropological treatises that explored how the particularities and historical trajectory of a given people manifested themselves in a distinctive cultural configuration.[4] Her first, most famous study, *Patterns of Culture*, published in 1934, established Benedict as an intellectual innovator within anthropology and Boasian ethnology. *The Chrysanthemum and the Sword: Patterns of Japanese Behavior*, published in 1946, was the culmination of her work in the Office of War Information.[5] It offered a penetrating examination of the Japanese for

---

of a Life (Philadelphia: University of Pennsylvania Press, 1983), pp. 110–125, 256–258; Virginia Heyer Young, *Ruth Benedict: Beyond Relativity, Beyond Pattern* (Lincoln: University of Nebraska Press, 2005), pp. 7–10, 39–52, 102–104.

  In addition to the biographies already cited, there are a number of other texts that explore aspects of Benedict's life and work: Margaret Mead, *An Anthropologist at Work: Writings of Ruth Benedict* (Boston: Houghton Mifflin Company, 1959); Margaret Mead, *Ruth Benedict* (New York: Columbia University Press, 1974); Lois W. Banner, *Intertwined Lives: Margaret Mead, Ruth Benedict, and Their Circle* (New York: Alfred A. Knopf, 2003); Virginia Heyer Young, "Ruth Benedict: Relativist and Universalist," in Jill B. R. Chernoff and Eve Hochwald, eds., *Visionary Observers: Anthropological Inquiry and Education* (Lincoln: University of Nebraska Press, 2006), pp. 25–54; Dolores Janiewski and Lois Banner, eds., *Reading Benedict/Reading Mead: Feminism, Race, and Imperial Visions* (Baltimore: The Johns Hopkins University Press, 2004), including four essays on *The Chrysanthemum and the Sword*; Clifford Geertz, "Us/Not-Us: Benedict's Travels," in *Works and Lives: The Anthropologist as Author* (Stanford: Stanford University Press, 1988); Richard Handler, "Vigorous Male and Aspiring Female: Poetry, Personality, and Culture in Edward Sapir and Ruth Benedict," in George W. Stocking, Jr., ed., *Malinowski, Rivers, Benedict and Others: Essays on Culture and Personality*, History of Anthropology, vol. 4 (Madison: University of Wisconsin Press, 1986), pp. 127–155; George W. Stocking, Jr., *The Ethnographer's Magic and Other Essays in the History of Anthropology* (Madison: University of Wisconsin Press, 1992), especially "The Ethnographic Sensibility of the 1920s and the Dualism of the Anthropological Tradition," pp. 276–341; Darnell, *Invisible Genealogies*.

[4] Benedict, *Patterns of Culture*, p. xvi.
[5] Ruth Benedict, *Patterns of Culture* (New York: The New American Library of World Literature, Inc., 1934); Ruth Benedict, *The Chrysanthemum and the Sword: Patterns of Japanese Behavior* (Boston: Houghton Mifflin, 1946); Modell, *Patterns of a Life*, pp. 267–293; Caffrey, *Stranger in This Land*, pp. 302–326.

an American audience then little inclined toward cross-cultural sympathy toward their former enemy or critical self-examination. Both books promoted a comparative approach to cultural analysis, encouraging readers to examine their own customs with the same anthropological eye they cast upon alien societies. Both remain in print. While *The Chrysanthemum and the Sword* is still widely read and continues to provoke debate about Benedict's depiction of the Japanese, it is *Patterns of Culture* that has had the most lasting effect on the field of anthropology and American society.

In *Patterns of Culture*, Benedict analyzed what Franz Boas in his introduction to the book called "the genius of the culture."[6] To "grasp the meaning of the culture as a whole," Boas said, one had to "understand the individual as living in his culture; and the culture as lived by individuals."[7] Making an astonishing break from typically descriptive Boasian ethnography, Benedict analyzed as a whole three widely disparate "primitive" cultures – the Zuñi of the Southwestern United States, whom she had studied and about whom there was a substantial literature; the Dobu of Melanesia, studied by Margaret Mead's husband, Reo Fortune; and the Kwakiutl of the North American Northwest coast, studied extensively by Boas – characterizing them as either Apollonian or Dionysian (concepts she borrowed from Friedrich Nietzsche's *The Birth of Tragedy*).[8] By examining each culture as an integrated whole, Benedict aimed to demonstrate the way they each "selected from the great arc of human potentialities certain characteristics," elaborating them in distinctive ways, a kind of cultural "personality writ large."[9] The possible combinations and elaborations of cultural features were nearly limitless. Some would prove more cohesive and functional than others, but all functioned according to an internal logic and consistency that meant no culture was superior or inferior to another, only differently configured.

Patterns of Culture has come to be seen as a key text in an anthropological narrative about the ascendance of the culture concept, and the idea of cultural relativity, in anthropology and American society.[10] In her

---

[6] Benedict, *Patterns* (1961), p. xvii.

[7] Ibid., p. xvi.

[8] Ibid., p. xi; Caffrey, *Stranger in This Land*, p. 54.

[9] Benedict, *Patterns* (1961), p. viii.

[10] Margaret Caffrey argues that *Patterns of Culture* first reached leaders in the social sciences and a relatively small number of the educated public in its first hardcover edition in 1934. The "new twentieth-century paradigm," cultural relativism, purveyed

1958 preface, when the book had already sold more than 800,000 copies, Margaret Mead – Benedict's student, friend, colleague, lover, executor, and biographer – wrote, "That today the modern world is on such easy terms with the concept of culture... is in very great part due to this book."[11] In her foreword to the latest 2005 edition, anthropologist Louise Lamphere called Benedict's book "a foundational text in teaching us the value of diversity" and, quoting Benedict, asserted that "recognition of cultural relativity will create an appreciation for 'the coexisting and equally valid patterns of life which mankind has created for itself from the raw materials of existence.'"[12]

But Mead also noted in 1958 that Benedict's work offered a novel reflection on a classically holistic Boasian conundrum, one that was not simply a question of culture. Mead wrote that Benedict's comparison of cultural configurations in *Patterns* was meant to illuminate "the relationship between each human being, with a specific hereditary endowment and particular life history, and the culture in which he or she lived."[13] Although Benedict was not a physical anthropologist, nor especially interested in the biological, bodily side of human life and history, she employed a fundamentally Boasian philosophy in which individuals and cultures could only be understood as simultaneously products of heredity, social relations, environments, and history. Indeed, for a text widely lauded as a foundation of cultural relativism, Benedict had quite a bit to say in *Patterns of Culture* about the nature of race and its relation to culture (Figure 6.1). In part, this reflects the era in which she wrote the text, and the long-standing and still much debated hereditarian and biologically determinist presumptions still widespread among both scientists and a wider American readership. But it also reflects a Boasian approach that both acknowledged race and rejected any sort of determinism. As in the work of Franz Boas and Harry Shapiro, Ruth Benedict carefully delineated the appropriate place of biology and race in a broad understanding of human culture and history, noting ways in which human heredity and

by *Patterns* only reached a mass public, according to Caffrey, with the release of a twenty-five cent paperback version in 1946. This Mentor edition, from Houghton Mifflin, was reprinted at least once and sometimes twice a year over the next decade. Caffrey, *Stranger in This Land*, pp. 213–214.

[11] Benedict, *Patterns* (1961), p. vii.

[12] Ruth Benedict, *Patterns of Culture*, preface by Margaret Mead, forward by Louise Lamphere (Boston: Houghton Mifflin Harcourt, 2005), pp. vii–xii.

[13] Benedict, *Patterns* (1961), p. ix.

FIGURE 6.1. Heads in the circle are red, white, dark brown and light brown. Four-color cover of paperback 1956 Mentor Book edition. *Source:* Ruth Benedict, *Patterns of Culture*, New American Library of World Literature, Mentor Book, 13th edition, [1934] 1956. Courtesy of Penguin Group.

constitution contributed to anthropological understanding and ways in which she regarded it as distinctly separate from her concern with cultural integration.

Benedict opened *Patterns* with a chapter on "The Science of Custom," offering readers an initial, orienting discussion of anthropologists' approaches to "human beings as creatures of society.... Those physical characteristics and industrial techniques, those conventions and values, which distinguish one community from all others that belong to a different tradition."[14] Her discussion of the anthropological notion of "culture" makes it clear that this was still a piece of technical jargon requiring extensive exposition, particularly for the wide general audience she anticipated. In particular, as the chapter title suggests, Benedict felt the need to explain that anthropologists were most interested in matters of everyday life and belief – of custom – rather than in an earlier definition of culture as arts or high civilization. In a theme that Benedict would find herself stressing to the public repeatedly as the years wore on, the anthropological notion of culture viewed the customs of "primitive" peoples and those of American or other more complex societies (or "civilizations") as neither better nor worse, neither higher nor lower, but simply as different but equally plausible schemes for dealing with common human problems.

Beyond explication of the culture concept, Benedict also devoted a large portion of the chapter to a discussion of race prejudice, foreshadowing some of the concerns she would address in her wartime work. Race prejudice, according to Benedict, stemmed from Europeans' lack of contact with other cultures, and their consequent overvaluing of their own cultural traditions. Rather than viewing them as the historically contingent configurations that they were, Benedict argued, Europeans, lacking a relativistic cultural consciousness, viewed their own culture as "necessary and inevitable."[15] Lacking a broader perspective on the varieties of human culture, Europeans read "white culture" as "Human Nature."[16] Thrown into contact with other cultures, they had responded with nationalism and "race-snobbery." In fact, Benedict argued that misguided chauvinistic attitudes toward cultural differences among members of the same race were more responsible for modern race prejudice in the United States and Europe than "intolerance directed against the mixture of blood of biologically far-separated races." "Traditional Anglo-Saxon intolerance"

[14] Ibid., p. 1.
[15] Ibid., pp. 5–6.
[16] Ibid., pp. 6–7.

and the cultural construction of "in-groups" and "out-groups" explained Americans' antipathy for the "Irish Catholic in Boston" and the "Italian in New England mill towns," as well as the estrangement between French and German in Alsace-Lorraine when "in bodily form they alike belong to the Alpine sub-race."[17]

Against this mistaken "gospel of pure race," Benedict argued first that "the cultural process of the transmission of tradition" was vastly more important than whatever might be "the small scope of biologically trans-mitted behavior."[18] Babies from one race adopted into another grew up evincing all the cultural traits of the adopted society and none from their original people. The culture of the "American Negro" in Northern cities, Benedict noted, had "come to approximate in detail that of the whites in the same cities."[19] Moreover, if one looked at the history of Western civilization, its leaders had at various points come from Semitic-speaking peoples, Hamites, "the Mediterranean sub-group of the white race," and recently, "the Nordic." Conversely, people of the same race, for example North American Indians, did not all display the same cultural behavior.[20] Clearly, Benedict viewed race as a meaningful way to distinguish among various groups of people.[21] But for Benedict, the racial reality of human "biological constitution" did not extend to hereditary cultural

---

[17] Ibid., p. 11.

[18] *Patterns of Culture*, p. 15. Benedict left room for the possibility of "organically deter-mined" instinctual behavior and "common" "human adjustments" that are "inevitable in mankind." She noted that such behaviors required proof that they were not merely culturally conditioned behaviors, pp. 16–17.

[19] Ibid., p. 13.

[20] Ibid., p. 234.

[21] Benedict offers a confused discussion of heredity and race. At some points she seemed to restrict "racial" heredity to "family lines" and small, isolated groups (such as an Eskimo village), rejecting the idea that "Nordics" could exist as a hereditary group across "a wide area." Yet elsewhere she employs without qualification terms such as "Mediterranean sub-group of the white race," and in the same discussion in which she rejects a broad Nordic hereditary group mentions "Alpine or Mediterranean communities." Despite writing *Patterns of Culture* in 1932, Benedict had little expertise in genetics of that era, relying on Franz Boas to guide her through the more technical aspects of race and heredity. At least one scholar has argued that Boas's own understanding of genetics was fairly limited (he explained heredity in terms of "family lines"). Benedict's discussion of "physical types," "intermixture," and the production of "local types," and "biological homogeneity" suggests no more than a rudimentary grasp of Mendelian inheritance. Her discussion of local mixing and family inheritance betrays a conception of inheritance closer to family genealogy than to the emerging science of population genetics. While her overall critique of the limits of racial typology in terms of genetics and heredity is sound, the particulars of her argument as well as her use of racial terms is sometimes unclear or inconsistent. *Patterns of Culture*, pp. 15–16.

determinism. The lesson for Benedict was clear: "Culture is not a biologically transmitted complex."[22] Appeals to "racial heredity" in "our cosmopolitan white civilization" were nothing but "slogans," meant to rally people with similar socioeconomic interests, something Benedict regarded as a dangerously misleading distraction that diverted Americans from genuine cultural understanding and unity.[23]

Ruth Benedict and Race: *Race: Science and Politics*

In 1939, at the behest of Franz Boas, Ruth Benedict spent her sabbatical composing a book for popular consumption exclusively on the topic of race and racism. Undertaken just as WWII began in Europe, *Race: Science and Politics* was Benedict's effort as a "citizen scientist" to intervene more forcefully to shape public attitudes toward race, culture, and discrimination.[24] The book was intended not only for a broad reading

---

[22] Ibid., p. 14.

[23] Ibid., p. 16.

[24] Ruth Benedict, *Race: Science and Politics*, with a foreword by Margaret Mead (New York: The Viking Press, 1959), p. vii.

Prior to its publication, Benedict engaged in a protracted debate with editors at Modern Age Books over the title of her book. From practically the outset, Benedict favored "Race and Racism," deeming it an accurate representation of her topics and intentions. Vice President Louis Birk at first went along with Benedict's suggestion, "until we get a better idea." Benedict rejected a series of "better" ideas that in her opinion not only failed to capture the point of her book, but which were likely to mislead readers: "Races of the World," "The Races of Man," and "The Races of Mankind." Manager David Zablodowsky told Benedict that although her preferred title was "perfectly descriptive" of her book, "it does somehow strike me as not attractive." Zablodowsky also urged Benedict to adopt a title that would reflect the large amount of "descriptive ethnography" in the manuscript. Ultimately, the early suggestions were each abandoned. The editors agreed that "Races of the World" was in fact a misnomer, although Birk continued to lament that it was nonetheless "more attractive," as it did not leave the dreaded impression that the book was a hard-to-sell monograph nor use the word "racism," which Zablodowsky argued connoted hatred. "The Races of Man" and "The Races of Mankind" had both already been used. Birk and Zablodowsky suggested "Race: Science and Myth," "The Science and Mythology of Race," "Race: Science and Prejudice," or, finally, "Race: Science and Politics." Benedict replied that everyone she had consulted thought "Race and Racism" was "a fresh and saleable title," but agreed "reluctantly" to accept "Race: Science and Politics," although she noted that it was hard to say and "doesn't suggest anything." See Louis P. Birk to Ruth Benedict, Nov. 14, 1939, Folder 51.3; Ruth Benedict to David Zablodowsky, Jun. 13, 1940, Folder 51.4; David Zablodowsky to Ruth Benedict, Jun. 14, 1940, Folder 51.4; Louis P. Birk to Ruth Benedict, Jun. 20, 1940, Folder 51.4; David Zablodowsky to Ruth Benedict, Jun. 21, 1940, Folder 51.4; Ruth Benedict to Louis P. Birk and David Zablodowsky, Jun. 27, 1940, Folder 51.4; and Ruth Benedict to David Zablodowsky, Jul. 1, 1940, Folder 51.4, "Race: Science and Politics: Correspondence," Race: Science and Politics

public, but also as a "handbook" for "tolerance education" in schools, churches, and clubs throughout the United States.[25] Although pushed by Boas to write on a topic outside her area of expertise, Benedict felt compelled to comply, not simply because it would have been difficult to spurn such a request from a revered mentor and colleague who had worked so tirelessly himself to fight racism, but because she also hoped to combat inaccurate, racist distinctions that promoted a divisive ideology and undermined society. "Racism," she wrote, "is an *ism* to which everyone in the world today is exposed; for or against, we must take sides. And the history of the future will differ according to the decision which we make."[26]

To this end, she emphasized the importance of history and culture in creating the differences commonly attributed to race, continuing to press the anthropological culture concept. Her argument against racism proceeded from two critical assumptions: first, that individual prejudice was the critical locus of racism; and second, that such prejudice could be eradicated by disabusing biased people of their misguided notions about the nature of race and the superiority of their own culture. From Benedict's perspective, anthropology offered a scientific approach to both bodies and cultures that allowed comparison across time and space. Through her comparative approach, Benedict hoped to disabuse readers of their complacent sense of racial and cultural superiority. In an era of great faith in the ability of science to proffer objective truth and clear-eyed solutions to social problems, Benedict, like many of her peers, believed an unvarnished exposition of the scientific facts purveyed by a leading authority would be effective in combating social ills; knowledge would cure the afflicted and inoculate the vulnerable.[27]

(1940), Series IV, Writings of Ruth Benedict, Ruth Fulton Benedict Papers, Vassar College Library Archives and Special Collections Library, Vassar College, Poughkeepsie, New York (hereafter RBP-VC).

[25] Ruth Benedict to Louis P. Birk, Modern Age Books, Oct. 27, 1939, Folder 51.3, "Race: Science and Politics: Correspondence," Race: Science and Politics (1940), Series IV, Writings of Ruth Benedict, RBP-VC.

[26] Ibid., 5.

[27] Zora Neale Hurston, who trained in anthropology under Franz Boas and Ruth Benedict and alongside Margaret Mead at Columbia University in the 1920s, was more skeptical about the salutary effect of an educational approach, arguing that people were impervious to facts, hanging onto what they wanted to believe. Hurston, cited in Francesca Sawaya, *Modern Women, Modern Work: Domesticity, Professionalism and American Writing, 1890–1950* (Philadelphia: University of Pennsylvania Press, 2003), p. 135. On Hurston's life and work, see, inter alia, Valerie Boyd, *Wrapped in Rainbows: The Life of Zora Neale Hurston* (New York: Scribner, 2003).

More surprising is Benedict's account of race itself. Like the laudatory praise given *Patterns of Culture* for its role in promoting culture consciousness, *Race: Science and Politics*, and later efforts with colleague Gene Weltfish to reach an even wider audience, have been praised for their message of human unity and cultural relativity, highlighting their assertions that all humans share the same blood types or that intelligence is shaped by the environment, not head shape. Such accounts create the impression that Benedict and Weltfish were embarked on a recognizably modern postwar project to "reject scientific racism," but a closer inspection of their texts reveals that although racism and invidious norms were rejected, they, like Shapiro or Boas, were a long way from casting race as a "dangerous myth," as Ashley Montagu shortly would, or as a "modern superstition," like Jacques Barzun.[28] Among Benedict's assertions in *Race: Science and Politics* that readers expecting a denunciation of the race concept and racial science would find surprising include:

- Race is biologically transmitted.
- Race is a scientific field of study.
- In general usage this [racial] classification corresponds to the division into Caucasoid (white), Mongoloid (yellow), and Negroid (black); these are the obvious and striking human varieties. Further physical characteristics, as we shall see, can be associated with these three groups. They have each a large geographical range, and represent real differentiations.
- The Mediterranean coast is characterized by short, dark, narrow-headed peoples – the Mediterranean sub-race.
- Race is not "the modern superstition" as some amateur egalitarians have said. It is a fact.
- To minimize racial persecution...it is necessary to minimize conditions which lead to persecution; it is not necessary to minimize race. Race is not in itself the source of the conflict.[29]

What becomes clear in reading Benedict's disquisition on race and racism is that her goal was to sharply distinguish race from culture, and both from racism. She had no interest in rejecting racial science and the race concept as viable means of delineating human variation. In her

---

[28] M. F. Ashley Montagu, *Man's Most Dangerous Myth*, with a foreword by Aldous Huxley (New York: Columbia University Press, 1942); Jacques Barzun, *Race: A Study in Modern Superstition* (New York: Harcourt, Brace & Co., 1937).

[29] Benedict, *Patterns*, pp. 13, 18, 26, 47, 97–98, 155.

analysis, both race and culture existed as subjects of legitimate scientific investigation. Indeed, it was her scientific authority and expertise as a professional anthropologist that authorized Benedict and her colleagues to address the public on the topic of race and pressing social matters such as racism. Throughout her text, Benedict again and again invoked the scientific nature and value of sound racial studies. She viewed her job as setting forth an accurate account of race (and culture), distinct from her analysis of racism. She repeatedly made the point that it was the misuse of race and racial science, whether misguided or deliberate, that defined racism. Not race itself. Thus, Ruth Benedict's form of antiracism, like her mentor's, was not the late-twentieth-century critique of race as an inherently pernicious concept. Like Boas, Benedict tried to tread a more perilous line, distinguishing a science of human racial and cultural variation from its exploitation.

Benedict, like Boas, attempted to move anthropology away from biological essentialism and hereditarianism while retaining the idea that morphological, biologically based classifications and investigations of human variation were meaningful and useful. She saw the methods of physical anthropologists as a scientific way to describe evident physical differences that arose via migration, isolation, and intermixing over the course of human history. "Race," she asserted, "is a subject which can be investigated by genealogical charts, by anthropometric measurements, by studies of the same zoological group under different conditions, and by reviews of world history."[30] Advising the Council Against Intolerance in America on the text of their publication *The American Answer to Intolerance* in the summer of 1939, Benedict urged them to distinguish race from racist "propaganda" and expand the bibliography to include "good and reliable works on the subject of racial differences," including Julian Huxley and Alfred Haddon's *We Europeans*, Herbert Seligmann's *Race Against Man*, and Franz Boas's *Anthropology and Modern Life*.[31] By the late 1930s and 1940s, most anthropologists rejected the idea once commonly held that particular constellations of physical traits provided a reliable basis for assessing individual or group mental, moral, or social capacities.[32] But

[30] Ibid., p. 97.
[31] Ruth Benedict to Mrs. McCormick, Aug. 20, 1939, Folder 19.11, Council Against Intolerance in America, 1939–1941, Organizational Correspondence, Series I, Correspondence, RBP-VC.
[32] There were significant exceptions among anthropologists, particularly physical anthropologists, including Carleton Coon and Stanley Garn. Among psychologists and geneticists were also a significant number who thought, at the very least, it was still an open

few argued that the concept of race itself and the practice of evaluating, describing, and categorizing human variation were inherently essentialist and racist. Most seemed to regard the broadest racial categories simply as common-sense acknowledgment of obvious differences. Even Ashley Montagu, renowned for his vigorous repudiation of race and racism, retained the idea of race in the 1940s, despite his growing discomfort with it.[33]

Race: Science and Politics deployed broadly accepted racial terminology common to anthropological classifications since the nineteenth century. Although Benedict noted that there was no consensus about how to classify human varieties nor how many races there were, and that many individuals were difficult to place because they displayed traits characteristic of more than one type, she nonetheless wrote, "No one doubts that the groups called Caucasoid, Mongoloid, and Negroid each represent a long history of anatomical specialization in different areas of the world."[34] Given her belief that a descriptive, classificatory racial science was a legitimate exercise, and her rhetorical strategy of presenting facts to disabuse Americans of their racial misconceptions, Benedict found herself rehearsing the various criteria anthropologists used to delineate racial types (including jargon devised to describe human features, such as leucodermi for the white-skinned; platyrrhine for flat, broad nostrils; and dolichocephalic for narrow-headed). These criteria included skin color, eye color and form, hair color and form, nose shape, height (or "stature"), cephalic index, and blood types. In the course of her presentation of the basic variations on these supposedly racial characteristics, Benedict also noted their limitations. For example, the wide range of skin color made it useful only in designating broad groups across wide geographical ranges, such as the three major racial divisions. Conversely, the cephalic index, a measure of head shape, which had been arbitrarily divided into three types (narrow, medium, and broad), was primarily useful in distinguishing small local variations and racial subgroups. The wide range of head shapes found in larger racial groupings made the cephalic index of little diagnostic or descriptive value for these broader classifications. Nose shape, eye color, and hair color and form were similarly problematic for distinguishing types, as they were very widely distributed and not

question. This became evident during debates over framing the UNESCO "Statement on Race" in 1950, discussed in Chapter 7 of this book.

[33] See Chapter 7 in this book for a discussion of Ashley Montagu's popular and professional disquisitions on race and physical anthropology.

[34] Ibid., pp. 23, 31.

uniform in any given group.[35] Benedict concluded from this, as had anthropologists since the late nineteenth century, that if there ever had been pure races comprised of mutually distinct constellations of characteristics, they had long since disappeared amid the history of human migrations and mixing.[36] Despite these shortcomings, Benedict nonetheless offered her readers descriptions of traits associated with "the major human stocks," Caucasian subgroups (Nordic, Mediterranean, Alpine), and a few other "distinctive races" such as the Polynesians. "The Caucasian," Benedict stated, "is relatively hairy, has smooth (wavy to curly) hair, minimum prognathism [the degree to which the jaw juts forward, like an ape] . . . a fairly high, thin nose, and straight eyes." Mongoloids, she wrote, "have very little facial and body hair, and their head hair is lank and straight. . . . The slant eye is local and among American Indians is absent except sporadically." And the "Negroid type has the darkest skin range, kinky hair, thick lips, frequent prognathism, a flat nose."[37]

Race: Science and Politics framed the problem of racism as a kind of category error, a failure to appropriately distinguish between race and culture. Race, Benedict argued, was exclusively an issue of bodily characteristics that were inherited and physically distinguished one group of human beings from another. Culture, the "socially acquired" traits displayed by an individual or a group, was ontologically and conceptually distinct from race. And yet they were constantly confused and conflated. To exemplify this distinction, at a time when Germans propounded a theory of Aryan supremacy, Benedict emphasized the difference between race and language: language was learned; "[r]ace is biologically transmitted." Drawing examples from Siberian Manchus who adopted Chinese, Africans who spoke Arabic, and members of the African diaspora who spoke Spanish, English, French, or Portuguese, depending on which part of the Western hemisphere they settled, Benedict argued that "race and language have had different histories and different distributions."[38] Benedict argued that the complexities of human history belied any causal relationship between the inheritance of distinctive somatic characteristics and the "primitive" or "civilized" quality of a given culture. The Manchus, Benedict wrote, "were a rude and unnoted nomadic Tungus tribe," which, through contact with the Mongols and conquest of China,

[35] Ibid., pp. 25–31.
[36] Ibid., pp. 33–34.
[37] Ibid., pp. 34–35.
[38] Ibid., p. 10.

ultimately ruled a dynasty "unsurpassed in riches and glory."[39] Such varied fortunes were typical, Benedict argued. "Wherever we look – to the Malays, the Manchus, the Mongols, the Arabs, or to the Nordics – the same story repeats itself over and over." She did not contest Nordic racial identity, but instead pointedly argued that "their participation in civilization" was due to "the universal processes of history" and not "their racial type."[40] "The lesson of history is that pre-eminence in cultural achievement has passed from one race to another," Benedict wrote, and those achievements were not "mechanically transmitted" or guaranteed "by any racial inheritance."[41]

That said, Benedict regarded the study of race as a legitimate and valuable science, one that had contributed "many important facts" to our understanding of human migrations, the results of racial mixing, and the differences between "a group of people who constitute a nation and a group who constitute a biological type." The anthropology of race had been "an integral part of anthropology from its earliest beginnings," Benedict wrote, "because it provided a record, written in the bones and other bodily characteristics of men, of the history of mankind." Benedict even thought it possible that sometime in the future science might devise a way to determine whether "some ethnic groups have identifiable emotional or intellectual peculiarities which are biological and not merely learned behavior."[42] But the scientific record had overwhelmingly demonstrated that "race did not correlate with superiority or inferiority."[43]

The racist "dogma" that imputed superiority and inferiority to individuals and whole peoples on the basis of hereditary differences was no scientific doctrine, according to Benedict, and it could not be explained by recourse to the history and conclusions of racial science. Benedict praised the work of Paul Broca and Jean Louis Quatrefages in the nineteenth century as "pioneers in accurate anthropometric measurement of great populations." They employed "actual measurements in their arguments," unlike racist propagandists such as Arthur Comte de Gobineau, whom Benedict denounced at length.[44] Broca and Quatrefages's "researches on

---

[39] Ibid., pp. 16–17.
[40] Ibid., p. 17.
[41] Ibid., p. 18.
[42] Ibid., p. 98.
[43] Ibid., p. 65.
[44] Gobineau is author of "Essay on the Inequality of the Human races," (1853–1855), translated and reprinted in the United States in an abbreviated version in 1912. Gobineau, along with Houston Stewart Chamberlin, author of *Foundations of the Nineteenth*

race advanced knowledge," she argued, but their "racist conclusions" were "a quite different matter"[45] (italics original). Racism, Benedict argued, was not motivated by the search for truth, nor constrained by adherence to "the facts" about race. It was a "belief," not subject to scientific investigation.[46] Modern hereditarian racism, she argued, developed in the nineteenth century, erected upon a body of scientific knowledge about human evolution and variation, as an expression of a much older human penchant for privileging one group over another, in-groups against out-groups. Drawing on anthropological and historical evidence extending from human prehistory through the ancient world to the history of Europe, Benedict reconstructed a long story of discrimination based on various rationales, none of them reliant on hereditary physical differences. Even more recent forms of prejudice and discrimination, which rooted their claims in hereditary superiority, upon closer examination revealed their use of race and heredity to be a mask for class and nationalist conflicts. Racism "stemmed not from the sciences – which have repudiated it, and which, indeed, racism has constantly distorted in its pronouncements," Benedict asserted, "but from politics."[47]

Although readers today might be surprised by Benedict's defense of race and racial science, her strenuous efforts to distinguish race from culture, and race from racism, suggest a shrewd strategy. In framing her discussion of race as an opposition of science and politics, Benedict crafted a rhetorical means to salvage a critical field of work in anthropology that

---

Century, first published in Germany in 1899 and then in the United States in 1911 (by which time it had gone through eight editions in Germany), and Madison Grant, author of *The Passing of the Great Race*, published in 1916, have frequently been credited with – or perhaps better, held responsible for – promoting and cementing the doctrine of Nordic/Teutonic/Aryan/Anglo-Saxon superiority. Gossett, *Race: The History of an Idea* (New York: Shocken Books, 1963), pp. 339–364, passim; Reginald Horsman, *Race and Manifest Destiny: The Origins of American Racial Anglo-Saxonism* (Cambridge, MA: Harvard University Press, 1981); David Roediger, *The Wages of Whiteness: Race and the Making of the American Working Class* (London and New York: Verso, 1991); Noel Ignatiev, *How the Irish Became White* (New York and London: Routledge, 1996); Matthew Frye Jacobson, *Whiteness of a Different Color: European Immigrants and the Alchemy of Race* (Cambridge, MA: Harvard University Press, 1999); Charles W. Mills, *The Racial Contract* (Ithaca, NY: Cornell University Press, 1999); Richard Dyer, *White* (New York and London: Routledge, 1997); Thomas A. Guglielmo, *White on Arrival: Italians, Race, Color and Power in Chicago, 1890–1945* (New York: Oxford University Press, 2003); Bruce Baum, *The Rise and Fall of the Caucasian Race: A Political History of Racial Identity* (New York: New York University Press, 2006).

45 Ibid., pp. 129–130.
46 Ibid., p. 98.
47 Ibid., p. 138.

threatened to be cast into disrepute by its association with Nazi racial doctrine, further popularize the anthropological culture concept, position anthropology as an efficacious social science, and counter dangerous racial dogma. *Race: Science and Politics* shifted the ground for discussion of race and race prejudice from bodies and heredity to culture. She moved from an emphasis on race (bodies) to a focus on prejudice (behavior and culture). By emphatically presenting the study of race as science, strictly limiting race to inherited somatic traits, and trafficking only in what she cast as broadly accepted "facts" about race, Benedict hoped to counter misinformation and remove from debate assertions that normative or evaluative judgments about human capacities or societies had any place in a discussion of race. Defending human variation as a legitimate field of scientific investigation, and race as an uncontested fact, was supposed to have the counterintuitive effect of minimizing or removing it from the debate. Benedict could then turn to the real crux of her argument, and employing her favorite method of comparing cultures and peoples across time and space, examine the nature of prejudice and discrimination. Race, given its due as a measure of genuine human diversity, was left aside as largely irrelevant. The problem was not race, but prejudice. And prejudice had been fomented under a wide variety of guises to meet equally varied human cultural ends.

Benedict's answer to the problem of prejudice was not only the knowledge and arguments provided in popular tracts such as hers, but also deliberate social engineering. "If civilized men expect to end race prejudice," she concluded, "they will have to remedy major social abuses.... Whatever reduces conflict, curtails irresponsible power, and allows people to obtain a decent livelihood will reduce race conflict. Nothing less will accomplish the task."[48] Benedict exhorted her readers to strive for a "better America," that would "make democracy work." Writing in the midst of the Great Depression, she warned:

[S]o long as there is starvation and joblessness in the midst of abundance we are inviting the deluge. To avert it, we must "strongly resolve" that all men shall have the basic opportunity to work and to earn a living wage, that education and health and decent shelter shall be available to all, that regardless of race, creed, or color, civil liberties shall be protected.[49]

Although Benedict had felt that spending her sabbatical writing *Race: Science and Politics* took her away from her own work, her approach

---

[48] Ibid., pp. 150, 160.
[49] Ibid., p. 160.

to the task bore the hallmarks of the Boasian anthropological questions she spent her career exploring: What makes a culture coherent? How does human psychology shape a culture?[50] Benedict's interest in "socio-psychological problems" and her comparative cultural methodology not only shaped the way she practiced her profession but were manifested in the way she communicated about race and culture with the public.

Popularizing Race and Culture, Fighting Racism: *The Races of Mankind*

In 1943, Ruth Benedict continued her efforts to reach a broad American public on the subjects of race and culture, this time with a shorter, more accessible version of *Race: Science and Politics*. Adopting the same title as the Field Museum a decade earlier, Benedict, with her colleague Gene Weltfish, crafted a thirty-one-page, ten cent pamphlet, *The Races of Mankind*, that distilled key facts and arguments from Benedict's book and paired them with evocative illustrations by Adolph Reinhardt, an artist known for cartoons and abstract expressionist paintings.[51] It was, according to *TIME* magazine, "designed to fit a serviceman's pocket and to fight Nazi racial doctrine."[52] The pamphlet was one in a large educational series published by the Public Affairs Committee, a non-profit organization in New York dedicated to disseminating "in summary and inexpensive form the results of research on economic and social problems to aid in the understanding and development of American policy." Although the organization professed to be purely educational, with "no economic or social program of its own to promote," the titles of the eighty-nine short pamphlets produced by 1944 implied a consistently progressive agenda. The organization published on a wide variety of topics, including *Schools for Tomorrow's Citizens*, *Pensions After Sixty?*, *Women at Work in Wartime*, *Freedom from Want: A World Goal*, *Who Can Afford Health Care?*, and *What's Happening to Our Constitution?*

---

[50] Young, *Ruth Benedict: Beyond Relativity*, p. 10.

[51] For more on Adolph Reinhardt's cartoons for Benedict and Weltfish's *Races of Mankind*, see Marianne Kinkel, *Races of Mankind: The Sculptures of Malvina Hoffman* (Urbana, IL: University of Illinois Press, 2011). For a consideration of his artwork and contemporaries, see Thomas B. Hess, "The Art Comics of Ad Reinhardt," *Artforum*, vol. 12, no. 8 (Apr. 1974), pp. 46–51; and Annika Marie, "The Most Radical Act: Harold Rosenberg, Barnett Newman and Ad Reinhardt" (PhD diss., University of Texas, 2007). His paintings reside in the collections of the Museum of Modern Art and the Guggenheim Museum, among others.

[52] "Education: Race Question," *TIME*, Monday, Jan. 31, 1944.

Some addressed questions of race, including *The Negro and the War* and *Why Race Riots?*[53]

During the 1940s, Benedict's vision of race and culture were widely circulated in a variety of popular forms. *The Races of Mankind* proved enormously popular. By 1945, more than 750,000 copies had been sold or distributed to community organizations and civic groups, schools, churches and synagogues, and individual buyers, and it had been translated into seven languages.[54] Within the first decade, almost 1 million copies were sold.[55] In 1944, the Cranbrook Institute of Science, in Bloomfield Hills, Michigan, a suburb of Detroit, opened an exhibition based on the book (Figure 6.2). Spurred in part by local racial tensions and violence, particularly a 1943 riot in Detroit in which thirty-four people died,[56] Director Robert T. Hatt enlisted the help of Columbia University anthropologist John Adams and the advice of other anthropologists including Weltfish, Harry Shapiro, and Ralph Linton, as well as advice from "experienced social workers," to construct a series of displays portraying Benedict and Weltfish's main arguments that would be "a contribution to better interracial understanding."[57] Hatt's rationale, in the face of local racial violence and a worldwide race war, invoked science and education: "We were convinced that since race is a biological concept and since science has much to contribute to an understanding of race problems, it was our mission to help promote understanding by a visual presentation of some scientifically sound conclusions about human races."[58]

[53] Ruth Benedict and Gene Weltfish, *The Races of Mankind*, Public Affairs Pamphlet No. 85 (New York: Public Affairs Committee, Incorporated, 1943), last page, back cover.

[54] Zoë Burkholder, "'With Science as His Shield': Teaching Race and Culture in American Public Schools, 1900–1954" (PhD diss., New York University, 2008), p. 378; Zoë Burkholder, *Color in the Classroom: How American Schools Taught Race, 1900–1954* (Oxford: Oxford University Press, 2011), pp. 74, 215, n121; Violet Edwards, "Note on *The Races of Mankind*," in Benedict, *Race: Science and Politics*, pp. 167–168. Burkholder discusses the extensive use of Benedict and Weltfish's work in American classrooms in the 1940s and early 1950s.

[55] Caffrey, *Stranger in This Land*, p. 298.

[56] Vivian M. Baulch and Patrical Zacharias, "The 1943 Detroit Race Riots," *The Detroit News*, Feb. 11, 1999. On the history of race and violence, see inter alia, Harvard Sitkoff, "The Detroit Race Riot of 1943," in *Toward Freedom Land: The Long Struggle for Racial Equality in America* (Lexington, KY: University Press of Kentucky, 2010); Thomas Sugrue, *The Origins of the Urban Crisis* (Princeton: Princeton University Press, 1996); Kevin Boyle, *Arc of Justice: A Saga of Race, Civil Rights, and Murder in the Jazz Age* (New York: Macmillan, 2007).

[57] Robert T. Hatt, "The Races of Mankind, Announcing an Exhibition," *Cranbrook Institute of Science News Letter* [sic], vol. 13, no. 5, Jan. 1944, Bloomfield Hills, Michigan, pp. 1–3; Robert T. Hatt, "Report of the Director," *Fourteenth Annual Report, July 1, 1943–June 30, 1944*, Cranbrook Institute of Science, Bloomfield Hills, MI, p. 12.

[58] Hatt, "Report of the Director," p. 13.

FIGURE 6.2. Cranbrook Institute of Science. "All Mankind Is One Family," opening panel of *The Races of Mankind* (1944). Courtesy of Cranbrook Archives, "Three children in front of Adam and Eve."

Drawing on the principle themes, and in many cases replicating illustrations, from *The Races of Mankind* pamphlet, the exhibition consisted of twenty-two displays:

1. All Mankind Is One Family
2. Our World Shrinks
3. What Is Race?
4. Early Concepts of Race
5. Physical Characters of Human Races
6. Why Are There Different Races?

7. No Race Is Mentally Superior
8. No Race Is Most Primitive
9. Nationalities Are Not Races
10. Culture Is Not Inborn
11. Art Forms Define Cultures, Yet Transcend Racial Bounds
12. All Races Enrich Architecture
13. Poetry Is Universal
14. The Foods We Cultivate Are a Gift From All Peoples
15. Our Inventions Have Come from Many Races
16. Love of War Is Taught
17. Negroes Are an Integral Part of Our Culture
18. Composition of the American Negro
19. The Jews Are Not a Race
20. Who Are the Aryans?
21. Blood Groups
22. Let Us Live in Peace[59]

Following its exhibition in its home institution, the Cranbrook *Races of Mankind* continued as a traveling exhibition of thirty-four panels that toured the country's public libraries, museums, schools, and other community organizations.[60]

---

[59] "The Races of Mankind," Cranbrook Institute of Science, Bloomfield Hills, MI, n.d., 2 p., Folder: Press Releases, etc., Races of Mankind Collection, Archives, Cranbrook Center for Collections and Research, Bloomfield Hills, Michigan (hereafter Cranbrook-ROM); "List of Exhibits," Folder 52.8, The Races of Mankind (1948), Series IV, Writings of Ruth Benedict, RBP-VC.

 #16, Love of War is Taught, can be viewed as a direct rebuttal to Arthur Keith's notorious argument that national antagonisms were natural and beneficial, and that war was nature's "pruning hook," an evolutionary tool for producing vigorous races. See Chapter 3 in this book for a discussion of Keith's views.

[60] M. F. Ashley Montagu, *Man's Most Dangerous Myth: The Fallacy of Race*, with a foreword by Aldous Huxley (New York: Columbia University Press, 1945), pp. 259–260; Joseph Klimberger, "The Races of Mankind Exhibitions at the Detroit Public Library," *Library Journal*, vol. 69 (Nov. 1, 1944), pp. 919–921. "List of Exhibits," Folder 52.8, The Races of Mankind (1948), Series IV, Writings of Ruth Benedict, RBP-VC.

 The exhibition boosted attendance at the Cranbrook Institute of Science from a previous annual attendance of 13,261 in 1943 to 15,508 for the 1944 fiscal year. The exhibition itself drew the "greatest single month's attendance" for the year, totalling 2,127 people, "four times the normal number for that time of year." The *Pontiac Press* noted that the Institute was keeping the exhibit open in the evenings from 7:30–10:00 PM "due to great demand," to accommodate those who could not come during the day. Hatt, "Report of the Director," p. 21; *Pontiac Press*, Feb. 3, 1944, Scrapbook, 1943–46, Cranbrook Foundation, p. 33, Archives, Cranbrook Center for Collections and Research, Bloomfield Hills, Michigan (hereafter Cranbrook Archives).

FIGURE 6.3. *In Henry's Backyard,* four-color cover (1948). *Source:* Ruth Benedict and Gene Weltfish, *In Henry's Backyard. The Races of Mankind* (New York: Henry Schuman, Inc., 1948).

In 1946, the United Auto Workers approached United Productions of America, an independent animation studio, to create an animated film based on the pamphlet with the hope that it would help ease racial tensions in recently desegregated union branches in the South.[61] The popular eleven-minute film, *The Brotherhood of Man*, released in 1947, was then transformed in 1948 into *In Henry's Backyard*, a small fifty-page book, heavily illustrated in the manner of a children's book, with illustrations adapted from the film[62] (Figure 6.3). Although the book reads and looks like a children's book, the inside cover insisted the book would be "enjoyed by every reader in your house," and my own copy spent time in the United States Army Finance Office in Lima, Peru.[63] *In Henry's Backyard* was the work of Gene Weltfish and Violet Edwards, director of education for the Public Affairs Committee, supervised by Benedict.[64]

[61] Christopher P. Lehman, *The Colored Cartoon: Black Representation in American Animated Short Films, 1907–1954* (Amherst: University of Massachusetts Press, 2007), p. 105.

    The film was for sale in cities around the country, including Boston, Providence, Rhode Island; Hartford, Connecticut; New York, Baltimore, Atlanta, Tampa, Detroit, Cleveland, Toledo, Columbus, Cincinnati, Lexington, Indianapolis, Milwaukee, Racine, WI, Minneapolis, St. Louis, Kansas City, Wichita, Austin, Dallas, Houston, Portland, OR, San Francisco, and Los Angeles. "Points of Service for Brotherhood of Man Throughout the United States," Folder 76.2, General Subject Files: In Henry's Backyard, Series X, General Subject Files, RBP-VC.

[62] Ring Lardner, Jr., Maurice Rapf, John Hubley, and Phil Eastman, "'Brotherhood of Man': A Script," *Hollywood Quarterly*, vol. 1, no. 4 (Jul., 1946), pp. 353–359 (p. 353); Ruth Benedict and Gene Weltfish, *In Henry's Backyard. The Races of Mankind* (New York: Henry Schuman, Inc., 1948), n.p.

[63] Benedict and Weltfish, *In Henry's Backyard*. My copy is repeatedly stamped:

War Department
Finance Office, U.S. Army
Lima, Peru
Official Business

[64] Ibid. In addition to the credited authors of these works, there were also a number of scholars involved as advisors. Products from the Public Affairs Committee were supervised by a committee of the American Association of Scientific Workers that included zoologist Leslie C. Dunn, psychologist Otto Klineberg, and anthropologist Marion Smith, all of Columbia University. The Cranbrook Institute *Races of Mankind* exhibition was advised by anthropologists John Adams, Ralph Linton, Gene Weltfish, and Harry Shapiro of Columbia University (Shapiro was an instructor in addition to his primary museum job at the American Museum of Natural History), and Carl Guthe, a Harvard-trained anthropologist and director of the University of Michigan Museum of Anthropology. L. C. Dunn was coauthor with geneticist Theodosius Dobzhansky of *Heredity, Race and Society* (New York: New American Library of World Literature, 1946); Benedict

The Races of Mankind and its derivatives were particularly widely circulated in American public schools for lessons in "tolerance education" and "intercultural relations." According to Zoë Burkholder, it was "the single most influential anthropological text produced for school use in the 1940s."[65] In Detroit, the school board Advisory Committee on Human Relations distributed the pamphlet and accompanying poster to all schools. The Detroit public library also hosted the traveling exhibit from the Cranbrook Institute, reaching a broad audience of adults as well as children. Joseph Klimberger, an Austrian immigrant and librarian in the Detroit Public Library Circulation Department, noted the exhibition attracted a larger-than-expected audience of groups and individuals, and sparked "heated arguments." According to Klimberger, a "factory worker of foreign stock" commented, "I wish I could bring all my co-workers from the shop to the library to see this for themselves. Too many are still prejudiced." In contrast, an "aristocratic-looking gentleman" commented, "It seems as though everyone is working very hard to put the nigger and the Chinese on the same basis with me."[66] The New York City Board of Education encouraged teachers to read *The Races of Mankind* and use it in their classrooms. Enterprising teacher Alice B. Nirenberg and her students at Public School No. 6 created "Meet Your Relatives," a musical version set to the popular tune "Pistol Packin' Mama." The musical featured twelve students in lab coats portraying "eminent scientists" conveying lessons about tolerance, including one that emphasized that most people had an intermediate "tan" skin color.[67] Prior to the creation of *The Brotherhood of Man*, New Tools for Learning, in New York, in cooperation with the Public Affairs Committee, created a thirty-minute filmstrip for schools entitled "We Are All Brothers – What Do You Know About Race?"[68]

The Cranbrook Institute *Races of Mankind* exhibition circulated not only in local Detroit schools and libraries but crossed the nation in

---

and Weltfish, *Races of Mankind*, p. 1; Hatt, "The Races of Mankind, Announcing an Exhibition"; John Ripley Forbes, "Valuable Racial Exhibit," *Hobbies: The Magazine for Collectors*, vol. 49 (Mar. 1944), p. 22.

[65] Burkholder, "'With Science as His Shield,'" p. 116.

[66] Klimberger, "Races of Mankind Exhibitions at the Detroit Public Library," p. 919.

[67] Burkholder, "'With Science as His Shield,'" pp. 384–385. Burkholder discusses many more examples of the way antiracist, intercultural materials were used in schools in the early 1940s. For another perspective on race and culture in the schools in the 1920s and 1930s, see Diana Selig, *Americans All: The Cultural Gifts Movement* (Cambridge, MA: Harvard University Press, 2008).

[68] Edwards, "Note on *The Races of Mankind*," p. 168; Montagu, *Man's Most Dangerous Myth* (1945), p. 268.

traveling versions. The original Cranbrook traveling exhibition was distributed to twenty sites along the Eastern seaboard, including Virginia, and in middle America, including Minnesota and Tennessee, between 1943 and 1945.[69] Growing demand spurred creation of three photographic reproductions that were distributed out of New York, Chicago, and Cleveland. The American Missionary Association then purchased the exhibition and created a series of fifteen posters, based on the Cranbrook exhibit, which further circulated throughout the country in schools, churches, and civic organizations, including the New York Public Library, where it was mounted in February 1945, "Brotherhood Month," sponsored by the Mayor's Committee on Unity.[70] The American Missionary Association version was distributed via the Race Relations Department at Fisk University in Nashville, Tennessee, a department founded by the Missionary Association in 1942 in response to concern about growing racial tensions, fears, and violent conflict following the outbreak of WWII. Led by sociologist Charles S. Johnson, the Race Relations Department sent *Races of Mankind* posters and a teacher's manual with discussion materials to schools.[71]

The Races of Mankind was not popular with everyone. The pamphlet was originally written at the request of the United Service Organizations for distribution to American soldiers encountering and serving alongside people of various races and cultures.[72] Although the War Department ordered 50,000 copies for officers in orientation courses, intended as background material to help instructors "counteract the Nazi theory of a super-race," they were never distributed to American servicemen.[73] In January 1944, United Service Organizations (USO) President Chester Barnard, calling the pamphlet "controversial" and "propaganda," ordered the YMCA to stop distributing it and removed it from

[69] "List of Exhibits," Folder 52.8, The Races of Mankind (1948), Series IV, Writings of Ruth Benedict, RBP-VC.

[70] "Education Notes," *New York Times*, May 13, 1945, p. E9; "Display Aids Tolerance. 'Races of Mankind' Exhibition Opened at Public Library," *New York Times*, Feb. 2, 1945, p. 17; Edmonia W. Grant, Director of Education, Race Relations Division, American Missionary Association, News Release, Feb. 1, 1945, Folder: Publicity, Cranbrook-ROM.

[71] Race Relations Department Archives, United Church Board for Homeland Ministries, 1942–1976, Amistad Research Center, Tulane University, New Orleans, Louisiana.

[72] Caffrey, *Stranger in This Land*, p. 298; Juliet Niehaus, "Education and Democracy in the Anthropology of Gene Weltfish," Jill B. R. Cherneff and Eve Hochwald, eds., *Visionary Observers: Anthropological Inquiry and Education* (Lincoln: University of Nebraska Press, 2006), pp. 87–118 (p. 95); Lehman, *The Colored Cartoon*, p. 104.

[73] "Army Drops Race Equality Book; Denies May's Stand Was Reason," *New York Times*, Mar. 6, 1944, p. 1; "Education: Race Question," *TIME*.

USO reading rooms around the world, a move that spurred black news-papers to characterize the USO as "Jim Crow."[74] The following March, the House Military Affairs Committee, led by Republican Andrew J. May from Kentucky, blocked army distribution of the pamphlet. May objected to the inclusion in the pamphlet of intelligence test scores that showed African Americans besting white Southerners. The scores, obtained from WWI army recruits in the American Expeditionary Force, showed African Americans from New York, Illinois, and Ohio scored higher on average on the Alpha tests for literate test takers than did whites from Kentucky, Arkansas, and Mississippi. May hoped the House Committee would "'expose the motive behind" publication and distribution of the book-let, which he described as promoting "tolerance by teaching the fun-damental unity of races and contending that economic differences were largely responsible for racial differences."[75] In April, a subcommittee report by Democrat Carl Thomas Durham of North Carolina contended that the pamphlet contained "statements ranging 'all the way from half-truths through innuendos to downright inaccuracies,'" and argued that "'wartime is no time to engage in the publication and distribution of pamphlets presenting controversial issues or promote propaganda for or against any subdivision of the American people.'"[76] The subcommittee had been charged with "investigating the 'infiltrations' into the Army and its training programs of teachings and philosophies 'inimical to the interests of the people and the Government.'"[77]

Participants on both sides of the flap surrounding *The Races of Mankind*'s controversial presentation of race and intelligence deployed the rhetoric of science, objectivity, and patriotism during war. Respond-ing directly to May's accusations, Gene Weltfish argued the congressman had misunderstood what *The Races of Mankind* contended, an argument that was proffered by other defenders of Benedict and Weltfish, such as Public Affairs Committee Editor Maxwell S. Stewart, who asserted that May's "objection is based on his failure to read the text carefully," as

---

74 "Education: Race Question," *TIME*, Monday, Jan. 31, 1944; David H. Price, *Threaten-ing Anthropology: McCarthyism and the FBI's Surveillance of Activist Anthropologists* (Durham, NC: Duke University Press, 2004), p. 114; "'The Races of Mankind,'" *The Commonweal*, vol. 39 (Mar. 17, 1944), p. 532.

75 "Army Drops Race Equality Book," p. 1.

76 "Hits 'Races of Mankind.' House Group Says Book Army Used Has Misstatements," *New York Times*, Apr. 28, 1944, p. 7.

77 Article from the Washington, DC, *Times-Herald*, Friday, Apr. 28, 1944, cited in Mon-tagu, *Man's Most Dangerous Myth: The Fallacy of Race*, fourth revised and enlarged edition (Cleveland and New York: Meridian Books, The World Publishing Company, 1965), p. 390.

"anyone who read the text carefully could not possibly misconstrue it." *The Commonweal* criticized the notion that the pamphlet was a form of propaganda, noting that its "thesis . . . is that of every competent scientist in the field: that race is no determinant of ability and that the similarities within the human family are far greater than the dissimilarities." It could hardly be deemed "propaganda," the writer continued, "since it limits itself to readily verifiable facts of our human existence."[78] Responding to May's accusations, Weltfish asserted that the pamphlet in fact offered no scientific explanation for the source of differences among the races, "such as skin pigmentation," but merely contended that "economic and educational advantage" made a difference in how people scored on intelligence tests, backing up her contention with reference to the work of noted Columbia University psychologist Robert S. Woodworth, author of a leading college textbook, *Psychology: A Study of Mental Life*, then in its fourth edition.[79] Ordway Tead, chair of the Public Affairs Committee that published the pamphlet, characterized the criticism as "smear tactics." In defending the work, he cited Benedict and Weltfish's high professional standing and noted that their research results were accepted in anthropology as "sound and unassailable." As evidence of the uncontroversial, wholesome nature of *The Races of Mankind*, Tead noted that the pamphlet had been bought by the Boy Scouts and the Junior League, as well as by churches and schools.[80] Defenders contrasted the ignorance, inattention, and bias of those objecting to the claims made in *The Races of Mankind* with the careful reasoning and scientific authority of Benedict, Weltfish, and other social scientists involved in the production of the pamphlet. At stake was not only the fate of the pamphlet and the reputations of its authors and advisors, but also the stature of the discipline of anthropology and its kindred fields as science.

Assertions of the sound, uncontroversial science underlying the pamphlet's claims were tied to its patriotic wartime purpose by many commentators. In a political salvo aimed directly at May and the House Committee, Weltfish argued, "[T]he pamphlet does nothing more than present the biological and psychological material in support of an ancient

78 "The Races of Mankind," *The Commonweal*, Mar. 17, 1944, p. 532.
79 "Army Drops Race Equality Book," p. 1; Clarence H. Graham, "Robert Sessions Woodworth, 1869–1962, A Biographical Memoir" (National Academy of Sciences, Washington DC, 1967), pp. 541–572; Robert S. Woodworth, *Psychology: A Study of Mental Life* (New York: Henry Holt & Co., 1940). Woodworth retired from Columbia University at seventy-five, in 1945, but continued to teach there until he was eighty-nine, when he was forced to quit. Graham, pp. 541–542.
80 "Tead Defends Race Book. Charges House Group With 'Smear' of Anthropological Work," *New York Times*, Apr. 29, 1944, p. 13.

principle that 'all men are created equal,' set out in the Bible, the Declaration of Independence, and the Constitution."[81] The Council of the
Society for the Psychological Study of Social Issues, an organization of
professional psychologists and other social scientists, issued a press release
defending *The Races of Mankind* as "not only good science; it is also a
contribution to true democracy." The psychologists defended Benedict
and Weltfish's booklet as "an able popularization of sound anthropological and psychological knowledge" and pointedly argued that "its central
thesis – that there is no real evidence for the belief in the innate superiority
of any one race over any other, and that racism is therefore superstition –
represents the consensus of opinion among the overwhelming majority of
American social scientists."[82] Placing May and the booklet's detractors
on the defensive, the psychologists asserted that efforts to interfere with
distribution of *The Races of Mankind* represented "a concession to prejudice," whereas its wider distribution would "help bring a little nearer
the kind of world for which the democracies are now striving."[83]

Like the psychologists, many of Benedict and Weltfish's defenders put
the dispute explicitly into a wartime context, arguing that the pamphlet
was making a valuable contribution to the war effort by undermining
Nazi racial doctrines. Some went so far as to accuse the War Department and other critics of *The Races of Mankind* of supporting white
supremacy, and questioned their patriotism in a time of war against an
enemy espousing such doctrines. Harold S. Sloan, director of the Alfred
P. Sloan Foundation, which supported the work of the Public Affairs
Committee, said, in the midst of the controversy, "It seems a pity that
the Military Affairs Committee of the House sees fit to withhold from
our armed forces the simple facts of science that completely refute the
enemy's contention of a superrace [sic]."[84] Constance Warren, president
of Sarah Lawrence College in New York, knit all three strains of argument
together in a strongly worded letter to *The New York Times*. Titled by
the *Times* "A Plea for Racial Truth," Warren's letter assailed the House
Military Affairs Committee and Congressman May for rejecting the work
of "recognized anthropologists," scientists in a "comparatively new field

---

[81] "Army Drops Race Equality Book," p. 1.
[82] Press release, Apr. 12, 1944, Folder 52.8, The Races of Mankind (1948), Series IV,
    Writings of Ruth Benedict, RBP-VC. The press release was signed by Gordon W. Allport
    as chairman, Steuart H. Britt, George W. Hartmann, Daniel Katz, Otto Klineberg, Kurt
    Lewin, Rensis Likert, Gardner Murphy, Ruth Tolman, and Goodwin Watson.
[83] Ibid.
[84] "Plans New Edition of Race Pamphlet. Public Affairs Group Differs With May's View
    That Led to Army Circulation Ban," *New York Times*, Mar. 14, 1944, p. 11.

of study" whose expertise was necessary to combat Nazi racial ideology, and "our widespread ignorance and consequent racial prejudice." Ending with a series of rhetorical questions aimed squarely at May and the House Committee, Warren concluded:

[T]he Representative from Kentucky thinks it fatal to let our soldiers know that tests show black men to be as capable of education as white men and that there are no inborn racial characteristics which mean inferiority of intelligence and spirit. Are we going to let Representative May and his colleagues keep our men who are fighting for democracy in ignorance of the fact that modern research proves Jefferson to have been right when he said "All men are created equal"? Do we deprive them of the knowledge that the democracy for which they are fighting is valid? What are we going to do?[85]

As critics of May and the House Military Affairs Committee eventually pointed out, the data to which he and his committee so strenuously objected had been gathered and published by the federal government more than two decades earlier.[86] What seems to have angered May and others was the selective presentation of the data in a manner that drew attention to the superior performance of Northern African Americans over Southern whites. The median AEF scores, as Benedict and Weltfish published them,[87] were:

Southern Whites		Northern Negroes	
Mississippi	41.25	New York	45.02
Kentucky	41.50	Illinois	47.35
Arkansas	41.55	Ohio	49.50

Benedict and Weltfish compounded this infraction by arguing that low Southern scores were the result of poor housing, diet, and schooling. Or as they put it, "Negroes with better luck after they were born got higher scores than whites with less luck. The white race did badly where economic conditions were bad and schooling was not provided, and Negroes living under better conditions surpassed them." It was differences in income, education, "cultural advantages, and other opportunities" that made the differences in test scores, not heredity, they argued. Their socioeconomic analysis also explained why, taken in aggregate, African Americans did more poorly on the tests than whites. A greater

---

[85] Constance Warren, "A Plea for Racial Truth. President of Sarah Lawrence College Protests Suppressing Army Pamphlet," Letter to the Times, *New York Times*, Mar. 14, 1944, p. 18.

[86] Montagu, *Man's Most Dangerous Myth* (1965), p. 387.

[87] Benedict and Weltfish, *The Races of Mankind*, p. 18.

proportion of blacks lived in the South, where conditions were poor.[88] Years later, Ashley Montagu pointed out that Benedict and Weltfish actually understated the results of the AEF Alpha tests. Computing median scores for blacks and whites, North and South, from Robert Yerkes' original Alpha test data, Montagu calculated that blacks in four states surpassed whites in the bottom three states, blacks from Illinois surpassed whites in five states, and blacks from Ohio performed better on average than whites from all eight Southern states recorded. Congressman May took the data presented in *The Races of Mankind* and the authors' analysis as an affront to white Southerners on the basis of both racial and social standing. Montagu's analysis supported not only the magnitude of the differences between black and white test takers, but reinforced Benedict and Weltfish's regional cultural analysis. For it was not only African Americans in the North who surpassed Southern whites, but Northern whites, as well. The lowest average attained by Northern whites, in Indiana, was still nearly twelve points higher than the highest average attained by whites in the South (in Tennessee); the highest white average, in Ohio, surpassed that in Tennessee by more than eighteen points.[89] And the

[88] Ibid.

[89] Below are the numbers provided by Ashley Montagu for "Median Alpha Scores of White and Negro Recruits, United States Army, Five Northern States and Eight South Central States." The data was derived from Robert Yerkes, ed., "Psychological Examining in the United States Army," *Memoirs of the National Academy of Sciences*, XV (1921), 690 and 691, Tables 205 and 206. Adapted from Frank Lorimer and Frederick Osborn, *Dynamics of Population* (New York: Macmillan, 1934), p. 140, Table 46.

	White Recruits	"Negro" Recruits
Five Northern States		
Ohio	62.2	45.3
Pennsylvania	62.0	34.7
Illinois	61.6	42.2
New York	58.3	38.6
Indiana	55.9	41.5
Eight South Central States		
Tennessee	44.0	29.7
Texas	43.4	12.1
Oklahoma	42.9	31.4
Kentucky	41.5	23.9
Alabama	41.3	19.9
Mississippi	37.6	10.2
Louisiana	36.1	13.4
Arkansas	35.6	16.1

According to Montagu, Benedict and Weltfish's numbers differ from his because they took them from Otto Klineberg in *Race Differences* (1935), who included Beta test

wide range of average scores, black, white, or combined, suggested that something other than inherited abilities was being assessed by the Alpha tests.[90]

The Races of Mankind pamphlet was vehemently condemned in some quarters outside Congress, as well. Senator Theodore Bilbo of Mississippi, author of a repeatedly introduced but unsuccessful bill "repatriating" American blacks to Liberia, brought Benedict and Weltfish's pamphlet to the attention of Earnest Sevier Cox, fellow separatist and a leading white supremacist from Virginia, asking him to review it.[91] By 1944, when Bilbo contacted him, Cox, an "elder statesman of white supremacy," was notorious for his vociferous advocacy of racial segregation.[92] In the 1920s, Cox cofounded the Anglo-Saxon Clubs of America, devoted to the separation of the races; published *White America*, an account of the dangers of integration and miscegenation buttressed with evidence from

scores (for illiterate and foreign test takers). Montagu, *Man's Most Dangerous Myth* (1965), pp. 388–389.

[90] The differences amounted to some 27 points among whites, 35 points among blacks, and an overall range from 10.2 points for blacks in Mississippi to 62.2 point for whites in Ohio. See footnote 89.

[91] Ethel Wolfskill Hedlin, "Earnest Cox and Colonization: A White Racist's Response to Black Repatriation, 1923–1966" (PhD diss., Duke University, 1974), pp. 156–157; Earnest Sevier Cox, *The Races of Mankind: A Review*, foreword by Arthur Daugherty (Jellico, TN: Arthur Daugherty, 1951), p. 1. My thanks to John P. Jackson for bringing Cox's review to my attention. On Bilbo's alliance with Marcus Garvey's Universal Negro Improvement Association and the Back to Africa movement and the Peace Movement of Ethiopia, see Hedlin, "Earnest Cox and Colonization." For an example of Bilbo's philosophy, see Theodore Bilbo, *Take Your Choice: Separation or Mongrelization* (Poplarville, MS: Dream House Publishing Co., 1947). See also William H. Tucker, *The Science and Politics of Racial Research* (Urbana and Chicago: University of Illinois Press, 1994), pp. 134, 177. In addition to Hedlin, Tucker offers an extended discussion of Cox and his allies in *The Funding of Scientific Racism: Wickliffe Draper and the Pioneer Fund* (Urbana and Chicago: University of Illinois Press, 2002); as does John P. Jackson, *Science for Segregation: Race, Law, and the Case Against* Brown v. Board of Education (New York: New York University Press, 2005).

Hedlin describes Bilbo's book thus: "With some pretense toward scholarship, Bilbo vents spleen against the American Negro and predicts either the separation of black from white or the 'mongrelization' of the U.S. population." Bilbo was open in his disregard for African Americans, and those who supported integration and civil rights. Hedlin quotes him describing his opposition as "Communists and nigger-lovers," and the Fair Employment Practices Commission, which he opposed, as "nothing but a plot to put niggers to work next to your daughters and to run your business with niggers." She notes that he was a member of the Ku Klux Klan, as was Cox. Hedlin, "Earnest Cox and Colonization," pp. 160, 241.

[92] Jason Ward, "'A Richmond Institution': Earnest Sevier Cox, Racial Propaganda, and White Resistance to the Civil Rights Movement," *Virginia Magazine of History and Biography*, vol. 116, no. 3, pp. 262–293.

his travels in Africa, Asia and the Americas;[93] and was a leading advocate of Virginia's 1924 Racial Integrity Act, the nation's strictest antimiscegenation statute, which enshrined the "one-drop" rule into law, as well as the Massenburg Bill, which outlawed racial mixing in all public places. All this was premised on his deeply held belief that racial mixing, whether social integration or intermarriage, would undermine the clear superiority of the white race and lead to the collapse of "Western civilization."[94]

Delighted with Cox's 1944 "Review" of *The Races of Mankind*, Bilbo intended to read it into the Congressional Record and circulate it among the American public, in an attempt to discredit the claims made by Benedict and Weltfish.[95] It was finally published in 1951, funded partially by Cox himself, by fellow segregationist Arthur Daugherty, who appended a brief explanation of its origins and his own anti-Semitic, anticommunist screed in a foreword.[96] *The Races of Mankind*, Daugherty warned, was an "inflammable [sic] and misleading pamphlet" whose "evil influence and false information" was leading the public astray, part of "the gigantic race mongrelizing propaganda campaign" propagated by Jews. ("I wonder if I am correct," he mused, "in assuming that the authors of The Races of Mankind are of the Jewish branch of the white race?") In an apparent effort to undermine their credibility as dispassionate scientists

---

93 Earnest Sevier Cox, *White America* (Richmond, VA: Mitchell and Hotchkiss, 1923), subsequent editions 1924, 1925, 1937, 1966. For an anthropologist's critical review of *White America*, in which the author described Cox's book as "fallacious in its assumptions, incompetent in its handling, and loose in its logic," see Melville J. Herskovitz, "Extremes and Means in Racial Interpretation," *The Journal of Social Forces*, vol. 2, no. 4 (May 1924), pp. 550–551. For a contemporary review in an African-American Virginia newspaper, see Walter J. Scott, *Norfolk Journal and Guide*, Aug. 22, 1925. For more on Cox, as well as a fuller account of the intersection of white supremacy and racial science, see Tucker, *The Science and Politics of Racial Research* (1994), and *The Funding of Scientific Racism: Wickliffe Draper and the Pioneer Fund* (2002); Jackson, *Science for Segregation* (2005); Jonathan Peter Spiro, *Defending the Master Race: Conservation, Eugenics and the Legacy of Madison Grant* (Lebanon, NH: University Press of New England, 2009); Gregory Michael Dorr, *Segregation's Science: Eugenics and Society in Virginia* (Charlottesville, VA: University of Virginia Press, 2008). See also pp. 335 n91.
94 J. Douglas Smith and the *Dictionary of Virginia Biography*, "Earnest Sevier Cox (1880–1966)," *Encyclopedia Virginia*, Virginia Foundation for the Humanities, Apr. 19, 2013, retrieved from www.encyclopediavirginia.org/Cox_Earnest_Sevier_1880-1966. See also, J. Douglas Smith, *Managing White Supremacy: Race, Politics and Citizenship in Jim Crow Virginia* (Chapel Hill: University of North Carolina, 2002); Hedlin, "Earnest Cox and Colonization"; Ward, "'A Richmond Institution"; Matthew Pratt Guterl, "The Importance of Place in Post-Everything American Studies," *American Quarterly*, vol. 61, no. 4, Dec. 2009, pp. 931–941 (931–932).
95 Hedlin, "Earnest Cox and Colonization," pp. 156–157, 157 n42.
96 Cox, *The Races of Mankind: A Review*, pp. 1–3.

proffering objective information, Daugherty portrayed the authors as parties to a Jewish conspiracy. "Should the Anglo-Saxon, who made the United States what it is today, become mongrelized, the Jew would have no competition in ruling America in every walk of life. A mongrel cannot compete with a thoroughbred. The Jew looks centuries ahead in planning world domination." According to Daugherty, Benedict, Weltfish, and their ilk bamboozled unwitting churches and educators into accepting a program of "Communistic race suicide" that promoted damaging race mixing, instead of "racial freedom," which he defined as the "right to be racially pure."

Cox, too, decried the "inflammable" propositions in *The Races of Mankind*, but in somewhat more measured, ostensibly more objective and scientific terms. Cox accused Benedict and Weltfish of "gratuitously" including material that would promote "racial discord," rather than "allaying racial prejudices" and "promoting racial harmony," goals "worthy of the support of every one of us." Like others in the wartime and postwar racialist right, Cox rhetorically positioned antiracist humanists as members of a leftist conspiracy to dupe the American public into believing what Cox and his brethren viewed as nonsense about equality, propaganda that directly contradicted what they viewed as the reality of innate racial differences and their consequences. He accused Benedict and Weltfish of deliberately misleading readers, calling their work "a studied effort to conceal from the untutored the simple fact that those who deny the influence of race and exaggerate the influence of environment in human achievements are a dwindling lot, in retreat." Much of the review is dedicated to undermining the scientific credibility of Benedict and Weltfish – and "the late Professor Franz Boas, a Jew" who "promote[d] general miscegenation" – by constructing an alternative account of human history and evolution in which whites, especially racially superior Anglo-Saxons, have been responsible for higher civilization. Asserting that Benedict and Weltfish littered their pamphlet with "unsupported dogmatic statements" and "questionable" data, Cox proceeded to build a history of "Teuton" success and "Negro" failure.[97] Cox asserted, for example, that "**blond** Russians" (his emphasis), not a more varied set of peoples, produced the "mighty Russian commonwealth," basing his claim on a 1943 article in *LIFE* magazine.[98] Several pages were devoted to arguing that Egyptians were not black Africans, and citing British imperialist Henry Hamilton Johnston, went so far as to assert that "Egyptians were

[97] Ibid., pp. 5–6.
[98] Ibid., p. 7.

a 'Caucasian' people" whose "early contact with the Negro imparted to that race all the arts of civilization they possessed up to the coming of the Persians, Greeks, Romans, Arabs, and modern Europeans to the continent of Africa."[99] Cox spent several pages of his twenty-two-page review excoriating Jews as a race bent on domination, "a composite specialized race stock historically given to ganging up at the cross roads of commerce and overflowing into the professions" who "pay, cajole, or seduce" gentiles into unwittingly abetting their schemes. Franz Boas and his "disciples" in the "hot-bed of miscegenation propaganda in the Department of Anthropology at Columbia University" were agents of this deception, luring foolish whites to their doom.[100] In Cox's view, the race mixing of black and white Americans, which he believed *The Races of Mankind* advocated, was a Jewish plot to weaken Anglo-Saxon control of the United States. Summed up as "Communism and Race-Mixing, '40 Acres of Land and a Mule and a White Wife,'" Cox ended his review arguing that sex between members of different races was abnormal – "atavistic" and "perverted" – and inevitably led to inferior "mixbreeds" who were "racial misfits."[101] His prescription, which he had been promoting for more than twenty years, was full separation of the races, each "a nation of their own," whites in the United States, blacks in Africa, "the land of his ancestors."[102]

Ruth Benedict shared Ernest Sevier Cox's concern that racial disharmony imperiled American society, but her understanding of its roots and prescription for its resolution were entirely different. *Race: Science and*

---

[99] Ibid., p. 11. Cox based this claim on the work of Sir Harry Johnston, "an eminent authority on African history." Cox provided neither footnotes nor bibliography, but presumably this refers to British colonial officer and explorer Sir Henry Hamilton Johnston (1858–1927). Johnston was the leader of an 1884 expedition sponsored by the Royal Geographic Society and the British Association for the Advancement of Science to Mount Kilimanjaro, in which treaties he brokered were later transferred to the British East Africa Company. As a colonial officer, he was involved in efforts to acquire land and access for imperial powers and trading companies across Africa. His publications included short novels, several accounts of his travels, and descriptions of Africa and its peoples, most written in later life, after his return to England. His contemporaries regarded him as a scholar, although he had no training in fields where he was cited as an expert, such as ethnology or linguistics, but only an honorary degree from Cambridge University in ornithology. The "authoritative" text to which Cox refers is likely Johnston's *The Backward Peoples and Our Relations with Them* (New York and London: Oxford University Press, 1920). See Roland Oliver, *Sir Harry Johnston & the Scramble for Africa* (New York: St. Martin's Press, 1959).

[100] Ibid., pp. 6, 11, 14–16.

[101] Ibid., pp. 20–22, 26.

[102] Ibid., p. 23.

Politics stressed the importance of understanding and tolerating both racial and cultural diversity. Like Boas, Benedict argued that human beings shared a fundamental psychic unity that underlay the wide variety of cultural solutions different groups of people devised for common problems of living. And, also like Boas, Benedict argued that although human beings could generally be sorted into broad groups on the basis of physical differences, those differences were neither uniform among any group of people nor fixed. Human beings had proved enormously malleable not only in their social and cultural formations, but in their physical constitutions. Looked at broadly, human variation was largely a matter of local variations on a theme, a matter of constant change in response to changing conditions, both human and environmental. Rather than essentializing racial and cultural differences, Benedict urged readers to focus on human commonalities and the possibility of change and improvement that a study of history and cultures offered.[103]

The Races of Mankind distilled this message into the idea of "brotherhood." In the face of a "steadily shrinking" world fraught with conflict at home and abroad, Benedict and Weltfish wrote their pamphlet to directly combat what they viewed as inaccurate, racist distinctions that promoted a divisive ideology and undermined society and the war effort.[104] Like Harry Shapiro, or Berthold Laufer, they hoped to use their anthropological expertise to convince their readers of the fundamental unity of all human beings. The pamphlet noted, for example, that diverse peoples from around the world were fighting fascism together, showing how "the whole world has been made one neighborhood."[105] "White men, yellow men, black men, and the so-called 'red men' of America, peoples of the East and West, of the tropics and the arctic, are fighting together against one enemy."[106] This shrinking world, with its modern perils, made harmony and tolerance a necessity. The world could not afford for "hard feeling" among "different races and nationalities" to leave the Allies defeated.[107] Robert T. Hatt, director of the Cranbrook Institute, captured Americans' sense that they were living in a rapidly changing world:

Though four centuries ago it took Magellan over 1,000 days to circumnavigate the globe, and only a century ago a clipper ship required 150 days, the airplane

---

[103] Benedict, *Race: Science and Politics*, pp. 37–38, 63.
[104] Lardner, "'Brotherhood of Man': A Script," p. 354.
[105] Benedict and Weltfish, *Races of Mankind*, p. 2.
[106] Ibid., p. 1.
[107] Ibid., p. 2.

has shrunk our earth to a size that requires but seven days for encirclement. This with radio and international commerce has indeed again made neighbors of all men and broken down all but the social barriers between races.[108]

The idea of a shrinking world and the humanist notion of a "brotherhood of man" was becoming commonplace in the early 1940s. For example, at the 1944 Social Studies Conference "Diversity Within National Unity," presided over by Alain Locke, a number of speakers argued for the importance of diverse people realizing they "are truly brothers," including Carey McWilliams, author of *Brothers Under the Skin* (1942), psychologist Otto Klineberg, and Howard E. Wilson, chair of the Department of Education at Harvard University. Following travels in Britain, the Middle East, Russia, and China, Wendell Willkie, whom Hatt cited, wrote the widely read *One World* (1943), an account of his meetings with leaders, citizens, and soldiers, which pressed the common interests of people around the world.[109] Closer to home, the *Detroit News* also picked up on the brotherhood theme in its February 1944 *Sunday Graphic* highlighting "National Brotherhood Month," featuring the Cranbrook *Races of Mankind* exhibition and a call by the Churches of Christ in America for all Christians to be "unprejudiced and wise enough to bridge and cross the chasms of racial isolation and segregation."[110] The opening panel of the Cranbrook exhibition, featuring Adam and Eve seemingly guiding children from the three principal races (see Figure 6.2), also framed the world's peoples as neighbors: "Early Man lacking means of wide travel and communication developed into varied types which we call races. People then needed to keep peace with few neighbors. Today all men are neighbors and with all men in time we can and must live at peace."[111] The shift in tone since the 1933 opening of *The Races of Mankind* exhibition at the Field Museum in Chicago is striking. In 1933, museum publicity and news accounts all marveled at Malvina Hoffman's "world tour" to far away, exotic locales. The Field Museum exhibition was touted as an opportunity for Americans to encounter exotic peoples

---

[108] Hatt, "The Races of Mankind, Announcing an Exhibition," p. 1; "Valuable Racial Exhibit," *Hobbies*, vol. 49 (Mar. 1944), p. 22.

[109] Alain Locke, et al., *Diversity Within National Unity: A Symposium* (Washington, DC: The National Council for the Social Sciences, Feb. 1945); Carey McWilliams, *Brothers Under the Skin* (Boston: Little, Brown, 1942); Wendell Willkie, *One World* (New York: Simon and Schuster, 1943).

[110] *Detroit Free Press*, Sunday Graphic, Feb. 13, 1944, Scrapbook, 1943–46, Cranbrook Foundation, pp. 37–38, Cranbrook Archives.

[111] "Script on panels in the exhibit 'The Races of Mankind' . . . ," p. 1, Folder: Press Releases Etc., Cranbrook-ROM.

they were not likely to otherwise meet, except perhaps along the midway of a world's fair. There was no sense that visitors were likely to travel to India, Japan, or Africa themselves, much less that the people displayed at the Field Museum might be found in visitors' "backyards." Rather, the Field Museum exhibition was constructed on an older anthropological model that viewed other, "primitive" people as a means for understanding "civilization." Diversity was read as a progressive hierarchy, offering white Americans the promise of hope for the future.

By 1943, the times and anthropology were changing. In the fantasy world of *In Henry's Backyard* and *The Brotherhood of Man*, the proximity of the world's peoples to middle America was made manifest. The book and animated film opened with Henry, an "ordinary, friendly person, who lived in an ordinary house," dreaming of travel – via propeller-driven arm chairs – to distant lands, and dreaming of the world's "odd people" becoming his neighbors. He awoke to find "It's really happened!" Bubbling over with curiosity and goodwill, he scampers out to meet his new neighbors, who include men from Africa, China, Mexico, Turkey, Europe, and the Arctic.[112] For Ruth Benedict and Gene Weltfish, cultural relativity alone was not enough. To effectively combat racism and prevent more violent conflict required more than an appreciation for the wide variation in human cultures; Americans needed a basis for bridging those differences, some kinds of commonalities to form the basis of unified action, some sort of unity in diversity. In the postwar period, many found this in the notion of brotherhood.

The Races of Mankind presented a radically simplified account of race, racial science, racism, and human unity. The nuances of Benedict's *Race: Science and Politics* were inevitably lost in paring down a 164-page book into a brief pamphlet, film, and booklet. Moreover, the shorter texts and film were produced after the United States had entered the war, making their messages more urgent and their context more openly political and polemical. Combining less nuanced prose with profuse illustrations also intensified both the message of cultural relativity and the message that race differences were real. These were combined with more obvious attempts to highlight points of commonality amid diversity and disrupt assumptions about racial or cultural superiority.

As in *Race: Science and Politics*, Benedict and Weltfish proclaimed the validity of racial science at the outset. Just as chemists, physicists, and

[112] Lardner, "'Brotherhood of Man': A Script," p. 355; Benedict and Weltfish, *In Henry's Backyard*, n.p.

engineers contributed to the war effort, so, too, did sciences that provided insight into the nature of racial differences, including not only anthropologists and biologists but also historians, psychologists, and sociologists.[113] In what seems to have been an attempt to link the scientific facts about race to another unassailable truth, Benedict and Weltfish also asserted that "The Bible story of Adam and Eve, father and mother of the whole human race, told centuries ago the same truth that science has shown today: that all peoples of the earth are a single family and have a common origin." This sort of claim also offered an evolutionary account of human unity in the guise of the biblical origin story. The accompanying illustration offered a biblical-evolutionary tree of race, one that evoked Blumenbach's taxonomic system of Negroid, Malayan, Mongoloid, and Caucasoid, although left aside his American race, "the so-called 'red men'"[114] (Figure 6.4). The Cranbrook Institute of Science crafted a similar image that joined scientific racial categories and Biblical imagery with the heading "All Mankind Is One Family," which they used to open the exhibition (see Figure 6.2). By employing Albrecht Dürer's famous sixteenth-century Adam and Eve, and calling its grouping of Renaissance imagery and children a "family," Cranbrook was able to drop the genealogical/evolutionary tree but still convey the same mix of biblical origins and racial taxonomy.[115] Using living children as models also suggested that the races (here three) were, if not the product of God, at least the progeny of an ancient evolutionary process. In *The Races of Mankind*, Benedict and Weltfish reinforced the idea of common evolutionary origins with anatomical evidence, such as the complexities of the foot and

---

[113] Benedict and Weltfish, *The Races of Mankind*, p. 3.

[114] Ibid., pp. 1, 3–4. Benedict noted in *Race: Science and Politics* that "'Three' is not a sacred number" in classifying the major racial divisions, noting that Polynesians "may be a well-marked variant" on the usual triumvirate: Caucasian, Negroid, Mongoloid. Interestingly, Adam and Eve were not illustrated as white, but rather as an intermediate tone, in keeping with Benedict and Weltfish's assertion elsewhere in the pamphlet that most people had an "intermediate" skin tone. This was also implied in *In Henry's Backyard* in an illustration of the peopling of the world that begins in Central Asia, not Africa, as it would today, nor in Mesopotamia, given the Adam and Eve genealogy.

[115] The Adam and Eve images were copied from Albrecht Dürer's Garden of Eden engraving, *Adam and Eve*, 1504. Cranbrook's image lacks the snake from the biblical narrative, which was included in the pamphlet illustration, although it retained the apple, which makes the narrative legible as Adam and Eve. Dürer was an ironic choice – or perhaps an intentional one? – as the Nazis embraced him as a pure Aryan, "the most German of all German artists," and displayed his self-portrait and art throughout the regime. Jane Campbell Hutchinson, *Albrecht Dürer: A Guide to Research* (Taylor & Francis, 2000), pp. 15–17.

FIGURE 6.4. "The Peoples of the Earth Are One Family." *Source:* Ruth Benedict and Gene Weltfish, *The Races of Mankind*, Public Affairs Pamphlet No. 85 (New York: Public Affairs Committee, Incorporated, 1943).

type and arrangement of teeth, to demonstrate fundamental similarities among human beings. They continued by pointing out that "racial" differences belied deeper commonalities "in what all races are physically fitted for":

All races of men can either plow or fight, and all the racial differences among them are in nonessentials such as texture of head hair, amount of body hair, shape of the nose or head, or color of the eyes and the skin. The white race is the hairiest, but a white man's hair isn't thick enough to keep him warm in cold climates. The Negro's dark skin gives him some protections against strong sunlight in the tropics, and white men often have to take precautions against sunstroke.... The shape of the head, too, is a racial trait; but whether it is round or long, it can house a good brain.

The races of mankind are what the Bible says they are – brothers. In the bodies is the record of their brotherhood.[116]

In *Race: Science and Politics*, Benedict repeatedly made the point that diverse races and cultures had reached great heights of invention and art, and these innovations were diffused from one group of people to another, so that no civilization could claim any special superiority. She illustrated her point with examples from world history. Although Americans might think of steel and gunpowder as Western innovations, they were wrong. Steel was invented in India or Turkestan, she noted, and gunpowder in China, as were paper and printing. Cultivation of grains and animals originated in Neolithic Asia, she continued, and corn and tobacco were first domesticated by Native Americans. Even our numerical system of mathematical notation was invented in Asia and introduced to the West by the Moors, she added.[117] In *The Races of Mankind* pamphlet and the Cranbrook exhibition, this idea was illustrated graphically, with an image of a round table set with food from a wide variety of world cultures including peoples from Asia, Africa, and the Americas. The scene employs a visual shorthand common to the *Races of Mankind* pamphlet and all its subsequent iterations, using stereotypical clothing and artifacts to direct readers' and viewers' assumptions about the identity of the "others" referred to in the text. In the pamphlet, "Our Food Comes From Many

---

[116] Benedict and Weltfish, *The Races of Mankind*, pp. 4–5.

[117] Benedict, *Race: Science and Politics*, p. 15. This was a favorite conceit of anthropologists, who could draw on their encyclopedic knowledge of world cultures. Otto Klineberg recounted a version attributed to Ralph Linton in "Cultural Diversity Within American Unity," Alain Locke, et al., *Diversity Within National Unity: A Symposium* (Washington, DC: The National Council for the Social Sciences, Feb. 1945), pp. 16–22 (p. 17–18). According to Klineberg, Linton illustrated how "'certain insidious ideas have wormed their way'" into American civilization despite strenuous efforts to preserve "Americanism at all costs," by following a typical man through his day. It started with him in bed in his pajamas, an East Asian garment. He drinks his morning coffee, an African plant discovered by Arabs in Ethiopia, and heading out into the rain, picks up an umbrella, invented in India. Before meeting his train, an English invention,

he pauses for a moment to buy a newspaper, paying for it with coins invented in ancient Lydia. Once on board he settles back to inhale the fumes of a cigarette invented in Mexico, or a cigar invented in Brazil. Meanwhile, he reads the news of the day, imprinted in characters invented by the ancient Semites by a process invented in Germany upon material invented in China. As he scans the latest editorial pointing out the dire results to our institutions of accepting foreign ideas, he will not fail to thank a Hebrew God in an Indo-European language that he is one hundred percent... American.

FIGURE 6.5. "Our Food Comes from Many Peoples." *Source:* Ruth Benedict and Gene Weltfish, *The Races of Mankind*, Public Affairs Pamphlet No. 85 (New York: Public Affairs Committee, Incorporated, 1943).

Peoples" showed a map of the world on the tabletop, identifying where each foodstuff originated. At the head of the table sat a white man, his hat resting on a coat rack behind him, fork and knife in hand ready to eat as the others stand beside the table, presenting him with food and drink[118]

---

[118] The text accompanying the Cranbrook display reads:

> The crab-apple and the hazelnut are the only important food contributed to our table by the ancient Europeans. Everything else we eat we have borrowed from someone else. If the contribution of the American Indian alone were removed we should have no kidney beans, lima beans, chocolate, corn, peanuts, pineapple, sweet potatoes, Irish potatoes, pumpkins, squash, avocado, tomatoes, tobacco or turkeys.

"Script on panels in the exhibit 'The Races of Mankind,'" p. 4, Folder: Press Releases Etc., Cranbrook-ROM. Foods represented included: apple, avocado, banana, bread, chicken, cocoa, cocoanut [sic], coffee, corn, crabapple, ham, hazelnut, lima beans, milk, orange, peanuts, pineapple, potato, pumpkin, rice, spaghetti, sugar cane, steak, sweet potato, tea, tobacco, turkey, watermelon, and wine. "Inventory. Races of Mankind Exhibit," Feb. 19 and Apr. 26 (revised), 1944, Folder: Press Releases, Etc., Cranbrook-ROM.

FIGURE 6.6. Cranbrook Institute of Science. "The Foods We Cultivate Are a Gift from All Peoples," *The Races of Mankind* (1944). Courtesy of Cranbrook Archives, "Six figures at table."

(Figure 6.5). The Cranbrook Institute replicated this scene, making the table homier, covered with a checkerboard tablecloth, the repast more sumptuous (much of it packaged and processed), and the racial identifications more specific. The seven nonwhite individuals had been pared down to five, and all six figures at the table were identified: (from left to right) African Negro, Black Race; American Indian, Yellow Race; North American, White Race; West Asiatic, Yellow Race; East Asiatic, Yellow Race; and South Asiatic, Yellow Race[119] (Figure 6.6). As in the booklet, the only figure seated and prepared to eat is the white man. Although this arrangement was meant to illustrate the intercultural origins of American meals, the display also equates the American viewer with a white man, the deserving recipient of other, racialized peoples' agricultural innovations.

[119] Note in this display the Native American has been subsumed into the "Yellow Race," perhaps reflecting broad anthropological opinion, reiterated by Benedict, that American Indians were of "Mongoloid" or Asian descent.

They are not portrayed joining him in the repast, but instead appear to be serving him, retaining implicitly the racial hierarchy the text disavows.

The original Cranbrook Institute *Races of Mankind* exhibition included a number of novel images intended to undermine various erroneous racial assumptions, inspired by Benedict and Weltfish's pamphlet, but not produced in it. "No Race Is Mentally Superior" attacked the longstanding notion that brain size was related to mental capacity by comparing average brain sizes for the three major racial divisions with the much larger brain of a "caveman," to demonstrate the irrelevance of brain size, as well as the similarity among the races[120] (Figure 6.7). "Who Are the Aryans?" one of four displays directed at specific races, attacked the claim that Aryans were a race. The display pointedly presented a photo of Adolph Hitler making the Nazi salute next to the statement, "The Nazi Ideal 'Aryan,'" mocking Nazi claims that Aryans were blond and blue eyed.[121] Another countered Nazi racial theory with a display titled "The Jews Are Not a Race." American race relations were addressed in "Negroes Are an Integral Part of Our Culture," a display that emphasized the valuable contributions blacks had made to American society.[122]

---

[120] The "cavemen" are a reproduction of Charles Knight's famous mural of Neanderthals, commissioned by Henry Fairfield Osborn for the American Museum of Natural History. For a recent discussion of such imagery, see Constance Areson Clark, *God – or Gorilla: Images of Evolution in the Jazz Age* (Baltimore, MD: The Johns Hopkins University Press: 2008); and also Ronald Rainger, *An Agenda for Antiquity: Henry Fairfield Osborn and Vertebrate Paleontology at the American Museum of Natural History, 1890–1935* (Tuscaloosa, AL: The University of Alabama Press, 1991). The panel appears to be the creation of Columbia anthropologist John Adams, who is pictured next to the display in archival photographs, and also in the small inset photo in the panel reproduced here. Two versions of this panel existed, one used in the institute exhibition, which used beakers filled with millet seed to display cranial volumes, and another, used for the traveling exhibition and depicted here, that illustrated cranial volume with bars. Pictures of both panels are available in the Cranbrook Institute Archives. "Inventory. Races of Mankind Exhibit," Feb. 19 and Apr. 26 (revised), 1944, Folder: Press Releases, Etc., Cranbrook-ROM.

[121] Lee A. White, et al., *Cranbrook Institute of Science: A History of Its Founding and First Twenty-Five Years* (Bloomfield Hills, MI: Cranbrook Institute of Science, 1959), Plate 21.

[122] The text on the "Negroes are an Integral Part of Our Culture" display reads:

The myth of Negro inferiority was a direct result of the effort of plantation owners to prove their slaves inherently unfit to take on the responsibilities of freedom. That their slaves and sons and grandsons of their slaves have in numerous instances, and in spite of tremendous adverse pressure, succeeded not only in proving their worth as artists, poets, scientists, and administrators, but in rising to the top ranks of those professions,

FIGURE 6.7. Cranbrook Institute of Science. "No Race Is Mentally Superior," *The Races of Mankind* (1944). Courtesy of Cranbrook Archives, "No Race Is Mentally Superior."

The depiction of African Americans was not, however, limited to a discussion of their cultural contributions. The racial identity of American blacks was addressed directly in a display devised for the Cranbrook version of *The Races of Mankind*, in which African Americans were

is ample evidence of the ridiculous nature of such assertions. The Negro has contributed greatly to modern society. Our enjoyment of his contributions is proof of his ability.

The individuals pictured included Marion Anderson, Roland Hayes, Coleridge Taylor, Richard B. Harrison, Paul Robeson, Lt. James Fowler, General Benjamin O. Davis, Joe Louis, Booker T. Washington, George Washington Carver, and Duke Ellington. Cranbrook-ROM.

FIGURE 6.8. Cranbrook Institute of Science. "Composition of the American Negro," *The Races of Mankind* (1944). Courtesy of Cranbrook Archives, "Composition of the American Negro."

presented as a "new race," developing through a combination of "each of the three great races" (Figure 6.8). The display depicted a young black man in a suit coat and tie holding a label – "American Negro" – flanked by images of an "American White," "American Indian," and "African Negro," each representing equal portions of the "American Negro's" heritage.[123] The text on the display reads: "The American Negro is developing as a new race, combining elements of each of the three great

---

[123] White, et al., *Cranbrook Institute of Science*, Plate 19.

races. These are represented by the African Negro, the American Indian and the European White." Despite this description, the photograph of the white man is labeled "American," revealing the underlying assumed conflation between Americanness and whiteness. It also offers a clue as to why Robert Hatt and John Adams felt it was important to create a display presenting American blacks as a racially distinct group. It seems to have been an effort, following Benedict's work, to combat antiblack prejudice by making the problematic argument that American blacks were a valuable part of American society because they were not really the same population as African blacks – "The American Negro is a unique type and is not the same as the African Negro" – instead combining their African roots with qualities found in whites and American Indians.[124] Hatt's description of the exhibition in his 1944 annual report for the Cranbrook Institute reveals this underlying racial logic, in which American blacks needed to be distanced from their African roots. In his report, Hatt noted that one of the displays accounted for the supposedly primitive nature of Africans by showing that "the backwardness of the central Africa Negroes was relative and explainable on the basis of isolation and absence of stable food resources," while the "Composition of the American Negro" display illustrated that "the American Negro is not an African transplanted but a blend of African, European and American Indian," although "African characters are dominant since they are genetically stronger."[125] The man portrayed as American/European white in this display is the same "North American" man seated at the table in "Our Food Comes From Many Peoples," linking the two displays and reinforcing the centrality of whiteness to the conception of Americanness offered to the public. Historian Matthew Jacobsen has argued persuasively that

---

[124] Printed text associated with this panel reads:

> The American Negro is a unique type and is not the same as the African Negro. Early in the history of this country the Negro mixed extensively with the Indian. Since that time a large amount of White blood has been added, so that today the American Negro is a fairly homogenous blend of White, Indian and Negro, and thus of the three races of mankind (Folder: Labels, Cranbrook-ROM).

[125] Hatt, "Report of the Director," p. 13. These comments reflect the development of the exhibition and changes in the displays. In February 1944, an exhibition inventory notes the preparation of a display entitled "A race loses a culture but gains another," intended to replace the original display, "Why were the Africans backward?" But in a later April version of the inventory, that entry has been crossed out altogether. "Inventory. Races of Mankind Exhibit," Feb. 19 and Apr. 26 (revised), 1944, Folder: Press Releases, Etc., Cranbrook-ROM.

displays such as these participated in the construction of broad racial categories along color lines, moving away from the earlier racialization of what came to be called "ethnic" types or nationalities. Thus, Jews and people the National Socialists classed as Aryans, as well as Poles, the Irish, and the rest of America's polyglot European immigrant populations, were scientifically subsumed under the Caucasian "white" category, erasing, or at least tempering, earlier racialized distinctions based on national, ethnic, or confessional origins. Similarly, despite efforts to offer a more nuanced vision of African Americans as a truly American mixture, the display on the composition of the "American Negro" reified the races presented and legitimated the practice of racial classification.[126] For African Americans, "Composition of the American Negro," and the *Races of Mankind* exhibition as a whole, offered an ambivalent embrace, at once encompassing them as part of the American melting pot, yet distancing them from the presumptively white "American" through the "strength" of their African genetic inheritance.[127]

The presence of African American social worker and community leader Beulah Whitby among the guest speakers at the exhibition opening reflected the institute's sincere desire to "create good will and interracial understanding" by counteracting "baseless notions of race superiority" that bred "ill will,"[128] and their efforts to reach out to African Americans to participate in that effort. In 1944, when the *Races of Mankind* opened, Whitby was executive secretary of the Office of Civilian Defense in Detroit. Daughter of a prominent Baptist minister, Edward Tyrrell, and his socially active wife, Elizabeth, who had helped organize Baptist conventions for women at the state and national levels, Beulah followed her parents' example in leading a life of service to her community. Graduated in 1926 with a degree in sociology from Oberlin College,

---

[126] Matthew Frye Jacobson, *Whiteness of a Different Color: European Immigrants and the Alchemy of Race* (Cambridge, MA: Harvard University Press, 1998), pp. 106–109. The display, like *The Races of Mankind* and the other brief treatments, hedges on the number of "major" races or divisions. The works generally proclaimed the usual triad of Negroid, Mongoloid, and Caucasian, but then frequently inserted additional apparently important racial types, most often American Indians.

[127] White, et al., *Cranbrook Institute of Science*, Plate 19. The "American Indian" is John Youngbear, a Fox Indian from Tama, Iowa, who earlier participated in a Cranbrook Institute recreation of an "Indian camp" at the 1940 Detroit Flower Show, advertised as "complete with Sauk Indians and a wild turkey." Folder: Photographs, Cranbrook-ROM.

[128] Comment made by American Museum of Natural History anthropologist Harry L. Shapiro at the opening ceremonies, and widely quoted in Detroit area newspapers. *Detroit Free Press, Sunday Graphic*, Feb. 13, 1944, pp. 37–38.

Beulah Tyrrell married dentist Charles Whitby and in 1931 began her
career as a caseworker in the Detroit Department of Public Welfare.
Following graduate work at the University of Chicago, New York University, and the University of Michigan, in 1941 she became the first
black director of the Alfred district, the city's largest, overseeing eighty
employees, including fourteen African Americans. In 1942, she resigned
that office to work for the Office of Civilian Defense. Whitby was also a
prominent clubwoman, including service as president of the black college
women's association, Alpha Kappa Alpha, vice president of the National
Council of Negro Women, membership in the American Association of
Social Workers and the National Urban League, and the Detroit Mayor's
Committee. Following WWII, Whitby testified before Congress on the
importance of the Fair Employment Practices Commission as a way to
fight racial discrimination in federally financed employment, and was a
staunch advocate of integrated housing. Whitby taught at Wayne University Graduate School of Social Work in Detroit, as well as at the University
of Michigan, eventually becoming head of the Department of Sociology
at Mercy College in Detroit in 1962.[129]

Following the 1943 Detroit race riot, Whitby was appointed executive
secretary of the Emergency Welfare and Evacuation Service in the Civilian
Defense Office, as well as assistant director of the Detroit Commission
on Community Relations, formerly the Interracial Committee, a post she
held for nineteen years. It was in this capacity that Whitby addressed
the public and press at the *Races of Mankind* exhibition opening in
January 1944. In her remarks, Whitby, "one of Detroit's most distinguished workers for the betterment of inter-racial relations" warned her
audience that Americans "must come to grips with this problem we call
race. Unless we do, we may lose a battle more important that those being

---

[129] Megan Taylor Shockley, *We, Too, Are Americans: African American Women in Detroit
and Richmond, 1940–54* (Urbana IL: University of Illinois Press, 2004), pp. 49–50, 60,
104, 129; Stephanie J. Shaw, *What a Woman Ought to Be and to Do: Black Professional
Women Workers During the Jim Crow Era* (Chicago: University of Chicago Press,
2010), p. 192; "Mrs. Beulah Tyrell Whitby," African American Biographical Database,
retrieved from aabd.chadwyck.com (accessed Jul. 26, 2013); A Handbook on the
Detroit Negro, pp. 128–129, African American Biographical Database, retrieved from
aabd.chadwyck.com (accessed Jul. 26, 2013); The Minister's Wife, pp. 21–22, African
American Biographical Database, retrieved from aabd.chadwyck.com (accessed Jul. 26,
2013); Who's Who in Colored America, 1950, p. 547, African American Biographical
Database, retrieved from aabd.chadwyck.com (accessed Jul. 26, 2013); Detroit African
American History Project, Wayne State University, retrieved from http://www.daahp
.wayne.edu/biographiesDisplay.php?id=7 (accessed Jul. 26, 2013).

fought on the war fronts." She stressed the importance of harmony on all fronts, home and war, among "members of the white, yellow and black races." The other panellists that day were anthropologists. In addition to Harry Shapiro, curator of physical anthropology at the American Museum of Natural History, the Cranbrook Institute also invited University of Michigan anthropologist Mischa Titiev, an expert on Hopi Indians, to speak at the opening.[130] Like Benedict, Weltfish, and Shapiro, Titiev stressed the "common fallacies" about race that anthropology and the exhibition were working to "explode." As if he were offering a rejoinder to Ernest Sevier Cox's paranoid review, Titiev rejected each of the claims Cox would later assert in print. In addition to the misconception that some races were superior to others, Titiev also rejected the idea that some races were more capable of high culture than others, that Jews were "marked by clannishness and fail to mingle with people around them," and that it was wrong to intermarry and have children. The themes of enlightenment and interracial harmony articulated at the exhibition opening were also illustrated in a photograph distributed by the Cranbrook Institute to publicize *The Races of Mankind*. The photo depicted local Boy Scout Billy Vigelius gazing at the opening display of cheerful black, white, and Asian schoolchildren in front of Albrecht Durer's Adam and Eve (see Figure 6.2) "[D]reaming of the common parentage," Billy "gains a new conception of brotherly feeling for children of other colors."[131]

The lessons Billy Vigelius was supposed to have learned from the Cranbrook Institute exhibition – people of many "colors" can live harmoniously – shows how the interracial accord promoted by *The Races of Mankind*, *The Brotherhood of Man*, and *In Henry's Backyard* reinforced a visual racial logic, even while emphasizing cultural diversity. In all three popularizations, racial types were presented coincident with cultural cues that suggested the persistence of a "common sense" notion that both race and culture were plainly evident facts of human variation. Both Benedict and Weltfish's pamphlet and the Cranbrook *Races of Mankind*

[130] "Mischa Titiev, 1901–1978," *American Anthropologist*, vol. 81, no. 2, Jun. 1979, pp. 342–344.
[131] "The Inter-cultural Exhibition at the Cranbrook Institute of Science Draws Large Number of Interested Visitors," *The Birmingham Eccentric*, Birmingham, MI, vol. 66, no. 45, Feb. 3, 1944, pt. 2, p. 4, Scrapbook, 1943–46, Cranbrook Foundation, p. 34, Cranbrook Archives; "Titiev Says Anthropology Explodes Theory of Race Superiority," *Jackson City Patriot*, Jan. 22, 1944, Scrapbook, 1943–46, Cranbrook Foundation, p. 33, Cranbrook Archives.

exhibition explicitly stated as a matter of established scientific fact that
there were three broad racial groups. At the Cranbrook Institute, "The
Physical Characters of Human Races" display proclaimed "There Are
Three Major Races," illustrated with pictures of skulls, noses, eyes, lips,
and beards arranged in columns for "Black," "White," and "Yellow"
races. This was not lost on readers. Alice Nirenberg's New York public school musical, "Meet Your Relatives," included the lyrics: "Any
one can notice/The color of a race/It's easily detected/By looking at a
face."[132] To later readers and viewers, Ad Reinhardt's cartoon figures and
the Cranbrook displays seem painfully stereotypical, often primitivizing
depictions of the world's peoples, presented from a Western, white perspective. Christopher Lehman has argued that *The Brotherhood of Man*
"participated in systemic racism by using racial stereotypes to illustrate
cultural differences," pointing to the "Eskimo" igloo and African hut
in Henry's backyard, next to his typical American frame house.[133] Zoë
Burkholder also found the use of blatant visual racial typing by skin color
at odds with the authors' purported intent to promote racial and cultural
tolerance and their insistence that all people were inherently the same.
Burkholder argues that *The Races of Mankind* was a "product of its
time," in which Benedict and Weltfish "unintentionally reified elements
of racialist thinking that they viewed as harmless."[134] Both Lehman and
Burkholder are essentially right. Neither Benedict nor Weltfish mounted
any kind of sustained critique of institutionalized racism, although they
approach it in their indictment of educational and other social conditions
that inhibited black students. And from the perspective of modern critics
who view the race concept itself as racist essentializing, the depiction of
racial types must inevitably be deemed inappropriate. It is also accurate
to say that Benedict and Weltfish thought the racial differences they delineated – the inherited physical variations by which science differentiated
among human beings – were not "harmless," but irrelevant to the problem of prejudice. On the other hand, Benedict and Weltfish's reification
of race was anything but unintentional, and their association of race with
culture was deliberate, accurately reflecting not only how they understood
those concepts but also their strategy for communicating their ideas and
undermining ingrained race prejudice.

---

[132] Burkholder, "'With Science as His Shield,'" p. 386.
[133] Lehman, *The Colored Cartoon*, p. 105.
[134] Burkholder, "'With Science as His Shield,'" p. 121; Burkholder, *Color in the Classroom*,
     pp. 76, 77.

Benedict and Weltfish were working in an era when most people assumed not only that race was an essential and immutable human characteristic, but that it was probably at the root of many differences among individuals and societies. Benedict and Weltfish distinguished themselves from these views, not in utterly rejecting the reality of race, but in granting it so little purchase. They counted as racial traits a very limited number, granted even fewer any effect on human behavior or culture, and rejected entirely the inheritance of the very broad range of beliefs, practices, abilities, and inventions collected under the umbrella of culture. Nonetheless, the proximity of race and culture in their work defies modern readers' expectations of antiracist literature and cultural relativists. Benedict did not write about race in ways that feel familiar. For example, Benedict herself expressed concern about the visual quality of racial distinctions. But to her, the "'visibility'" of a characteristic such as skin color was problematic not because of differences themselves – in other words, not because of the way race itself was constructed – but because she viewed racial differences as heritable, and therefore transmitted across generations. People could escape discrimination based on cultural attributes such as religion or language by changing their behavior, but a victim of racial prejudice could not change. "[T]oo dark a Negro cannot 'pass' and not even his children's children may be born light enough to do so," she wrote. Benedict argued that this was "a problem of relative permanence of distinctions, not specifically of 'visibility.'"[135] From her perspective, the problem was "*conflict, not race*" (italics original). This dichotomy was most graphically illustrated in *The Brotherhood of Man* and *In Henry's Backyard*, when the "Green Devils" of fear and prejudice in each man induce them to attack each other (Figure 6.9). Once they have been enlightened by the omniscient narrator about the profound similarities among all people in mental abilities, blood, life goals, and cultural achievement, none of them due to race, they realize "we're not *born* haters" and resolve to work together, "no matter where they were born or what the color of their skin," for a "better world"[136] (italics original) (Figure 6.10).

By making race visible to readers, first through exhaustive exposition in *Race: Science and Politics* and then with illustrations in the briefer distillations, Benedict first acknowledged race on her terms and then undermined its importance by crafting cross-cultural equivalencies. The racial

---

[135] Benedict, *Race: Science and Politics*, pp. 149–150.
[136] Benedict and Weltfish, *In Henry's Backyard*, n.p.

FIGURE 6.9. Henry's ugly Green Devil inhibits his neighborly humanist impulse. "It had...slithered...out...of him. And it whispered, 'Don't speak to these people, Henry! You won't like them. They're DIFFERENT!'" (ellipses and capital letters in the original). *Source:* Ruth Benedict and Gene Weltfish, *In Henry's Backyard. The Races of Mankind* (New York: Henry Schuman, Inc., 1948).

stereotypes that Lehman lamented were conflated with equally one-dimensional cultural stereotypes, with their ethnic costumes and distinctive architecture. By showing a racially and culturally diverse set of characters easily legible to a broad public who shared with Henry, the middle-class white protagonist, a life constructed around the same basic concerns – home, family, religion, education, work, health, and the stages of life – Benedict and Weltfish evacuated the "other" of its exoticism and replaced racial determinism with cultural relativity. Their approach prefigured possibly the most famous humanist exhibition of the postwar era, the 1955 *Family of Man* photographic exhibition at the Metropolitan Museum of Art. Like Benedict and Weltfish, *Family of Man* portrayed

FIGURE 6.10. Henry speaking to his neighbors from his stoop. *Source:* Ruth Benedict and Gene Weltfish, *In Henry's Backyard. The Races of Mankind* (New York: Henry Schuman, Inc., 1948).

people from around the world arranged to highlight the common stages of life – birth, growth, marriage, parenthood, family, work, death – and common human experiences – joy, sadness, hunger, faith, and politics.[137]

---

[137] Edward Steichen, *The Family of Man; The Greatest Photographic Exhibition of All Time* (New York: Museum of Modern Art by the Maco Magazine Corp., 1955), a catalogue published after the exhibition opened.

Cultural constructs such as *The Races of Mankind* (in all its incarnations), *In Henry's Backyard* and *The Brotherhood of Man*, and the *Family of Man*, created over a span of more than twenty years, pose a sociological, intellectual, and historical conundrum. They seem to have accomplished their humanist, relativist intentions with a large public, while still reifying racialized and exoticized "others." They seem to accomplish both, simultaneously.

In *Race: Science and Politics*, Ruth Benedict stressed understanding and tolerance for both racial and cultural diversity because nearly all observed differences were simply the result of human beings, who all shared a fundamental psychic unity, approaching the same problems of living with varied solutions. *The Races of Mankind* translated that message more pointedly into the idea of "brotherhood." It offered the same message, but stressed even more the smallness of the world, the increasing proximity of others, the need to get along, and the idea that all human beings were brothers who flourished when differences were accepted and opportunity and prosperity were shared. None was better or worse than another and all shared a common fate. Although diversity was appreciated, even celebrated in this period, race was also accepted. This perspective became pervasive among postwar humanists, although the matter-of-fact acceptance of race diminished by the mid-1960s, as ethnicity and cultural diversity took its place in humanist rhetoric. Like *Race: Science and Politics*, *The Races of Mankind*, *The Brotherhood of Man*, and *In Henry's Backyard* did not reject race but rather repudiated divisive, normative distinctions based on evident physical and cultural differences. *The Races of Mankind* and the other short, illustrated texts reified race more than *Race: Science and Politics* did. The brief, less-nuanced, heavily illustrated approach lost Benedict's point that most physical differences were a matter of small local variations, and that race was not fixed but changed with changing conditions, human and environmental. In *The Races of Mankind*, *The Brotherhood of Man*, and *In Henry's Backyard*, human variation was no longer a threat, but it still seemed immutable.

Conclusion

Ruth Benedict's efforts to undermine racism employed a strategy that did not involve repudiating race itself or the science of race. Rather, she dismissed race, rightly understood as merely heritable physical variations, as *irrelevant* to the proper appreciation of human cultural diversity. Although Benedict, as a Boasian anthropologist, viewed bodies and social

formations as two entwined elements in the broad picture of human existence, in combating racism for public consumption she was careful to
stress that they were also analytically and, more importantly, ontologically and causally distinct realms of life. Benedict's strategy had several virtues from the perspective of mid-century anthropologists. From
the perspective of geneticists and physical anthropologists, it helped preserve the study of human variation as a legitimate scientific topic, distinct
from various fallacious practices, especially Nazi racial theory. By carefully limiting the reach of "race" to hereditary somatic differences, and
divorcing those from any causal relationship to historical and cultural
change, Benedict also helped shift the grounds of the debate over racism
to the cultural realm of behavior and belief and away from heredity and
its conflation with cultural attainments. Her arguments also successfully
promoted a notion of brotherhood rooted in a culturally relativist tolerance for cultural diversity. But from a longer perspective, we can also see
that Benedict's argument and its popularization for a broad public reified
the race concept for another generation of Americans. Furthermore, the
ontological and causal barrier between race and culture that she and others worked so hard to erect was much harder to maintain than Benedict's
arguments suggested. The intermingling of race and culture remained a
problem for anthropologists and other students of human variety well
into the postwar period. In Chapter 7, I explore other leading voices on
the problem of race and racism in postwar America, and the persistence
of race into the 1960s in anthropology and natural history museums.

7

Alternatives to Race? Ethnicity, Genetics, Biology

Introduction

In 1942 in his popular book *Man's Most Dangerous Myth*, anthropologist M. F. Ashley Montagu famously argued that race was nothing more than a historically contingent, socially constructed fallacy. It was "nothing but a whited sepulchre... utterly erroneous and meaningless."[1] This chapter continues to examine the degree to which anthropologists and other scientists repudiated racism and the concept of race during and after World War II. I examine a range of works that illustrate the way anthropologists struggled to rethink race in this period. I analyze collective statements issued by disciplinary associations and organizations, most famously the UNESCO Statements on Race, whose pronouncements on the concept of race and racism reflected divisions between biological and social scientists. Ashley Montagu's struggle with the race concept, as well as Harry Shapiro's efforts to redefine race and racial science in the postwar period – including his popular text *Race Mixture*, published by UNESCO, and his 1961 exhibition for the American Museum of Natural History, *The Biology of Man* – exemplify the persistent intermingling of cultural and racial explanations for human diversity. In contrast to the impression created by Montagu's widely cited denunciation of the race concept, scientists in the decades following WWII were not united in the rejection of race. Deeply ingrained racialized ways of comprehending human diversity persisted even as disciplines such as physical anthropology,

[1] M. F. Ashley Montagu, *Man's Most Dangerous Myth*, with a foreword by Aldous Huxley (New York: Columbia University Press, 1942), p. 28.

once defined by its construction of race, shifted increasingly toward framing humanity in terms of evolution and biology.

Race, Anthropology, and War

In the midst of WWII, Harry Shapiro traveled to Bloomfield Hills, Michigan, to give the opening address for a new exhibit on the races of man at the Cranbrook Institute of Science.[2] The title of his talk, "Anthropology's Contribution to Interracial Understanding," signaled his discipline's aspirations more than its history in 1944. Indeed, Shapiro noted with bitter irony that physical anthropology had been criticized for years for its lack of practical use, castigated for being immersed in the pointless pursuit of abstractions "remote from everyday living." During a war in which his field's "academic concepts" had been used to justify and provoke conflict and mass murder, Shapiro took the opportunity of his address to expound on what he viewed as the correct understanding of race. "How practical it is then," he concluded ruefully, "to keep these concepts free from distortion and to expose the fallacies which they engender!"

Shapiro opened his address with cautionary tales about pitfalls of "progress" and the dangers of "practical" science run amok. "We have grasped eagerly at the fruits of science regardless of their price. Now we are discovering that they have a price," he warned. New technologies may have benefits, but they profoundly "alter the organization of our societies" and "may also serve evil purposes." Given current violent "racial" antagonisms, Shapiro argued that the history and character of American society took on new and urgent dimensions. Unlike Europe, with its waves of invasion in the remote past, the United States had been formed by a combination of successive waves of immigration, conquest of indigenous peoples, and forced introduction of Africans as slaves, all in a very short period of time. Like Ruth Benedict, Shapiro was concerned about the ability of groups within society to exaggerate their differences, foster prejudice, and engender conflict. Immigrants clung to their "group identities and group traditions," Shapiro said, and soon "there grew up a system of hierarchies, local and national, which excludes whole sections of the population and erects barriers to assimilation and participation." The coming together in one nation of people from different national

---

[2] Harry L. Shapiro, "Anthropology's Contribution to Interracial Understanding," *Science*, New Series, vol. 99, no. 2576 (May 12, 1944), pp. 373–376. See Chapter 6 in this book for a discussion of the Cranbrook Institute *Races of Mankind* exhibition.

cultures and religious faiths, as well as masses of "Negroes and other non-European peoples" who "bear with them the mark of their difference which neither cultural nor religious assimilation can efface" meant that the United States needed to muster "tolerance and understanding" if it was to avoid very "serious dangers"[3] to its society.

Shapiro abjured "evangelization" and emotional appeals as effective methods to build the tolerance, "mutual understanding and respect" required for "inter-racial harmony." The "unregenerate" would be "converted" only through education and "the use of reason," he argued. Shapiro implied that anthropologists themselves offered the world an example of the tolerant perspective that might be gained through the education of experience. Anthropology was a discipline that saw "mankind as a whole" and could scrutinize American society "with some degree of objectivity," he argued. Armed with anthropological insights, Americans might "escape the micro-culture of the specific group with which they are identified and achieve a larger perspective" on the nature of world cultures and human variation. Although Shapiro retained vestiges of social evolutionary theory in his suggestion that humans were universally "struggl[ing] towards civilization," he also argued that no culture developed in isolation, but instead borrowed from many other traditions in creating its unique way of life. Furthermore, echoing Benedict's comparative approach, Shapiro argued that "though we must admit the superiority of Western civilization in technology and science, anthropology is decisive in disclaiming any equivalent supremacy in the social organization of the nations of the Western world. Indeed, it would be easy to enumerate examples among non-European people with more complicated social systems or with more efficient ones."[4] Like Ruth Benedict, Shapiro stressed a clear distinction between culture and race. Echoing the principal messages in the Cranbrook Institute exhibition, Shapiro took traditional elements of physical anthropology and delimited or redefined them in ways that emphasized their independence of culture. Bodies, Shapiro pointedly noted, were studied by anthropologists and other scientists as a "biological phenomenon like any other organism." Racial classifications, based strictly on physical criteria, were fluid and, to some degree, conventional. Humans had migrated all over the world, "intermingling" from the beginning, with the result that "'pure' races, if they ever existed, are no longer to be found in nature." According to Shapiro, widespread intermixture had created populations with overlapping physical

[3] Ibid.
[4] Ibid., p. 375.

characteristics, making it practically impossible to divide human variation into anything but more or less arbitrary types. Moreover, he pointed out, as had Benedict, no nation was composed of only one race "or breed," but rather was constituted from various mixed "strains."

But even as he took pains to distinguish social relations from bodily formations and exclude mind and behavior from the purview of racial science, he retained the most foundational racial classification. People at "geographic extremes," Shapiro noted – the "northwest European, the Chinese, and the Negro of Central Africa" – did show "pronounced differences in physical criteria." Just as Ruth Benedict and Gene Weltfish reified broad racialist categories while simultaneously decrying invidious racialized hierarchies, so, too, Harry Shapiro reinscribed the key racial categories he had spent the bulk of his address undermining.

Shapiro concluded his address with his most critical anthropological insight. The "doctrine that a racial hierarchy exists based upon physical and psychological superiorities" he argued, was "one of the most pernicious breeders of ill-will among the various races of mankind." He observed that those who created the hierarchies reserved the most exalted position for themselves, arguing that the same "smugness" that characterized the "city slicker's airs of superiority" animated the supposed distinctions between the Nordic and Mediterranean, or "the white races from the colored." Like Benedict, Shapiro laid much of the blame for pernicious racial hierarchies at the feet of ideologues such as Arthur Count de Gobineau and Madison Grant who had "encrust[ed]" the "zoological concept" of race "with psychological attributes and assignments of value." Unlike most of his peers, Shapiro also displayed a remarkable sensitivity to the complexities of racial constructions and the way even those who contest them are implicated in their creation and perpetuation. Racialized normative hierarchies were a "monstrous doctrine," Shapiro said, that had been "elevated to a credo" and used to "inflame and manipulate masses of men . . . insidiously . . . calculated to make even those who attack it disseminate its seeds."[5]

---

[5] Julian Huxley made a similar observation in *We Europeans: A Survey of "Racial" Problems* (New York and London: Harper, 1936). More recently, Barbara Fields, Ann Laura Stoler, and Virginia Dominguez have addressed this idea. Barbara Fields, "Ideology and Race in American History," in *Region, Race, and Reconstruction: Essays in Honor of C. Vann Woodward*, J. Morgan Kousser and James M. McPherson, eds. (New York: Oxford University Press, 1982), pp. 143–177; Ann Laura Stoler, "Sexual Affronts and Racial Frontiers: European Identities and the Cultural Politics of Exclusion in Colonial Southeast Asia," *Comparative Studies in Society and History*, vol. 34, no. 2 (1992), pp. 514–551; Virginia Dominguez, "A Taste for 'the Other': Intellectual Complicity in Racializing Practices," *Current Anthropology*, vol. 35, no. 4 (1994), pp. 333–348.

Neither culture nor race was a stable concept in 1944. The relativism of the anthropological culture concept was still novel to the broad American public, and even ethnologists themselves slipped in and out of older vocabulary and concepts. In his address, Shapiro had abandoned the terminology of primitivism, but vestiges of such thinking were apparent in his discussion of past barbarous ages and the movement of people toward "civilization." Race was under even greater pressure, as an anthropological and biological concept, and as a subject of "objective" science. By 1944, under the pressures of world events and their close ties to racial science, Shapiro was already shifting toward an explicitly and purportedly exclusive emphasis on biology (and, in some respects, back to the former "zoological" basis of physical anthropology) that would increasingly characterize his subfield.

Scientists Unite

The concern Harry Shapiro voiced about "racial misconceptions fostered for evil purposes" at the Cranbrook Institute opening of *The Races of Mankind* in 1944 was shared by other scientists in the early years of the Second World War. The rise of Adolph Hitler and Nazi racial hygiene programs provoked similar alarm and action, if not a uniform or consistent response among a range of scientists in the United States from the late 1930s through the 1960s. Like Ruth Benedict in 1939, many scientists felt compelled to register their concerns about and opposition to events in Europe.

American Anthropological Association

Anthropologists were among the first to pass a resolution condemning Nazi racial doctrine. In December 1938, at the annual meeting of the American Anthropological Association in New York, they formally weighed in on questions of freedom and equality that had become increasingly pressing as events in Europe unfolded. They felt compelled to counter German National Socialist racist dogma with a stance that distanced them from both Nazi racial ideology and its increasingly violent, undemocratic application. Anthropologists were also concerned to position their work as genuine science, countering Nazi distortions and misinterpretations with a "statement of facts" about race and their role as scientists dedicated to an "unbiased search for truth." Anthropologists attacked Nazi racism not only as repugnant but also as fundamentally unscientific. Scientists seemed concerned not only with what Nazis were doing to their innocent victims, but that they were doing it under the banner of government-promoted science. At the same national meeting where anthropologists first passed a resolution repudiating Nazi racial science, they also elected members to represent their discipline on a raft of cross-disciplinary agencies, funded by the likes of the Rockefeller and Carnegie Foundations, whose support had become critical to anthropology: the Social Science Research Council, American Council of Learned Societies, National Research Council, and American Association for the Advancement of Science.[6] American social science was itself becoming more and more of a government-supported enterprise, and it was clear by the late 1930s that this trend away from private individual philanthropy would continue.[7] Since the 1910s, anthropologists, psychologists, sociologists, and others who appealed to organizations such as the National Research Council for funding found that they increasingly had to frame their research proposals in ways that addressed broad social issues or offered a possibility of contributing to social or government policy (as well as fitting within the particular preferred research orientations of their disciplines). The escalation of international conflict in the 1930s and the vigorous misapplication of racial science by the Nazis to undermine equality and freedom provoked scientists to look for both ways to assist beleaguered populations and nations (and later, help in the U.S. war effort) and protect their vision of science and its relationship to government and democracy.

---

[6] "The New York Meeting of the American Anthropological Association," *Science*, New Series, vol. 89, no. 2298 (Jan. 13, 1939), pp. 29–30.

[7] George W. Stocking, Jr., "Ideas and Institutions in American Anthropology," *The Ethnographer's Magic and Other Essays in the History of Anthropology* (Madison: University of Wisconsin Press, 1992), pp. 130–131; Regna Darnell, *Invisible Genealogies: A History of Americanist Anthropology*, Critical Studies in the History of Anthropology (Lincoln and London: University of Nebraska Press, 2001); *Dorothy Ross, The Origins of American Social Science* (Cambridge: Cambridge University Press, 1991); Dorothy Ross and Theodore M. Porter, *The Modern Social Sciences*, vol. 7, The Cambridge History of Science (Cambridge: Cambridge University Press, 2003). The changing institutional and funding context of the social sciences in the twentieth century is also addressed in a variety of recent work on various social science fields, including Elazar Barkan, *The Retreat of Scientific Racism: Changing Concepts of Race in Britain and the United States between the World Wars* (Cambridge: Cambridge University Press, 1992); John P. Jackson, Jr., *Social Scientists for Social Justice: Making the Case against Segregation* (New York: New York University Press, 2001); William H. Tucker, *The Funding of Scientific Racism: Wickliffe Draper and the Pioneer Fund* (Urbana: University of Illinois Press, 2002); John P. Jackson, Jr. and Nadine Weidman, *Race, Racism and Science: Social Impact and Interaction* (Piscataway, NJ: Rutgers University Press, 2005); Nadine Weidman, *Constructing Scientific Psychology: Karl Lashley's Mind-Brain Debates* (Cambridge: Cambridge University Press, 2004).

The Anthropological Association resolution itself, although brief, high-lighted a set of concerns that would occupy scientists' discussions of race and culture throughout WWII and the postwar period. The resolution stated:

Whereas, The prime requisites of science are the honest and unbiased search for truth and the freedom to proclaim such truth when discovered and known; and

Whereas, Anthropology in many countries is being conscripted and its data distorted and misinterpreted to serve the cause of an unscientific racialism rather than the cause of truth;

Be it resolved, That the American Anthropological Association repudiates such racialism and adheres to the following statements of facts:

> (1) Race involves the inheritance of similar physical variations by large groups of mankind, but its psychological and cultural connotations, if they exist, have not been ascertained by science.
> (2) The terms "Aryan" and "Semitic" have no racial significance whatsoever. They simply denote linguistic families.
> (3) Anthropology provides no scientific basis for discrimination against any people on the ground of racial inferiority, religious affiliation, or linguistic heritage.[8]

Not only did the resolution not repudiate the idea of race, it hedged on what, if any, "psychological or cultural connotations" might exist. Like the later UNESCO "Statement on Race" by physical anthropologists and geneticists, discussed later in this chapter, anthropologists in 1938 equivocated on the existence of mental, behavioral, or social manifestations of race by arguing that these were questions science had not yet resolved. As Franz Boas had been striving to do for decades, anthropologists made a distinction between the physical variations among human beings and the consequences, if any, of those differences for individuals and groups. Thus, the anthropologists noted that race was the "inheritance of similar physical variations by large groups of mankind," while "Aryan" and "Semitic" were labels for language families that bore no relationship to physical differences. What anthropologists repudiated was *"unscientific racialism"* (italics mine), including co-opting language terms, conflating race and religion, and asserting the inferiority of whole groups on the basis of hereditary physical differences. Positioning themselves as "honest and unbiased" defenders of truth, they rejected politically, socially,

---

[8] "The New York Meeting of the American Anthropological Association," p. 30.

and religiously motivated discrimination promoted under the authority of racial science, but attempted to preserve what they viewed as a genuinely scientific body of knowledge about race.

Psychologists

The Society for the Psychological Study of Social Issues, an organization representing 400 professional psychologists, also crafted a statement condemning fascist racial doctrines, issuing a formal proclamation at their national meeting in December 1938.[9] Like the anthropologists, psychologists directly challenged racially divisive hierarchies and hereditarian assertions linking physical differences with mental capacities. While the anthropologists declined to name Germany or the Nazis directly, the psychologists not only cited Germany as well as Italy, they framed their arguments in response to particular Nazi claims. In naming no particular racialisms, the anthropologists potentially cast their net beyond Germany and Italy; the psychologists were again more direct, asserting at the outset that they feared not only divisive racial rhetoric in fascist Europe, but also its increase in the United States and elsewhere.

Marshalling a set of authoritative researchers and their texts (Frank S. Freeman's *Individual Differences*, Anne Anastasi's *Differential Psychology*, Otto Klineberg's *Race Differences*, T. R. Garth's *Race Psychology*),[10] the psychologists argued that Nazi racial theories were scientifically groundless. The overall tenor of the statement was skeptical about the ability of scientists to distinguish "so-called 'races'" on the basis of their mental traits, arguing that no experimental evidence suggested any "characteristic, inherent psychological differences." They asserted that there was "absolutely no" scientific support for the notion "that people must be related by blood in order to participate in the same cultural or intellectual heritage," and "no conclusive evidence" for "racial or national differences in native intelligence and inherited personality

---

[9] Ruth Benedict, *Race: Science and Politics* (New York: Viking Press, 1959), pp. 196–198. Benedict published excerpts of a statement prepared for the Executive Council of the Society for the Psychological Study of Social Issues. *Race: Science and Politics* was originally published in 1940, revised in 1943, and issued in a new edition in 1945, that also included *The Races of Mankind*.

[10] Frank S. Freeman, *Individual Differences: The Nature and Causes of Variations in Intelligence and Special Abilities* (New York: H. Holt and Company, 1934); Anne Anastasi, *Differential Psychology: Individual and Group Differences in Behavior* (New York: The Macmillan Company, 1937); Otto Klineberg, *Race Differences* (New York: Harper, 1935); T. R. Garth, *Race Psychology: A Study of Racial Mental Differences* (New York: Whittlesey House, McGraw-Hill Book Company, Inc. 1931).

characteristics." Like anthropologists, the psychologists distinguished their scientific practice and theorizing from the Nazis, whose race theories, they argued, "have been developed not on the basis of objective fact, but under the domination of powerful emotional attitudes." Although the psychologists didn't categorically rule out the possibility that evidence of innate mental differences between racial groups might emerge, they noted that "the many attempts to establish such differences have so far met with failure." Moreover, they argued, "Racial and national attitudes are psychologically complex, and cannot be understood except in terms of their economic, political and historical backgrounds." Repudiating Nazi arguments about racial heritage and social fitness, the psychologists argued that Nazis were laboring under "a well-known psychological tendency [that] leads people to blame others for their own misfortunes." Under the pressures of economic and political "disabilities," Germans scapegoated Jews and then rationalized their aggression and depredations as a response to the Jews' own "plots and machinations."

Although psychologists' persistent experimental attempts to locate innate or racially inherent mental characteristics or "tendencies" suggest they suspected some link among heredity, mind, and behavior, their 1939 statement also pointedly argued that those tendencies and characteristics were manifest only within complex social and cultural environments. Indeed, the psychologists ended their statement by contrasting Nazi Germany, where race, nation, and culture had so grievously broken down, with the United States, where "members of different racial and national groups have combined to create a common culture."[11]

International Congress of Geneticists

On the eve of Hitler's invasion of Poland in August 1939, biologists gathered in Edinburgh, Scotland, for the Seventh International Genetics Congress. There, a group of distinguished geneticists circulated their own "manifesto" in response to *Science Service* editor Watson Davis's query, "How could the world's population improve most effectively genetically?" The geneticists, including J. B. S. Haldane, Lancelot Hogben, Julian Huxley, Hermann J. Muller, and F. A. E. Crew, regarded human genetics – and human improvement – as a problem that could only profitably be approached by recognizing that human potential and success was a matter of both biology and culture. "The ultimate genetic improvement

---

[11] Benedict, *Race: Science and Politics*.

of man," they argued, would remain elusive without "major changes in social conditions and correlative changes in human attitudes."[12] Among these major changes, according to the geneticists, were the provision of equal opportunities regardless of social rank or status; the elimination of economic exploitation and political conditions that led to "antagonism," including the "unscientific doctrine that good or bad genes are the monopoly of particular peoples or of persons with features of a given kind"; and the guarantee of "economic security," including medical and educational services, supportive employment conditions, adequate housing, and municipal services to aid families in raising children. According to the geneticists, steps taken toward realizing this ambitious program of "economic reconstruction" were necessary not only to facilitate the genetic improvement of humans "to a degree seldom dreamed of," but also to master "those more immediate evils which are so threatening our modern civilization." Although their statement decisively rejected biological determinism and stratification and expressed skepticism about the nature of race (the single use of the term was placed in quotes), the geneticists still retained the conviction that under the right conditions the genetic improvement of humanity was possible. Their statement suggested a vision of human biology and culture reminiscent of anthropologists such as Franz Boas and Harry Shapiro who saw human beings as products of both heredity and environment.

Ashley Montagu: Rejecting Race?

In the preface to the 1964 edition of Ashley Montagu's famous text, *Man's Most Dangerous Myth: The Fallacy of Race*, he asserts that his book is indeed "designed to expose the most dangerous myth of our age,

---

[12] H. Gruenberg, "Men and Mice at Edinburgh: Reports from the Genetics Congress," *The Journal of Heredity*, vol. 30, no. 9 (Sep. 1939), pp. 371–374. Excerpts of the biologists' statement from the Seventh International Genetics Congress published in Benedict, *Race: Science and Politics*, pp. 198–199.

The 1939 meeting in Edinburgh occurred after two years' postponement of the 1937 Congress, originally scheduled to be held in Moscow. Political controversy over genetics led the Soviet Union to cancel – some said ban, others merely postpone – the international Congress, which was ultimately held in Scotland without the participation of Soviet geneticists, as tensions mounted in Europe. During the controversy, Western scientists received reports of Soviet geneticists being threatened and punished as "enemies of the people." For an account of this episode, see Valery N. Soyfer, "Tragic History of the VII International Congress of Genetics," *Genetics*, vol. 165 (Sep. 2003), pp. 1–9.

the myth of 'race.'"[13] *Man's Most Dangerous Myth* is among the most prominent antiracist and anti-race treatises in American literature, often cited as an early, "classic" example of scientists' rejection of racial science and the biological validity of race.[14] Like Ruth Benedict, Montagu offered an extensive critique of the ways race had been made a social and political tool to divide and oppress, an overt critique of the "feeble and unjustified proceedings" in the work of contemporaries such as Carleton Coon.[15] However, upon closer inspection, Montagu's position looks a good deal more complicated than a blanket rejection of the race concept and racial science. A close look at his earliest edition, in 1942, reveals that even Ashley Montagu had difficulty fully abandoning traditional racial categories and ways of thinking about human difference, something some of his early reviewers noted. By the fourth edition of his popular text, in 1964, Montagu had substantially altered the way he talked about human variation. But even at that late date, Montagu had not, in fact, utterly rejected a science of human variation. Like Boas, Benedict, and Shapiro, Montagu was attempting to erect a barrier between race as he understood it as a physical anthropologist and race as a folk concept that had created untold suffering. An examination of Montagu's evolving formulation of the race concept and his rejection of it helps further illuminate mid-century debates about the science of race, the race concept, and the significance of human variation.

M. F. Ashley Montagu (later, simply Ashley Montagu) was born Israel Ehrenberg to working-class Jewish parents in the East End of London in 1905. At the age of fifteen, given a skull unearthed by a local workman from a site on the Thames River, Montagu took it in a paper sack to England's premier evolutionist and physical anthropologist, Arthur Keith, to learn what he could.[16] Thus began Montagu's career as a

[13] Ashley Montagu, *Man's Most Dangerous Myth: The Fallacy of Race*, fourth revised and enlarged edition (Cleveland and New York: Meridian Books, The World Publishing Company, {1942} 1965), p. 14.

[14] Barkan, *The Retreat of Scientific Racism*, p. 3.

[15] Ruth Benedict to Ashley Montagu, October 22, 1941, Folder 27.6, "Ashley-Montagu, Montague Francis. See Montagu, Ashley, 1934–1945," (formerly restricted) Correspondence with Students and Colleagues, Series I, Correspondence, Ruth Fulton Benedict Papers, Archives and Special Collections Library, Vassar College, Poughkeepsie, New York (hereafter RBP-VC). Benedict cites, in particular, Carleton Coon's book, *The Races of Europe* (New York: The Macmillan Company, 1939).

[16] Andrew P. Lyons, "The Neotenic Career of M. F. Ashley Montagu," Larry T. Reynolds and Leonard Lieberman, eds., *Race and Other Misadventures: Essays in Honor of Ashley Montagu in His Ninetieth Year* (New York: General Hall, Inc., 1996), pp. 3–22 (p. 3).

physical anthropologist, as well as one of several longstanding friendships and close professional relationships with men whose philosophies differed dramatically from his own.[17] Montagu studied anthropology at leading universities in Britain and the United States, doing his undergraduate work first at University College London, then at the London School of Economics, followed by graduate training under Franz Boas and Ruth Benedict at Columbia University in New York. This education acquainted him with many of the leading figures in anthropology, biometrics and statistics, and psychology, including Grafton Elliot Smith, Bronislaw Malinowski, Charles Seligman, Margaret Mead, Charles Spearman, and Karl Pearson.[18] Despite a solid pedigree, Montagu had difficulty securing a faculty position in any of the leading anthropology departments, spending much of his career at Hahnemann Medical College in Philadelphia, followed by a stint at Rutgers University. In 1955, he left the academy to pursue his career as a public intellectual devoted to combating racism. Montagu became one of the most renowned anthropologists in postwar America, despite spending most of his professional career on the disciplinary margins, in part because of his tireless fight against forms of biological determinism.[19]

In both the first and fourth editions of *Man's Most Dangerous Myth*, Montagu began, like Benedict, with a discussion of what race was, and what it was not. Montagu was forceful in stating at the outset that racial classifications had been complete failures because they relied on a misguided belief in fixed characters and made arbitrary distinctions.[20] He was equally explicit about the nature of race. In 1942, he stated, "In the biological sense there do, of course, exist races of mankind."[21] He explained that "mankind is comprised of many groups which are often

---

[17] These included Earnest Hooton in later years. Montagu also had an unfortunate tendency to lose friends and mentors as a consequence of his uncompromising stands. These included Grafton Elliot Smith and eventually Earnest Hooton, whose work on criminality Montagu harshly critiqued in print. One of his biographers, Susan Sperling, described him as "outspoken to the point of contentiousness." Susan Sperling, "Ashley Montagu (1905–1999)," *American Anthropologist*, New Series, vol. 102, no. 3 (Sep. 2000), pp. 583–588 (pp. 584, 585).

[18] Sperling, p. 584. Grafton Elliot Smith was, along with Arthur Keith, one of Britain's foremost experts on human prehistory. Spearman and Pearson were biometricians, also involved in eugenic movement in Britain; Spearman is the namesake for spearman's "g," the subsequent key to much later hereditarian work on IQ.

[19] Sperling notes that he was a regular on the *Tonight Show* with Johnny Carson in the 1960s. Sperling, pp. 583, 585, 586.

[20] Montagu, *Man's Most Dangerous Myth* (1942), p. 4.

[21] Ibid.

physically sufficiently distinguishable from one another to justify their being classified as separate races," although not all groups of people were easily classified. But, he wrote, "All the people of Western Europe belong to the same 'race,' the White race, and such differences as some of them may exhibit simply represent small local differences arising from either circumscribed inbreeding or crossbreeding with members of a different racial group."[22] Montagu recognized not only the "White race," but "five or six great 'races' or divisions," which he listed as "Mongolian, Caucasian, Negro, Australo-Melanesian, and Polynesian."[23] He argued, however, that racial scientists had been overly ambitious in their taxonomies, claiming the existence of various numbers of subraces, attempting to fit local populations into the "obvious" major divisions with little justification.[24] His critique of racial classifications was not based on a rejection of racial typology as a purely social convention, but rather the more pragmatic complaint that human beings were of such mixed racial origins that hard-and-fast distinctions among them were very difficult. He noted, for example, that many Russians "are obviously of Mongolian origin," and many Americans had "Negro" origins, whereas the majority of Americans and Russians "belong to the white division of mankind." These kinds of mixed origins made it impossible to say that such individuals belonged in one racial division or another; they had physical characteristics of both. "It is for this reason," Montagu argued, "that it is difficult to draw up more than a very few hard and fast distinctions between even the most extreme types."[25] As one of his reviewers, physical anthropologist Wilton Krogman, noted, Montagu "may deny race, but he hasn't legislated it out of his thinking."[26]

In *Man's Most Dangerous Myth*, Montagu did not go much beyond the major racial divisions in his popular exposition on the pitfalls of racial classification. We get a clearer picture of his thinking about race and taxonomy – and a window into the tensions in anthropological racial science in the 1940s – by looking at *An Introduction to Physical Anthropology*, an instructional guidebook for students and others interested

---

[22] Ibid., p. 5.
[23] Ibid., p. 7. In 1964, Montagu dropped this to four (dropping the Polynesian), although he argued that it was "arbitrarily possible" to divide human beings into "distinctive major groups," p. 28.
[24] Montagu, *Man's Most Dangerous Myth* (1942), p. 6.
[25] Ibid., p. 5.
[26] W. M. Krogman, Review, *Man's Most Dangerous Myth: The Fallacy of Race* by M. F. Ashley Montagu, *American Anthropologist*, New Series, vol. 45, no. 2 (Apr.–Jun., 1943), pp. 292–293.

in a more technical discussion of physical anthropological theory, data, and practice that Montagu published in 1945 after its contents were vetted by fifteen of his colleagues.[27] In a chapter on "The Divisions and Ethnic Groups of Man," Montagu set out to "present the systematist's view . . . before it finally bows gracefully out of existence to make way for the genetical analysis of mankind." Montagu was clearly uncomfortable with the extant racial classifications, noting that they were only "temporary," "arbitrarily distinguished groups" that were "merely labels for convenient abstractions which help us to appreciate broad facts." He had, he avowed, "no desire to perpetuate this type of classification" or "perpetrate another along similar lines."[28] He noted that classifiers had latched onto various external characteristics in an effort to "reduce the great variety presented by mankind to some sort of comprehensible system," but that the selection of characteristics was so arbitrary that one could just as logically recognize a "red-headed race" or a "deaf-mute race" as those erected upon the usual traits, such as skin color, head shape, and hair form.[29] Moreover, because they were predicated on fixed types, they actually obscured the dynamic reality of constant change that was better represented by frequencies of genes and traits among populations.[30]

And yet, despite all these criticisms and hesitations, Montagu then set out to present a "highly debatable," "probably nonsensical," "simple scheme, combining both the vices and virtues" of classifications based on the usual external characters.[31] He began by reiterating the three major divisions, and then got down to particulars. He proceeded for twenty pages thus:

Among the Negroid the skin is typically dark brown, but is often black. . . . The head hair varies from tightly curled to pepper-corn in form. . . . As specialized characters may be reckoned the long head, the various types of Negroid head hair, galbrousness (paucity of body hair), small ears, and thick, everted lips.[32]

The Nordic is nothing but a color-isolate of Mediterranean stock, its strongest affinities being with the Atlanto-Mediterranean sub-group of that stock. The hair may be either blond, yellow, very light brown, or reddish, and is slightly wavy to

---

[27] M. F. Ashley Montagu, *An Introduction to Physical Anthropology* (Springfield, IL: Charles C. Thomas, 1945); Ashley Montagu to Ruth Benedict, April 1, 1943, Folder 27.6, "Ashley-Montagu, Montague Francis, See Montagu, Ashley, 1934–1945," (formerly restricted) Correspondence with Students and Colleagues, Series I, Correspondence, RBP-VC.

[28] Montagu, *An Introduction to Physical Anthropology*, p. 156.

[29] Ibid., p. 157.

[30] Ibid., pp. 157–158.

[31] Ibid., p. 158

[32] Ibid., p. 159.

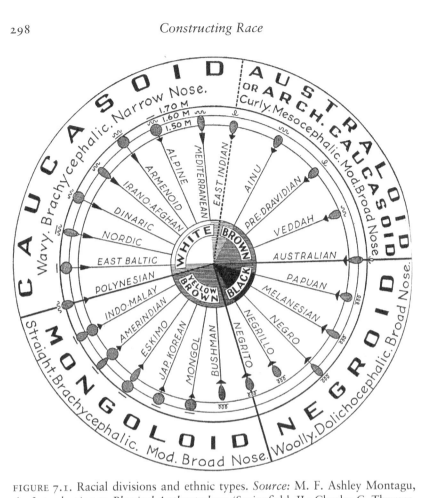

FIGURE 7.1. Racial divisions and ethnic types. *Source:* M. F. Ashley Montagu, *An Introduction to Physical Anthropology* (Springfield, IL: Charles C. Thomas, 1945).

curly in form ... skin is florid or pinkish white, eyes blue or gray; the head is long but of medium breadth in relation to its length (mesocephalic 76–79).[33]

The ideal type of the American Indian may be described as follows: Brachycephalic, large face, with large flaring cheekbones, a moderate degree of prognathism [forward projection of the upper jaw], prominent convex or aquiline nose with high bridge, depressed tip and flaring wings.... The Eskimo nose in general appears lower in the root and bridge, and approaches more closely to the ideal Mongoloid form in its soft parts than most other American groups.[34]

Montagu offered readers a diagram (Figure 7.1) that captured the division of humanity by stature, nose shape, hair form, head shape, and

33 Ibid., p. 177.
34 Ibid., p. 187.

skin color. Montagu's caption notes that his reductionist diagram was "simply intended to provide a useful approximation to facts which are at present not capable of accurate description."[35] Another figure offered a racial, ethno-national schematic of the various groups that had populated the United States (Figure 7.2). Like all Montagu's other discussions of racial/ethnic groups, and like Benedict and Shapiro, he hewed to the three-race triumvirate in parsing American society. Smaller divisions were depicted in the orbit of their primary "stock." In the concluding text, Montagu reasserted his argument that there were no "'pure races,'" saying the "divisions and ethnic groups of mankind are not, and never have been, so many separate streams and tributaries flowing within well-defined banks, but currents and eddies in one stream, in which there has been a constant interchange of what each has carried."[36] Yet his image suggested islands of bounded types, converging on the United States, not a stream of genetic mixing and recombination. As commentary, Montagu offered a surprisingly equivocal celebration of American diversity. "This diagram," he said,

may perhaps serve to suggest the probable truth that the United States stands out in high relief as the example of what, upon a lesser scale, has occurred throughout the history of man, the mingling of peoples to produce greater strength in almost every way, to illustrate the truth that diversity of genes as well as of culture is the basis of collective achievement.[37]

As Montagu himself seems to have been uncomfortably aware, his extensive descriptions of minutely defined racial characteristics ("The average height [of Armenoids] is 1.67 cm or 5 feet 6 inches") combined with diagrams reducing complex processes and highly variable groups to segments on a chart could only contribute to the reification of racial types.[38] For Benedict, this approach had been a strategy for distinguishing racial science from racist dogma and limiting the reach of "race." Montagu seems to have struggled more with the anthropological race concept itself. The constellation of traits upon which it was erected was at odds with a

---

[35] Ibid., p. 178.
[36] Ibid., p. 189.
[37] Ibid., p. 191.
[38] *An Introduction to Physical Anthropology*, p. 182. The figure Montagu cited as the average height of Nordics, 1.72 m, is very nearly the same figure given by Joseph Deniker for "the very lofty stature" of his "Northern Race," detailed in his widely cited elaboration of ten contemporary European racial types, *The Races of Mankind. An Introduction to Physical Anthropology*, p. 179; Joseph Deniker, *The Races of Mankind: An Outline of Anthropology and Ethnography* (London: Walter Scott Publishing Co., Ltd.; New York: Charles Scribner's Sons, 1913), p. 326.

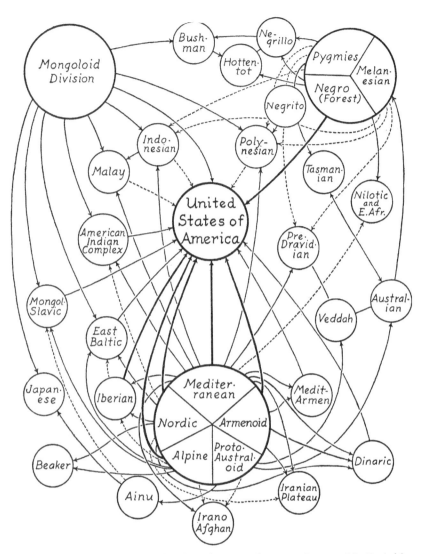

FIGURE 7.2. Varieties of mankind in the United States. *Source:* M. F. Ashley Montagu, *An Introduction to Physical Anthropology* (Springfield, IL: Charles C. Thomas, 1945).

genetic understanding of variation and change, yet he still had difficulty giving it up entirely.

By 1964, Montagu had adopted more of the language of population genetics, injected a stronger sense of racial categories as arbitrary groupings, and distanced himself further from the term "race," frequently replacing his earlier use of "race" with "population," "distinctive population," "varieties," or "ethnic groups."[39] But he nonetheless continued to assert the reality and importance of distinctive variations among groups of people. As he did in 1942, he stated that it was "easy to see that an African Negro and a white Englishman must have had a somewhat different biological history, and that their obvious physical differences would justify the biologist in classifying them as belonging to two different races." In both 1942 and 1964, Montagu explained that in biology, race is a subdivision of a species based on such marked differences in physical characteristics. In 1942, Montagu followed with the flat assertion that "in this sense there are many human 'races.'"[40] In 1964, he hedged a bit, telling his readers that "a number of human races have been arbitrarily recognized."[41] Montagu noted that not all groups differed sufficiently to qualify as "separate races," citing Germans and Englishmen as "distinctive populations" that were not distinct races. Rather, he claimed, all Europeans "belong to the same major group, the Caucasoid." He had replaced his 1942 language – "the same 'race,' the White race" – with a statement about a "major" racial group, "the Caucasoid."[42]

Moreover, he was quite explicit about just what sort of "race" was mythological. In new language introduced after 1964, Montagu explained, "The myth of 'race' refers not to the fact that physically distinguishable populations of man exist. Such populations are often called races. Distinctive populations of this kind are not myths, but neither are they races in the sense in which that term is usually employed." According to Montagu, "the myth of race" is "the belief that physical and mental traits are linked, [and] that the physical differences are associated with rather pronounced differences in mental capacities...and...cultural achievements."[43] Throughout his book Montagu argued, as Ruth Benedict had, that the myth was that distinctive physical differences could be any sort of guide for political or social "action," arguing instead

[39] Montagu, *Man's Most Dangerous Myth* (1965), pp. 25, 28, 29.
[40] Montagu, *Man's Most Dangerous Myth* (1942), p. 4.
[41] Montagu, *Man's Most Dangerous Myth* (1965), p. 25.
[42] Ibid., p. 25. See also, Matthew Frye Jacobson, *Whiteness of a Different Color: European Immigrants and the Alchemy of Race* (Cambridge, MA: Harvard University Press, 1999); Bruce Baum, *The Rise and Fall of the Caucasian Race: A Political History of Racial Identity* (New York: New York University Press, 2006).
[43] Ibid., p. 24.

that such variation was irrelevant to "the practical universe of human relations."[44]

But, like his teacher Ruth Benedict, Montagu believed such physical differences were worth studying. In the preface to his 1964 edition, Montagu warned readers not to misinterpret his critique. Although he was calling out racists for linking physical traits with mental abilities and alleging differences that did "not in fact exist," he was not arguing "that there are no genuine differences between the various groups of mankind." On the contrary, Montagu claimed, "The fact is that there are numerous differences between ethnic groups, and even regional segments of such groups, in many bodily traits. These differences are real enough, and they are of the greatest interest to the student of variation." Montagu noted that studies, such as his own on teeth, musculature, and skull structure, had shown how remarkably variable human beings were, how the human constitution was constantly changing. The meaning of such differences was "of the greatest significance to the student of the evolution of man."[45]

Montagu's substitution of the term "ethnic" or the phrase "ethnic group," for "race" was an attempt to abandon "race," as he had urged others to do. Montagu, like Benedict and other critics of racism and the misuse of racial science, was trying to distinguish between what he viewed as a legitimate definition and use of the race concept and one that was fallacious and dangerous. Obviously, using the term "race" for both notions was bound to perpetuate confusion. So Montagu chose the term "ethnic" as a substitute to discuss groups of people formerly described by the term "race." But, as Wilton Krogman pointed out in his review of the first edition, Montagu was inconsistent in his usage, sowing more confusion rather than clarity. Krogman noted, for example, that Montagu seemed to use the terms "race" or "racial" and "ethnic" interchangeably, sometimes using one, sometimes another, and at other points seemed to be making a distinction between them, as in a statement about "racial and ethnic psychology."[46]

After discussing the anthropological concept of race, Montagu devoted his third chapter to an exposition of genetics, and the formation and dissemination of hereditary physical differences. His principal point was that

---

[44] Ibid., p. 15.

[45] Montagu, *Man's Most Dangerous Myth* (1965), p. 14. The scientific study of human variation continues to provoke controversy. For a study of the Human Genome Diversity Project, see Jenny Reardon, *Race to the Finish: Identity and Governance in an Age of Genomics* (Princeton: Princeton University Press, 2004).

[46] Krogman, "Review," p. 61.

human characteristics were a product of continual mixing and mutations, a constant process of change that did not lend itself to tidy taxonomy.[47] His discussion of race and ethnicity exemplifies the confusion still present in this terminology. Montagu's definition of "ethnic group" is worth quoting at length to show just how fully his new terminology captured the ongoing intermingling of bodies and cultures in a way quite reminiscent of Franz Boas:

An ethnic group represents part of a species population in process of undergoing genetic differentiation; it is a group of individuals capable of hybridizing and inter-grading with other such ethnic groups to produce further genetic recombination and differentiation.... An ethnic group represents one of a number of populations, comprising the single species *Homo sapiens*, which individually maintain their differences, physical and cultural, by means of isolating mechanisms such as geographic and social barriers. These differences will vary as the power of the geographic and social barriers, acting upon the original genetic differences, vary. Where these barriers are of low power, neighboring groups will intergrade, or hybridize, with one another. Where these barriers are of high power, such ethnic groups will tend to remain distinct, or replace each other geographically or ecologically.[48]

Despite the injection of "ethnic group" where "race" once stood, and the application of a population genetics model to the process of human variation and differentiation, Montagu had not rejected all claims that human physical differences mattered. He certainly was not deploying "ethnic" in any kind of opposition to "race"; it was not a matter of culture versus heredity. Indeed, the collapse of meaning between race and ethnic group was even more pronounced in Montagu's *An Introduction to Physical Anthropology*. Writing about "the physical differences existing between the ethnic groups of men," Montagu asserted that "racial or ethnic characters" were only those "conditioned by heredity," and "insofar as nationality is determined by the accident of birth in a particular country, nationality has no connection with ethnic group."[49] Like Boas and Harry Shapiro, Montagu rejected biological determinism, and viewed

---

[47] Montagu, *Man's Most Dangerous Myth* (1965), pp. 37–43.

[48] Ibid., pp. 44–45.

[49] Montagu, *An Introduction to Physical Anthropology*, p. 118; see also pp. 151–152.
    In modern use, the Oxford English Dictionary notes at least three distinct definitions for the term "ethnic" since the mid-1940s: to denote religious affiliation; to denote national, linguistic, and/or "cultural" origins; and to denote "racial" origins. In the United States, it became common as a way to point to national or cultural origins rather than racial heritage, although this usage has never been exclusive, nor did it ever effectively fully exclude racial connotations.

physical and cultural variation as separate but interlinked processes that both had to be studied if the vast diversity of humanity was to be understood. The difficulty that Montagu and the others faced, however, was how to talk about the relationship of hereditary and cultural processes without supporting the fallacious links that fueled racist arguments about superiority and inferiority. Using the term "ethnic" to denote "distinct" groups was problematic, as it imported the very cultural connotations that elsewhere Montagu and others, such as Benedict, were trying so hard to exclude from any relation to physical variation.[50]

As Benedict had more than twenty years earlier, Montagu argued in 1964 that there was no need to deny human differences. Like Benedict, he argued that for decades the privileged had exaggerated human differences to justify their exploitation of the less fortunate, and to bolster the socially stratified status quo.[51] As did so many in the postwar era, Montagu believed that exposing racist exaggerations and fallacies would sap racism of its power. "[I]n the light of the facts," Montagu thought, reasonable people would come to see that human beings were more alike than different. Indeed, he argued that the scientific study of human variation had demonstrated "the fundamental unity of all mankind." "The very nature of the variations," he said, "provides the completest evidence of that truth, of the basic likeness in difference." In what now can only seem like astonishing faith in science and "the facts" in the face of entrenched, multiply-motivated hatreds and institutionalized socioeconomic and political discrimination, Montagu claimed that "all that should be required is to state the case against those who have endeavored to magnify the differences. This I have done." Like Benedict, Montagu had faith that race itself was not the problem and might be left out of any analysis of racism once people understood its irrelevance to the resolution of social problems. "If we can learn to understand the nature of the insignificant differences which remain, we shall then be happier in their presence, and find them in every way acceptable as we do those which

---

[50] This is a point also made nearly twenty years later by another reviewer, Gary W. Martin, of Yale University, this time of the collection *The Race Concept*, where Montagu repeated his definition of "ethnic group." Gary W. Martin, "Concerning Ashley Montagu and the Term 'Race,'" *American Anthropologist*, New Series, vol. 65, no. 2 (Apr., 1963), pp. 402–403.

[51] Ashley Montagu, *Man's Most Dangerous Myth: The Fallacy of Race*, fourth revised and enlarged edition (Cleveland and New York: Meridian Books, The World Publishing Company, 1965), p. 15. This argument was not new for Montagu in 1964; see, for example, in the 1942 edition, pp. 21–22.

exist within our own immediate group."[52] That "the facts" had not had such an effect in more than twenty years of reiterating them did not seem to lessen his faith in their power.

UNESCO and the Problem of Race

Perhaps the most famous repudiation of race and racism comes from UNESCO and its "Statement by Experts on Race Problems," issued by social scientists in July 1950. In subsequent decades, their declaration that "'race' is not so much a biological phenomenon as a social myth" became a rallying cry for a wide variety of scholars and activists eager to reject hereditarian, biologically essentialist conceptions of humanity.[53] Preceded in the 1940s by Ashley Montagu's widely read denunciation of race and racism, *Man's Most Dangerous Myth*, the 1950 UNESCO Statement has been seen by many as a turning point.[54] Galvanized by the calamity of racialized conflict in WWII, scientists from around the world and across disciplines came together to denounce pernicious and erroneous racial distinctions that had "taken a heavy toll in human lives and caused untold suffering."[55] In this view, the Statement captured an emerging and powerful consensus among scientists that racial distinctions and the essentializing biological, anthropological, and social sciences that had purveyed them were invidious remnants of a hereditarian philosophy best rejected by a postwar world seeking peace and greater equality. The

---

[52] Ibid., p. 15 for all quotations in this paragraph.

[53] Ibid., p. 364.

[54] See, for example, Michelle Brattain, "Race, Racism and Antiracism: UNESCO and the Politics of Presenting Science to the Postwar Public," *The American Historical Review*, vol. 112, no. 5; Anthony Q. Hazard, Jr., "Postwar Anti-Racism: The United States, UNESCO and 'Race', 1945–1968" (PhD diss., Temple University, 2008); Barkan, *The Retreat of Scientific Racism*, Elazar Barkan, "The Politics of the Science of Race: Ashley Montagu and UNESCO's Anti-Racist Declarations," in Larry T. Reynolds and Leonard Lieberman, eds., *Race and Other Misadventures: Essays in Honor of Ashley Montagu in His Ninetieth Year* (Dix Hall, NY, 1996), pp. 96–105; Donna Haraway, "Universal Donors in a Vampire Culture: It's All in the Family – Biological Kinship Categories in the Twentieth-Century United States," Modest-Witness@Second-Millennium.FemaleMan©-Meets-OncoMouse™: Feminism and Technoscience (New York: Routledge, 1997), pp. 213–266; Nancy Stepan, *The Idea of Race in Science: Great Britain, 1800–1960* (Hamden, CT, 1982). Jenny Reardon presents a different view in *Race to the Finish: Identity and Governance in an Age of Genomics*.

First published in 1942, Montagu's book went through seven editions (1942, 1945, 1952, 1964, 1974, and then reissued again in 1996 and 2007), each with substantial revisions, additions, and changes to appendices.

[55] Montagu, *Man's Most Dangerous Myth* (1965), p. 364.

1950 UNESCO Statement was seen as a harbinger of a new approach to human diversity that would emphasize common humanity amidst variety.

The UNESCO project was an extension of the prewar and wartime educational efforts of anthropologists such as Benedict and Montagu. Founded on the same premise – that "race hatred and conflict thrive on scientifically false ideas and are nourished by ignorance" – UNESCO meant to educate people around the world, to "show up these errors of fact and reasoning" by promoting the scientific pronouncements on the "facts" about race. UNESCO set out to untangle race and culture, whose "long-standing confusion" had "produced fertile soil for the development of racism."[56] The UNESCO constitution declared that WWII had been "made possible by the denial of the democratic principles of the dignity, equality and mutual respect of men, and by the propagation, in their place, through ignorance and prejudice, of the doctrine of the inequality of men and races." The UNESCO Statements and subsequent publications on race were meant to be a unified, institutionalized, scientific response to the "fundamentally anti-rational system of thought" at the root of racism, one "in glaring conflict with the whole humanist tradition of our civilization."[57]

The reality of UNESCO's effort to issue a ringing denunciation of racism and a unified statement about the "facts" was rather more complicated, however. As a number of scholars have noted, the first UNESCO Statement, issued in 1950, was largely the product of cultural anthropologists and other social scientists and met a great deal of criticism from physical anthropologists and geneticists, particularly its assertion that race was a social myth.[58] The result was a revised second Statement, "Statement on the Nature of Race and Race Differences," issued in 1951 by these natural scientists, presenting their much more heavily qualified conclusions about race, and its relationship to human abilities and cultural achievements and the study of human evolution. Not only were most physical anthropologists and geneticists unprepared to jettison race

---

[56] UNESCO, United Nations Education, Scientific and Cultural Organization, *The Race Concept: Results of an Inquiry*, The Race Question in Modern Science (Westport, CT: Greenwood Press, 1970), p. 5.

[57] Ibid.

[58] Most of the participants were sociologists. *Race Concept*, p. 6; Brattain, "Race, Racism and Antiracism"; John P. Jackson, Jr. and Nadine M. Weidman, eds., *Race, Racism, and Science: Social Impact and Interaction*, Science and Society Series (New Brunswick: Rutgers University Press, 2006).

as a biologically meaningful concept, many were quite unhappy about attempts to restrict its significance to a descriptive, classificatory device based solely on heritable physical characteristics.

There were several points of broad agreement on the nature of race and the problem of racism among the social and natural scientists gathered together by UNESCO. By 1949, when scientists began their work on a joint statement under Arthur Ramos, head of UNESCO's Department of Social Sciences, population genetics and an evolutionary view of human variation and change had become a consensus view. Both the first and second Statements defined race in terms of the frequency and distribution of genes across populations, subject to the forces of evolutionary change.[59] Both social and natural scientists noted that this evolutionary process had produced populations sufficiently distinct to be sensibly divided into three "major groups" or "divisions." Indeed, it was the social scientists who actually named those groups as "divisions," echoing traditional typologies, and enumerated them as Mongoloid, Negroid, and Caucasoid.[60] The physical anthropologists and geneticists merely noted that "the greater part of existing mankind" could be classified in "at least three large units, which may be called major groups."[61] Both Statements also noted that classifications varied, groups were neither fixed nor tightly bounded, and such taxonomies served a largely instrumental purpose (facilitating evolutionary studies or other "scientific purposes").[62] Both Statements also explicitly rejected the conflation of nation, religion, language, geography, or other cultural markers with race. They both flatly stated that "cultural traits" had "no demonstrated connection with racial traits"; the social scientists pointedly noted that it was not a "genetic" connection.[63]

Moreover, both Statements, as well as many individual scientists who offered comments on earlier drafts, stressed the importance of civil and social equality irrespective of any physical, cultural, or hereditary differences that might exist or be demonstrated in the future. "Equality as an ethical principle in no way depends upon the assertion that human beings are in fact equal in endowment," the first Statement declared.[64] Geneticists seconded this sentiment, even while stressing the extent of

---

[59] *The Race Concept*, p. 6; Montagu, *Man's Most Dangerous Myth* (1965), pp. 362, 367.
[60] Montagu, *Man's Most Dangerous Myth* (1965), p. 363.
[61] Ibid., p. 368.
[62] Ibid., pp. 362, 367.
[63] Ibid., pp. 363, 367.
[64] Ibid., p. 365.

"striking" differences among people. Ernst Mayr stated this forcefully in his comments on a proposed second Statement, writing:

> Equality of opportunity and equality in law do not depend on physical, intellectual and genetic identity. There are striking differences in physical, intellectual and other genetically founded qualities among individuals of even the most homogenous human population, even among brothers and sisters. No acknowledged ethical principle exists which would permit denial of equal opportunity for reason of such differences to any member of the human species.[65]

The reaction of many geneticists to the problem of equality reflected the same concern that Ruth Benedict had articulated a decade earlier. Most were convinced that there were demonstrable genetically based differences among human beings in a variety of traits, physical and possibly mental. But most were equally convinced that those differences, whatever they were or might be proved to be, were irrelevant to the problems of harmony, equality, and justice in modern society. Geneticist Walter Landauer forcefully expressed just this perspective in his comments on the draft of the second Statement. "It would surely make no difference," he wrote, "to the ethical standards of the Unesco [*sic*] group or to mine if, for instance, an unequal distribution of genes for certain mental traits were demonstrated. The declaration that 'all men are created equal['] was a fine one and remains so, even though and in the best sense *because* it is untrue in the biological sphere"[66] (emphasis in original). Comments such as Mayr and Landauer's reflected the sharp difference between the social and natural scientists in their approach to the "facts" about race and their sociopolitical context. While the social scientists acknowledged that individuals "varied greatly among themselves in endowment," their Statement repeatedly minimized those differences, especially group differences, arguing that they had been exaggerated and used for malign purposes. Rather than emphasize differences among human beings, only to disavow them as a basis for social relations, as natural scientists did,

---

[65] *The Race Concept*, p. 18.

[66] Ibid., p. 19. Others making similar points about the irrelevance of racial differences to the question of equality in society included Eugen Fischer, pp. 31–32; Kenneth Mather, p. 26; C. D. Darlington, pp. 26–27; and Carleton Coon, p. 28. Even German anthropologist Hans Wienert inveighed against the kind of inequities erected upon supposed racial differences by the Nazis, despite his invocation of classic white supremacist arguments, the most infamous of which was his query regarding the advisability of racial mixing: "I should like to ask which of the gentlemen who signed the statement would be prepared to marry his daughter for example to an Australian aboriginal." Weinert managed to raise issues of race, sex, and social standing that were largely evaded in the debate, all in one offensive swipe at his colleagues; p. 35.

social scientists emphasized biological unity. They argued that whatever the variations among individuals, or identifiable groups, the differences were small compared to the "vast number of genes common to all human beings." All were members of *Homo sapiens*, more alike than different. Although the varied "biological histories" of human populations created distinctive distributions of traits, the social scientists argued that it was not this acknowledged evolutionary process, but the arbitrary way people perceived these apparent differences that created trouble. Human variation was apprehended differently by different people and groups. "What is perceived is largely preconceived," the social scientists argued, "so that each group arbitrarily tends to misinterpret the variability which occurs as a fundamental difference which separate the group from all others."[67] Given persistent popular misconceptions about what race was – conflating national, religious, linguistic, and other cultural formations with physical variation – social scientists advocated replacing the term "race" with "ethnic groups."[68]

One area of particularly marked difference between social and natural scientists was the contentious question of mental abilities, temperament, and personality. In the first Statement, social scientists stressed the "essential similarity in mental characters among all human groups." They argued that it was very difficult to tease apart the relative influence of "innate capacity" and "environmental influences, training and education," but argued all the evidence suggested that given similar opportunities to realize their potential, "the average achievement of the members of each ethnic group is about the same."[69] They asserted that there was no "definite" evidence of inherent differences in temperament or personality among groups, again stressing the wide variation among individuals and the important influence of "environmental factors."[70] Although the first UNESCO Statement acknowledged physical and other "inborn" differences, it consistently framed them as less pertinent than the enormous variability and plasticity of humanity, and the critical role of cultural and other environmental factors.

This is the context in which the Statement famously asserted that race was a "social myth." In 1950, what caused a stir, especially among physical anthropologists and geneticists, was the then-radical suggestion that

---

[67] Montagu, *Man's Most Dangerous Myth* (1965), p. 362.
[68] Ibid., p. 363. It was no coincidence that Ashley Montagu, who was the rapporteur for the group crafting the first Statement, had advocated the same approach; see pp. 49–54.
[69] Ibid., p. 363.
[70] Ibid., p. 364.

race might be nothing more than a socially constructed, invidious invention. But the full assertion actually makes a less radical claim, one that left room for race as a natural phenomenon. The relevant section began by arguing that "the biological fact of race and the myth of 'race' should be distinguished," alluding to Ashley Montagu's earlier distinction between a legitimate, purely descriptive zoological definition of race and a dangerous one that erroneously linked physical diversity to character and culture. Following this critical distinction, the Statement continued, "for all practical purposes 'race' is not so much a biological phenomenon as a social myth."[71] As Benedict and Montagu had done earlier, the first Statement reiterated both the reality of often racialized differences and their irrelevance to the structure and quality of human relations. "Such biological differences as exist between members of different ethnic groups have no relevance to problems of social and political organization, moral life and communication between human beings," they wrote.[72] Stressing humanity's natural "cooperative spirit" and an "ethic of universal brotherhood," readers were urged to focus on "the unity of mankind from both the biological and social viewpoints."[73] For the social scientists crafting the first Statement, "To recognize this and to act accordingly is the first requirement of modern man."[74]

Physical anthropologists and geneticists took exception to the social scientists' environmentalist slant, the emphasis on unity and likeness, and the proscriptive tenor of the first UNESCO Statement. The lingering polygenism in the study of human variation was reflected in persistent questioning among commentators about the extent to which human variation was evidence of polyphyletic development.[75] Some of the most vigorous opposition came from scientists, such as German geneticist Fritz Lenz, who not only believed humans could be intelligibly distinguished as distinct populations on the basis of hereditary variations (as "races," "varieties," or "sub-species"), but questioned whether those distinctions might be sufficiently profound to justify regarding some groups as distinct species. "If an unprejudiced scientist were confronted with a West-African Negro, an Eskimo and a North-West European," Lenz argued in response to a draft Statement, "he could hardly consider them to belong to the

---

71 Ibid.
72 Ibid., p. 366.
73 Ibid., pp. 365, 366.
74 Ibid., p. 364.
75 *The Race Concept*, p. 39.

same 'species.'"[76] Although Lenz was among the more extreme critics of the UNESCO Statements (and among the few Germans invited to comment on the Statement at all), many physical anthropologists and geneticists insisted that "race" was more than just a "classificatory device."[77] Brazilian geneticist Oswaldo Frota-Pessoa argued that the race concept "results chiefly from the recognition of a natural fact, namely, that human populations differ in the incidence and frequency [*sic*] of certain hereditary characters." As such, the statement should have emphasized "its importance for the understanding of the actual biological texture of mankind."[78]

The greatest point of disagreement between social and natural scientists, as well as among the natural scientists themselves, was over how to discuss variation in mental and temperamental characteristics. Where the social scientists' Statement categorically asserted that anthropologists "never" included mental characteristics in making their racial classifications and dismissed the efficacy of intelligence tests in discerning the relative influence of "innate capacity" and "environmental influences," the natural scientists were much more circumspect. Their Statement noted that "most" anthropologists left aside mental characteristics in their taxonomies, and that although studies had shown that the results of intelligence and temperament tests were due to both capacity and environment, their relative influence was disputed.[79] Like the social scientists, the final Statement by natural scientists stressed the equalizing effect of cultural environments on intellectual performance, highlighted the salience of individual variation over differences between groups, and noted that heterogeneous human groups displayed "approximately the same range of temperament and intelligence." But they also gave much more credence to the possibility that "some types of innate capacity for intellectual and emotional responses are commoner in one human group than another," noting that much was still unknown about "the part played by heredity in the mental life of normal individuals."[80]

[76] Ibid., p. 37.
[77] See Robert Proctor, "From *Anthropologie* to *Rassenkunde* in the German Anthropological Tradition," George W. Stocking, Jr., ed., *Bones, Bodies, Behavior: Essays in Biological Anthropology*, History of Anthropology, vol. 5 (Madison: University of Wisconsin Press, 1988), pp. 138–179, esp. pp. 173–174.
[78] *The Race Concept*, p. 38.
[79] Montagu, *Man's Most Dangerous Myth* (1965), p. 368.
[80] Montagu, *Man's Most Dangerous Myth* (1965), p. 369.

Among the natural scientists, a great many were dissatisfied with the Statement's presentation on mental capacities, and their comments reveal the unsettled state of the field. Biologist Joseph Birdsell was concerned that the "multiple qualifications" might leave lay readers with the impression that differences in innate intelligence between racial groups would eventually be proved. He suggested the Statement clearly note that "so-called intelligence tests are not a measure of differences of innate or biologically endowed intelligence between groups which differ culturally," no psychological test was free of cultural content, and no method had been devised that could measure the innate intelligence of an individual.[81] Geneticist Alfred Sturtevant insisted that while environmental forces were a critical factor in the development of human behavior, "the argument that environment is the *sole* determinant is something every geneticist must protest against"[82] (emphasis original). Other geneticists, such as Fritz Lenz, maintained that contrary to statements minimizing the importance of mental factors in defining race, "psychical hereditary differences are much more important that physical differences" in racial classifications.[83]

Biologist Hermann J. Muller took exception to the implication that racial differences could be confined to bodily manifestations. "To the great majority of geneticists," Muller argued, "it seems absurd to suppose that psychological characteristics are subject to entirely different laws of heredity or development than other biological characteristics."[84] Among the foremost geneticists of his generation, a Nobel Prize winner for his studies of heredity and genetic mutation, Muller urged his colleagues to be candid about both the absence of scientific data on questions of innate mental capacities and what could realistically be inferred. "[W]e do have every reason to infer that genetic differences, and even important ones, probably do exist between one living racial group of men and another, and our statement should not imply the contrary." Muller also argued that such differences were no rationale for differential treatment. The enormous influence of culture and environment, the extent of genetic similarity among humans, and the variation among individuals of all groups, meant that all "racial groups" were "capable of participating and co-operating fruitfully in modern civilization." Like

[81] *The Race Concept*, p. 49.
[82] *The Race Concept*, pp. 55–56.
[83] Ibid., p. 50.
[84] Ibid., p. 53.

virtually all UNESCO commentators, Muller argued that all individuals should be granted opportunities and rights as "individuals without reference to their racial origins." But Muller, like many of his colleagues, was concerned about erecting a platform for fair treatment that rested upon the "spurious notion that [racial groups] are identical in the genetic basis of psychological traits."[85]

The natural scientists were also concerned about what physical anthropologist Carleton Coon described as "slanting scientific data to support a social theory." Coon cast the UNESCO Statements as no different than "what the Russians are doing," referring to recent controversy over the practice of genetics in the Soviet Union, "and what Hitler did."[86] Several geneticists made the same argument.[87] German anthropologist Walter Scheidt averred that he wanted "no part in attempts to solve scientific questions by political manifestoes [*sic*]" like Soviet Russia.[88] Kenneth Mather complained that even the draft of the natural scientists' "Statement on the Nature of Race and Race Differences" "was bending over backwards to deny the existence of race in the sense that this term has been used for political purposes in the recent past." Citing his own sympathy with the effort to reject Nazi racial theory, Mather argued that he did not believe "the case against it is strengthened by playing down the possibility of statistical differences in, for example, the mental capacities of different human groups."[89] British biometrician Ronald Fisher, an early proponent of eugenics and a statistician responsible for much of the mathematics foundational to the development of population genetics, objected even more strongly that human groups differed "profoundly 'in their innate capacity for intellectual and emotional development.'" He argued that the real international problem was how to "share the resources of this planet amicably with persons of materially different nature," a goal which was "obscured by entirely well intentioned efforts to minimize the real differences that exist."[90]

German physical anthropologist and eugenicist Eugen Fischer exemplified the complexities of racial science and international politics in these years. Fischer is a more complicated character than often portrayed. The leading German physical anthropologist in the interwar years, his

---

[85] Ibid., p. 54.
[86] Ibid., p. 28.
[87] Ibid., p. 7; see Eugen Fischer, pp. 31–32; Enrique Beltran, p. 21; Curt Stern, p. 30.
[88] Ibid., p. 32.
[89] Ibid., p. 26.
[90] Ibid., pp. 27, 31–32; Proctor "From *Anthropologie* to *Rassenkunde*."

work on racial mixing among Boers and indigenous South Africans – the "Rehobother Bastards" – was widely cited in the interwar period, although many are now unaware that his conclusions supported the idea that racial mixing was not deleterious. As Robert Proctor has argued, his work during the interwar period was fully within the mainstream of international racial science in that period. His support of the Nazi regime and elevated status in German institutions during that period as head of the Kaiser Wilhelm Anthropology Institute and rector of the University of Berlin, as well as the vigorous condemnation of his complicity following the war (notably at the Nuremburg trials), have, not surprisingly, resulted in a historical account of him that often neglects the longer trajectory of his career and those of others whose work became the justification and tool for murderous social and political policies. In the vein of Robert Proctor's efforts to situate Fischer's anthropology within a broader context of racial science in the mid-twentieth century, it is worth noting his comments on the draft Statement on race circulated to natural scientists in 1951. There Fischer commented:

In so far as the [UNESCO] Statement condemns any defamation of races and emphasizes the appalling nature of the recent abuse of racial theory, it has my full and unqualified approval. I wholeheartedly agree, also, with its explicit and implicit finding that anthropology and racial studies afford no justification for the assumption that the members of any particular race are not entitled to the enjoyment of all fundamental rights, or for any form of racial discrimination. And I am very glad that, after all the horrors that have been perpetrated, these principles should have been enunciated clearly and publicized widely by an organization of such standing and by such distinguished men as the authors of this Statement. . . . I recall the National Socialists' notorious attempts to establish certain doctrines as the only correct conclusions to be drawn from research on race, and their suppression of any contrary opinion. . . . The present Statement likewise puts forward certain scientific doctrines as the only correct ones, and quite obviously expects them to receive general endorsement as such. I repeat that, without assuming any attitude towards the substance of the doctrines in the Statement, I am opposed to the principle of advancing them as doctrines. The experiences of the past have strengthened my conviction that freedom of scientific inquiry is imperiled when any scientific findings or opinions are elevated, by an authoritative body, into the position of doctrines.[91]

---

[91] Ibid., pp. 31–32. For an account of human genetics, eugenics, and the Nazi regime, see Robert Proctor, "From *Anthropologie* to *Rassenkunde* in the German Anthropological Tradition," George W. Stocking, Jr., ed., *Bones, Bodies, Behavior: Essays in Biological Anthropology*, History of Anthropology, vol. 5 (Madison: University of Wisconsin Press, 1988), pp. 138–179; and Sheila Faith Weiss, *The Nazi Symbiosis: Human Genetics and Politics in the Third Reich* (Chicago: University of Chicago Press, 2010).

Both the agreements and the disagreements between social and natural scientists involved in UNESCO's educational project reveal the complexities of race and culture in the postwar period. Virtually all the scientists were united in their desire to undermine or "exterminate" invidious distinctions, prejudice, and discrimination. Even the biological scientists who clung to hereditarian views of human variation and its implications for society voiced support for civil rights and fair treatment. Certainly, there was no clear alignment of racial views along political or social lines. Moreover, the neat divide social scientists such as Ashley Montagu or Ruth Benedict claimed to erect between a genetic, physically based definition of race and a separate, culturally defined realm of human development and interaction never fully materialized. Bodies and cultures continued to intermingle, and race persisted.

Whither Race? From Physical to Biological Anthropology

The effort to find consensus about race among cultural anthropologists, sociologists, physical anthropologists, and geneticists revealed shifts in the study of human variation that preceded the events that had catalyzed UNESCO. By the early 1950s, neither anthropologists' culture concept nor the Evolutionary Synthesis and the concept of population genetics had fully penetrated the human and natural sciences. For physical anthropologists, concepts and methods that had been applied to the study of human variation for a century were under intense social, political, and scientific pressure. By the 1940s, physical anthropologists such as Harry Shapiro, who had been trained under a fundamentally typological conception of race, found themselves working through the implications of world events for the work to which they had dedicated their lives. Some such as Shapiro shifted to an increasingly populational genetic understanding of human variation, signaled by a new name for their discipline – biological anthropology – while an increasingly marginalized few, such as Carleton Coon and Stanley Garn, clung to the hereditarian essentialism of the old typologies.[92]

---

[92] For an incisive analysis of the racial politics of Carleton Coon's racial science in the postwar period, see John P. Jackson, Jr., "'In Ways Unacademical': The Reception of Carleton S. Coon's *The Origin of Races*," *Journal of the History of Biology*, vol. 34 (2001), pp. 247–285. See also, Alan Goodman and Evelynn Hammonds, "Reconciling Race and Human Adaptability: Carleton Coon and the Persistence of Race in Scientific Discourse," in Jonathan Marks, ed., *Racial Anthropology: Retrospective on Carleton Coon's* The Origin of Races (1962), *Kroeber Anthropological Society Papers* (2000),

By the 1950s, much had changed in the world of biological and human sciences, and in the wider world. In science, the Evolutionary Synthesis had transformed not only the study of evolution, but over the course of the 1940s and 1950s, transformed virtually all the biological sciences. In the first half of the 1940s, decades of work in biometrics, genetics, studies of natural selection, and Darwinian evolution were finally consolidated into a unified approach.[93] Work combining Darwin's mechanism with population biology created a new conceptual category, "population," and gave not only biology but anthropology a new approach through which to study difference and its origins, be it among plants, animals, or humans. The methodological shift in the biological study of evolution occasioned not only the ascendance of "populational thinking" in biology, but a concomitant shift in anthropological studies of human evolution, development, and difference. Physical anthropology became biological anthropology, and the once-dominant view of race as a static set of somatic characteristics associated with mental capacities and level of civilization – a view that had been questioned and qualified in earlier decades – became increasingly widely marginalized in favor of the view that physical differences existed only as characteristics distributed across

no. 84, pp. 28–55. For an account of the political, social, and economic context in which the Left and New Right contested racial narratives in this period, see also Jacquelyn Dowd Hall, "The Long Civil Rights Movement and the Political Uses of the Past," *The Journal of American History*, vol. 91, no. 4 (Mar. 2005), pp. 1233–1263. For Stanley Garn's postwar racial science, see, for example, *Human Races* (Springfield, IL: Thomas, 1961). Coon and Garn employed population genetics terminology and theory to better define racial types; they did not view it as a model of variation that undermined the idea of racial typology. For an example of this work, see Stanley Marion Garn and Carleton S Coon, *On the Number of Races of Mankind*, Fels Research Institute, *Publications*, no. 3 (Yellow Springs, OH: Antioch College, 1955), pp. 100–105.

[93] A handful of texts are generally regarded as the foundation of the Modern Synthesis, including Theodosius Dobzhansky, *Genetics and the Origin of Species* (New York: Columbia University Press, 1937); Ernst Mayr, *Systematics and the Origin of Species* (New York: Columbia University Press, 1942); Julian Huxley, *Evolution: The Modern Synthesis* (London: G. Allen & Unwin Ltd., 1942); and George Gaylord Simpson, *Tempo and Mode in Evolution* (New York: Columbia University Press, 1944). Secondary literature on the Synthesis includes Betty Smocovitis, *Unifying Biology: The Evolutionary Synthesis and Evolutionary Biology* (Princeton: Princeton University Press, 1996); William B. Provine, *The Origins of Theoretical Population Genetics* (Chicago: University of Chicago Press, 2001); William B. Provine, *Sewall Wright and Evolutionary Biology* (Chicago: University of Chicago Press, 1989); Donna Haraway, *Primate Visions: Gender, Race and Nature in the World of Modern Science* (New York and London: Routledge, 1990). On the relationship between studies of race mixing and evolutionary biological studies in the mid-twentieth century United States, see Paul Farber, "Race-Mixing and Science in the United States," *Endeavor* (2003), vol. 27, no. 4, pp. 166–170.

populations (and geography) that did not cohere into discrete types. Race, if it could be said to exist, had truly become an abstraction in the minds of many evolutionary biologists, anthropologists, and others. The physical differences in appearance that had traditionally characterized racial types were regarded as of questionable, or at least undetermined, biological significance, and whatever the import of evident external differences, there was no evidence that they were related to character, personality, intelligence, or behavior. Biological anthropology defined itself in terms of the new genetic and populational turn in wider biological studies, and shifted from the former preoccupation of physical anthropology with the study of human evolution and race to the study of human evolution and human growth, development, and disease among world and domestic populations. Race did not disappear entirely, but it was no longer one of the two defining areas of study for biological anthropologists.

It was no coincidence that the transformation of physical anthropology into biological anthropology coincided not only with the conceptual consolidation among the biological and evolutionary sciences, but also, and as importantly, with the social and political rejection of racism and colonialism in the postwar era. In the postwar period, Harry Shapiro began to construct a history of physical anthropology and racial science that distanced the emerging biological emphasis from its disreputable origins in essentialist racial types and positioned the "new" physical anthropology as a progressive, evolutionary science. In the early 1950s, Shapiro participated with other anthropologists and geneticists from around the world in drafting the UNESCO Statements on Race. In 1953, as part of a UNESCO series entitled "The Race Question in Modern Science," Shapiro authored *Race Mixture*, published as one of several companion pieces to the official UNESCO Statement on Race authored by leading scientists on their areas of expertise.[94] In *Race Mixture*, Shapiro revisited his studies of Pitcairn Islanders and the mixed populations of Hawaii, in a wide-ranging discussion of the various historical, political, social, psychological, international, biological, environmental, and methodological contexts for the study and evaluation of race mixing, understood as hybridization. Shapiro again asserted, as he had in his original studies, that race mixing was common, not deleterious, and sometimes beneficial. And he again issued a harsh critique of past studies that contributed to invidious racial comparisons, most especially Charles Davenport and Morris Steggerda's *Race Crossing in Jamaica*.

---

[94] Harry L. Shapiro, *Race Mixture, The Race Question in Science* (Paris: United Nations Educational, Scientific and Cultural Organization, [1953] 1965).

The most notable aspect of Shapiro's contribution to the UNESCO effort was his translation of his earlier work into a new analytical framework. Shapiro's argument in *Race Mixture* was explicitly framed in terms of contemporary genetic explanations of heredity, variation, populations, and change, although he didn't discuss these mechanisms in detail. He used his UNESCO platform to reposition physical anthropology as biological anthropology, allied to evolutionary biology and population genetics as much, if not more, than to the rest of anthropology. The transition of physical anthropology to biological anthropology was one Shapiro was plainly still working through. For example, Shapiro asked, "How much should be made genetically of a Negro strain that no longer is recognizable in a man who can pass as white or of a white strain in one who appears to be pure Negro." Shapiro had raised a question that went to the heart of American racial (il)logic. But even recognizing this fundamental conundrum, Shapiro could not let go of typological strains. He passed over the problem of "passing" by asserting that "phenotypes" were unreliable guides to determining race mixture, reiterating the persistent confusion over whether external appearance did or did not reveal inner essence. In terms of the sort of "zoological" classification he advocated, such a position also posed potentially fatal problems for human taxonomy if the external criteria upon which the classification was to be founded were an unreliable guide to the genes where true type resided. Shapiro's discussion of hybridization remained predicated on the typological notion of mixing "strains" that characterized his earlier work, which he had trouble reconciling with the new genetic/populational explanations of variation and heredity. Nevertheless, he self-consciously and consistently employed the terminology of genetics and population biology: genes, phenotype, populations, adaptation, recombination, and fitness.

By 1958, when Shapiro presented a paper on the history of physical anthropology to the American Anthropological Association (AAA), he had much more fully embraced the new biological anthropology of the postwar period. In his AAA paper, Shapiro presented a history of physical anthropology and Franz Boas's work in terms of Darwinian evolution, genetics, and population biology, obscuring the profoundly cultural perspective that both he and Boas stressed in the pre–WWII period. Although Shapiro continued to regard Boas as the seminal figure in shaping the methods and conceptual foundations of the field, by 1958 he either had cast aside the old physical anthropology with its racial taxonomies and studies of hybridization in which he had been trained, or had translated it into a new evolutionary genetic framework. This shift to a more biological perspective reflected larger changes, and perhaps was an attempt to revise and reclaim a discipline whose primary preoccupation for decades – race – was an increasingly problematic field of inquiry. Shapiro lamented the persistent "stereotype of the physical anthropologist with calipers in hand busily measuring heads" – something he did as recently as the late 1930s along with virtually all of his colleagues – and suggested that even in 1958 cultural anthropologists might not realize the "biological" revolution that physical anthropology had undergone, beginning, in Shapiro's chronology, as early as the mid-nineteenth century.[95]

Shapiro viewed Darwin's *Origin of Species* as a pivotal text not only in the discipline of biology, but in physical anthropology, which Shapiro was effectively redefining as a branch of biology.[96] Shapiro described a centuries-long interest among ethnologists and others in the problem of human difference, of "racial variation," and argued that physical anthropology originally arose to deal with the "comparative morphology" of "populations" whose cultures were being described and compared. Both ethnologists and physical anthropologists were interested in the origins of the people they studied. But in Shapiro's formulation, it was the theory of evolution, described by Darwin, which gave physical anthropology a "fundamental orientation," spurring studies of primates, growth, skeletal function, and hybridism by the latter half of the nineteenth century. These "tentative developments. . . began to flower" only after developments in biological fields, especially genetics, but also in serology, the comparative studies of blood groups,[97] as well as in statistics. Grouping physical anthropology with the "biological sciences," Shapiro claimed the field used statistics earlier and with more sophistication than psychology "or any of the strictly biological sciences." He regarded this "revolution" as "a natural extension of [physical anthropology's] concern with the human species into the various biological aspects of man that current

---

[95] Harry L. Shapiro, "Symposium on the History of Anthropology. The History and Development of Physical Anthropology," *American Anthropologist*, vol. 61 (1959), p. 375.

[96] This is, of course, not completely without reason. Disciplinary boundaries are obviously permeable, and as Shapiro notes, many disciplines have an interest in humans as organisms and in their changes over time and/or space, including both physical anthropology and biology.

[97] An interest Shapiro shared with Hooton. See "Genes, Populations, and Disease, 1930–1980: A Problem-Oriented Review," Kenneth M. Weiss and Ranajit Chakraborty, pp. 371–404, in Frank Spencer, ed., *A History of American Physical Anthropology 1930–1980* (New York: Academic Press, 1982), pp. 373–375, for a brief discussion of the study of blood type as a stable character for elucidating racial groups.

scientific developments permit," and redefined the discipline's field of
study, echoing both Hooton and Boas, as "a vast range of subject"
summed up "as a study of the human species in time, place, and culture, and the investigation of the factors underlying human variation."[98]
The reference to human variation pointed to his continuing interest in
questions of differences in human appearance that had earlier been conceptualized as questions of racial types and classifications, a field of study
that would rapidly disappear in the 1960s and 1970s except in certain corners of biological anthropology and by individual researchers who were
more often than not scorned by other anthropologists. Indeed, the subject of human variation and issues of methodology became questions that
estranged cultural and biological anthropology to a degree that would
have no doubt dismayed both Hooton and Boas, and defied Shapiro's
hopeful vision that a union of cultural and physical anthropology was
"still capable of mutual satisfaction."[99]

By 1958, Shapiro had shifted to an even more biologized version of the
sociocultural effects that he and Boas had stressed in an earlier period.
Rather than regarding humans as creatures whose appearance and heredity were incomprehensible without recourse to their culture and history,
Shapiro seems to have shifted to a view of humans as primarily evolved
organisms – as a species – whose culture was encompassed in a biological
notion of environment. Thus, culture became an element of the biological
history of man, rather than biology being understood through the lens of
particular histories and cultures.

Race in the Postwar Natural History Museum: Harry Shapiro and the *Biology of Man* Hall

In 1961, Harry Shapiro, as a curator of anthropology at the American
Museum of Natural History, opened a major exhibition entitled *The
Biology of Man*. Shapiro had been planning the exhibition for some fifteen years, hoping to finally be able to mount a major presentation of
his discipline inside the museum. Delayed first by the Depression then by
war, Shapiro was finally able to begin seriously planning for his hall in the
mid-1940s. By the time it opened to fanfare and publicity in New York
in 1961, the exhibition reflected profound changes in the field of physical
anthropology since the interwar period. The exhibition marked a stark

---

[98] Shapiro, "Symposium on the History of Anthropology," pp. 373–375.
[99] Ibid., p. 374.

contrast with its pre–WWII counterpart, *The Races of Mankind,* at the Field Museum of Natural History in Chicago. The Field Museum emphasized a hall filled with racial types, with scant attention paid to human biology; Shapiro inverted that scheme. *The Biology of Man* featured large displays on the evolution, development, growth, and physiology of the human body, broken down into discreet organ systems and described through graphical demonstrations of comparative and statistical studies of various populations. Only a small but notable portion of the exhibition was devoted to presenting the familiar racial divisions and types. Thus, for a brief few years before the Field Museum dismantled *The Races of Mankind* in 1967, an enterprising student of anthropology could travel to Chicago and New York and see at their natural history museums a history of anthropological racial science written across two exhibit halls.

The *Biology of Man* reflected broader efforts of the American Museum to reorient itself after WWII. During the war, newly appointed museum Director Albert Parr convened a Committee on Plan and Scope to examine the research and exhibition efforts of the museum. Under Parr's direction, the committee developed an ambitious program of renovation, reconstruction, and reorientation directed toward updating long-neglected exhibition halls and crafting new exhibitions for the postwar world.[100] Echoing Berthold Laufer's aspirations in the 1920s, Parr argued that natural history museums could offer a valuable service to "a troubled world":

In a world beset by hostility and want, the natural history museums have an opportunity, never before equaled, to serve the development of peace and of a better life for all by bringing their educational facilities and their scientific knowledge to bear upon the task of creating a better understanding of our own problems in relation to the country that surrounds us and supports us, and of the problems of other nations in relation to their natural circumstances, to one another, and to us.[101]

Noting that the American Museum received 2 million visitors a year, Parr argued that anthropology could contribute to international understanding by helping the American public appreciate the "various ways in which

[100] *The American Museum of Natural History, Seventy-Fourth Annual Report, for the Year 1942* (The City of New York, May 1, 1943), pp. 19–20.

[101] A.[lbert]. E. Parr, "Towards New Horizons," *The American Museum of Natural History, Seventy-Eighth Annual Report, July, 1946, Through June, 1947* (The City of New York, 1947), p. 10.

the peoples of the world have adjusted themselves, their customs, and social organizations to the natural conditions they have had to contend with." By fostering "approval or sympathy instead of jealousy or scorn," visitors might develop a "fairer judgment of [their] fellow men and lay a better foundation for lasting peace."[102]

Harry Shapiro's plans for a series of halls addressing major themes in anthropology – what Parr referred to as the "Epic of Mankind" – was at the center of efforts to reconceive the American Museum's role in public science education. During the war, Shapiro first proposed a radical restructuring of exhibit space in anthropology. He argued for a series of halls "entirely devoted to the dynamics of anthropology, illustrating general principles rather than culture areas."[103] Shapiro argued for subordinating the prewar practice of organizing exhibitions around geographical and tribal variations, shrinking and streamlining the old ethnological halls and installing new halls that would present current anthropological accounts of the natural and cultural world to a diverse public. The old halls neglected "the more dynamic aspects of the subject," Shapiro argued. They highlighted the museum's rich collections as "storehouses of treasures" in exhibits "barren of ideas," offering "the innocent visitor . . . little analysis and little explanation."[104] Overstuffed displays that provoked "mental indigestion" and illogical spatial orientations – why lodge European archaeology between Mexican archaeology and the South American hall? – along with an overemphasis on objects over ideas had made the American Museum anthropology halls "obsolescent" in both subject matter and presentation.[105]

Adopting the urgent rhetoric of wartime civic duty, Shapiro argued that the museum's responsibility in a radically altered postwar world would be to educate the American people about "the significance of this 'changing world,'" their "responsibilities" and the "vistas it opens up." Echoing the concerns of scientists engaged in combating racism and Nazi dogma, Shapiro invoked the perils of leaving American citizens "ill-informed, mis-informed, and weather vanes for every narrow, parochial interest that blows." Current conflicts demonstrated "the bitter

[102] Ibid., p. 13.
[103] H.[arry] L. Shapiro, "Suggestions for the Improvement of the Anthropological Halls" p. 3, Exhibition Plans – Dept. of Anthropology, Harry Shapiro Collection, Special Collections, Library, American Museum of Natural History, New York (hereafter AMNH-HLS).
[104] Shapiro, "Suggestions," p. 1.
[105] Ibid., p. 2.

consequences of a world left to languish in its 19th century illusions," he wrote. "It is our basic business to prepare the people for the universe they must live in."[106] Shapiro hoped to "revitalize" the Museum's presentation of anthropology through a sequence of five halls, starting with the evolution and biology of man, followed by a second hall devoted to the principles that underlie human behavior, a third hall on human social organization, a fourth hall presenting "the majestic sweep of human civilization" and material culture from its beginnings, and a final hall devoted to the "epic of America."[107] Shapiro argued that these halls, unlike older object-centered exhibits organized around culture areas, would help postwar visitors "discover the unities that bind all cultures and to comprehend the reasons for their differences," offering them "a knowledge of the rest of the world as well as . . . an understanding of the way we, in particular, tick."[108] Shapiro argued that these halls would present man's "origins, his development, the nature and growth of his civilization, the manner in which our society functions and the relationship of man to his environment."[109] Diversity was consistently presented as variation on a set of common themes; human beings were unified by common physical attributes (a large brain, upright posture and a bipedal gait, opposable thumbs) as well as common social structures and material cultures (family groups, spiritual beliefs, economic practices).[110]

The *Biology of Man* was conceived as the first installment in a comprehensive exposition encompassing the range of anthropological practice. Its privileged position as the first (and ultimately only) hall was undoubtedly due in part to Shapiro's position as chair of the anthropology department and author of the exhibition scheme. But it was also because Shapiro regarded human physiology and biology as the root of civilization, or as he put it, "Human society and human culture are highly evolved expressions of fundamental biological drives."[111] In his mind, the story of mankind progressed logically from the evolution of humankind to an examination

---

[106] H.[arry] L. Shapiro, "Report to Men's Committee, October 6, 1943," p. 2, Committee – Exhibition AMNH (formerly Committee on Plan and Scope), AMNH-HLS.

[107] "Master Plan for Proposed Revision of Anthropological Exhibits," (1951), p. 1; "Memorandum for Exhibition Committee, Plans for the Anthropology Halls," (1951), p. 2, Exhibition Plans – Dept. of Anthropology, AMNH-HLS.

[108] "Men's Committee," p. 5.

[109] Ibid.

[110] Ibid., pp. 5–7.

[111] "Presenting the New Hall of Man, Proposed Exhibit Arrangement and Architectural Design in Collaboration with the Department of Anthropology and the Department of Architecture," p. 7, Exhibition Plans – Dept. of Anthropology, AMNH-HLS.

of universal human behaviors, and finally to their diverse manifestations in social and cultural formations. Although Shapiro's orientation became increasingly evolutionary and genetic over the thirty years he spent working on elements of his hall, he retained his conviction that human biology, culture, and environment were deeply interwoven. In his view, population genetics and its synthesis with Darwinian natural selection had provided solid ground for understanding both biological and cultural diversity. In both cases, Shapiro argued, human diversity was an adaptive response to varying conditions of life. Population genetics provided an explanation, through the isolation and adaptation of small bands of hunter-gatherers, for how "mankind became differentiated into races."[112] Similarly, culture was a means humans had "devised to deal with the full and direct impact of the natural environment for our benefit." Living cultures were "all experiments in man's adjustment to nature," creative social and material ways of "meeting the demands of his environment." As cultures became more complex, Shapiro argued, they "became more of a selective agent in determining the course of evolution."[113] Thus, for Shapiro it made perfect sense to begin his sequence of anthropological expositions with a hall devoted to human evolution, biology, and variation.

The *Biology of Man* hall was conceived in three parts: evolution and phylogeny, physiology and the functions of the human body, and a section variously referred to as "aggregates of man" or "group biology." The last section, on human variation and grouping, encompassed racial classifications, body types, and a shifting set of issues related to human migrations, demography, and population.[114] After nearly a decade languishing in the planning stage due to lack of funds, Shapiro's *Biology of Man* finally got underway with a $95,000 gift of stock from trustee John D. Rockefeller III in 1953.[115] Subsequent funds from the James Foundation and the city of New York enabled the museum to open the hall, with the first two sections completed in 1961.[116] The exhibition was, as promised, a

[112] Harry L. Shapiro, "Anthropology: The Races and Cultures of Man," p. 4, AMNH-HLS.
[113] Ibid., pp. 6–7.
[114] "Suggestions"; "Master Plan"; H.[arry] L. Shapiro, "Notes on Proposed Installation of Collections on First Floor," "Memorandum on the Biology of Man Hall," (1952), Exhibition Plans – Dept. of Anthropology, AMNH-HLS.
[115] "Plans for a New Hall of Biology of Man," Natural History, vol. 62, no. 4 (Apr. 1953), p. 192. The exhibition ultimately cost $450,000. *The American Museum of Natural History, Ninety-Second Annual Report, July, 1960 through June, 1961* (The City of New York, 1961), p. 4.
[116] "Newest Permanent Exhibition, the Hall of the Biology of Man Opens at American Museum of Natural History" (press release), 1961, p. 1, AMNH-HLS.

radical departure from prewar ethnological dioramas and glass cases full of artifacts. In retrospect, Shapiro's approach was a harbinger of now-common science museum techniques. It was perhaps appropriate that it was Harry Shapiro who superceded Franz Boas's ethnological habitat groups, which had been in their day, at the turn of the century, as innovative and critical a response to earlier synoptic, functionalist artifact cases as Shapiro's were to Boas's own life groups and culture area presentations.[117]

The *Biology of Man* resembled nothing so much as an illustrated and animated human evolution and biology textbook. Visitors encountered a guided tour, first through a demonstration of the anatomical evidence of human evolution and the fossil evidence from nonhuman ancestors to modern man, and then through a detailed exposition on the major bodily components and functions of the "human organism." The emphasis was squarely upon human beings as "a form of organic life," no different to the biologist or evolutionary scientist than horses or fruit flies.[118] The exhibition hall opened with a fourteen-foot-high spiral, a visual metaphor for the two billion years life had been evolving on earth, showing human evolution as "a tiny, almost imperceptible dot."[119] The first section located human beings in the broad evolutionary phylogeny, explaining which features people share with vertebrates, mammals, primates, and all humans.[120] Elements of the human body, sometimes enlarged to massive proportions, and details of their functioning were explained through textual panels or sound recordings – cells, mitosis and meiosis, fetal specimens, heart and lungs, locomotion, nervous system, circulation, digestion, and more. For example, an enlarged model of a cell proclaimed it "the basic unit of life," with labels and supplementary texts explaining the chemical and biological functions of the cell membrane, protoplasm, organelles, and nucleus. Later in the exhibit, a neuron enlarged to "ten thousand times life size" was displayed, accompanied by text explaining

[117] Alison Griffiths, *Wondrous Difference: Cinema, Anthropology, & Turn-of-the-Century Visual Culture* (New York: Columbia University Press, 2002); Ira Jacknis, "Franz Boas and Exhibits: On the Limitations of the Museum Method of Anthropology," George W. Stocking, Jr., ed., *Objects and Others: Essays on Museums and Material Culture*, History of Anthropology, vol. 3 (Madison: University of Wisconsin Press, 1985), pp. 75–111.

[118] "Supplement to the Hall of the Biology of Man," n.p., AMNH-HLS.

[119] Museum press release announcing the opening of the Hall of the *Biology of Man* (1961), p. 3, AMNH-HLS.

[120] "Supplement."

FIGURE 7.3. Harry Shapiro in front of the neuron display, *The Biology of Man* (1961). Courtesy of American Museum of Natural History Library, #327768.

the function of dendrites, axons, and synapses (Figure 7.3). The complexities of humans' upright gait – the relationship of bones, joints, tendons, and muscles – was explained as a "living leverage system" via a panel of aluminum silhouettes of a leg in motion. The section on the human organism ended with the wildly popular "Transparent Woman," a life-size clear-plastic model whose detailed internal organs lit up in sequence

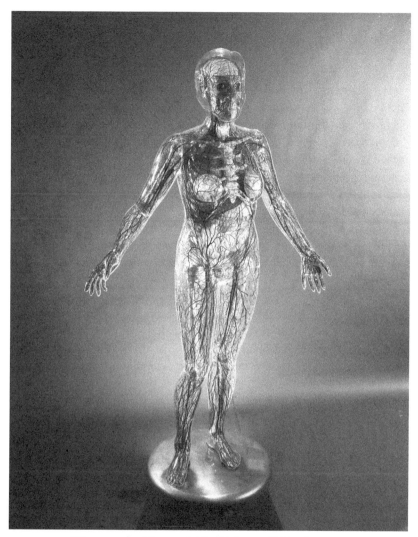

FIGURE 7.4. "Transparent Woman," *The Biology of Man*, American Museum of Natural History (1961). Image # 324639, American Museum of Natural History Library. Courtesy of American Museum of Natural History Library, #324639.

as a tape-recorded voice described their functions and relationships[121] (Figure 7.4).

The enormous detail provided in the *Biology of Man* about organs, structures, and functions of the human body emphasized the basic

---

[121] Press release, 1961, p. 4. The first "transparent woman" was imported from the German Hygiene Museum in 1936 and presented at the Museum of Science and Industry in New York City, "Medicine: Museum Piece," *TIME*, Monday, Aug. 31, 1936.

similarity of all human beings without saying it. The nervous system, bone structure, digestive system, blood – all were explained without any reference to the types of variation emphasized in racial classifications. People were referred to exclusively as "organisms" and placed in the context of nonhuman ancestors and evolution of the entire animal kingdom. Most of the exhibition did not present entire human bodies, but instead deconstructed human beings and their ancestors into their parts. Only the "Transparent Woman" was presented an integrated human body, but she lacked crucial markers of difference – no skin, facial features, hair. In that sense, she represented a raceless human organism, stripped of all racialized traits. She stands in stark contrast to the Field Museum figure that transcended the racialization in *The Races of Mankind*, "Nordic Man." The absence of racializing markers in the "Nordic Man" sculpture was not, in fact, the absence of race, but rather the presence of privileged whiteness and the absence of "otherness" or primitiveness. And unlike the Field Museum's story of human prehistory centered in Europe, the *Biology of Man* approached the evolutionary history of mankind with the advantage of a great deal more fossil evidence and a markedly changed theory of human evolution. In 1961, Harry Shapiro located human origins in Africa, pushed back in time millions of years to a prehominid ancestor, *Australopithecus*. In the *Biology of Man*, references to the development of intelligence, technology, and culture were described as part of universal human evolution, not the province of particularly gifted groups or racialized competition, as they had been thirty years earlier at the Field Museum and by Arthur Keith.

When *Biology of Man* opened in 1961, it superceded exhibits that had focussed on race and racial science throughout the 1920s and 1930s. In the 1920s, Louis Sullivan addressed what was seen then as a glaring absence of displays devoted to the "racial constitution and history of man" by drafting plans for a *Races of Mankind* hall that was to include an "exhibit of race types." Proposed elements included "typical face casts, detailed studies of the long bones, hair form, eye color and form, ear, stature, brain capacity, etc. . . . demonstrated by special models" as well as "a number of life-sized [painted] figures" including "a Dakota man, a lifecast of an Iroquois Indian, and of a northern European."[122] By 1923, in what became the *Races of Man* hall, sections devoted to Polynesia, Indonesia, Malaysia, "Eskimo," American Indian and "American white"

[122] The American Museum of Natural History, *Fifty-Fourth Annual Report for the Year 1922* (The City of New York, 1923), pp. 2, 101.

were complete or nearly so. Fullsize, painted figures arrayed as a "racial series" included Hawaiian, "Eastern Indian," "Eskimo," Bushmen, and "North European White."[123] In the 1930s, after Harry Shapiro had replaced Sullivan as the curator of physical anthropology, another series of painted, life-size figures "representing the chief racial types of the world" and a "large pictorial map of the world representing the spread of the primary races of man over the earth" were added to the hall. In 1941, a new exhibit "showing world distribution of the living types of man,"[124] was the last display installed in an American Museum hall devoted explicitly and exclusively to racial science.

At the same time that the Anthropology Department was augmenting the Museum's *Races of Man* displays, Harry Shapiro was also engaged in a collaboration with the comparative anatomy department on an exhibition that foreshadowed the biological turn physical anthropology would take after the 1940s. In 1931, Shapiro had developed sketches for figures intended to represent "chief racial types" for the new *Hall of the Natural History of Man*, a joint project between the departments of anthropology and comparative anatomy. Like the later *Biology of Man* hall, the *Natural History of Man* portrayed human evolution and variation, but did it primarily via comparative anatomy.[125] Anatomist and evolutionary theorist William King Gregory constructed displays that presented human evolution in a racialized context. His evolutionary tree showing "Man Among the Primates" featured human beings divided into four racialized figures highly reminiscent of those being exhibited at the Field Museum in the same period ("red" and African figures in loincloths holding a spear and bow; a "yellow" figure in Chinese garb; and a "white" figure posed like interwar bodybuilder Charles Atlas). The implication

[123] The American Museum of Natural History, *Fifty-Fifth Annual Report for the Year 1923* (The City of New York, 1924), p. 127.

[124] The American Museum of Natural History, *Sixty-Seventh Annual Report for the Year 1935* (The City of New York, 1936), pp. 13–14. The American Museum of Natural History, *Sixty-Eighth Annual Report for the Year 1936* (The City of New York, 1937), p. 35. The American Museum of Natural History, *Sixty-Ninth Annual Report for the Year 1937* (The City of New York, 1938), p. 18. The American Museum of Natural History, *Seventieth Annual Report for the Year 1938* (The City of New York, 1939), p. 23. The reports also note that the displays were being created with the help of Works Progress Administration (WPA) workers throughout the latter part of the 1930s. The American Museum of Natural History, *Seventy-Third Annual Report for the Year 1941* (The City of New York, 1942), p. 22.

[125] *Sixty-Third Annual Report of the Trustees for the Year 1931: The American Museum of Natural History* (The City of New York, 1931), pp. 46–47; American Museum *Annual Report*, 1935, p. 9; American Museum, *Annual Report*, 1936, p. 13.

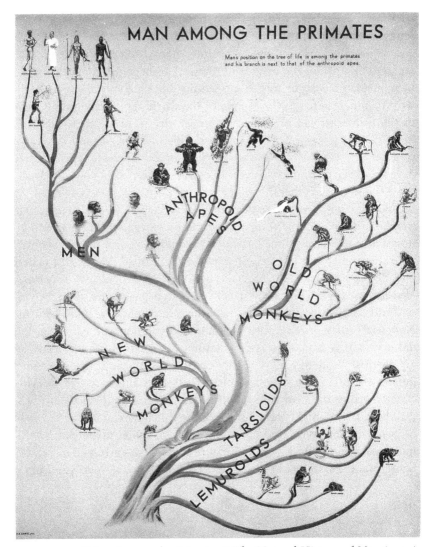

FIGURE 7.5. "Man Among the Primates," *The Natural History of Man* (1931).
Courtesy of American Museum of Natural History Library, #315595.

of Australian primitiveness was even more explicit in Gregory's family tree than at the Field Museum; Gregory positioned his boomerang-
wielding man between Charles Knight's Neanderthal figure and the four
primary racial stocks (Figure 7.5). Elsewhere, Gregory employed the
head of Apollo Belvedere to stand in for modern *Homo sapiens*, again
separated from an Australian figure interposed between the white race

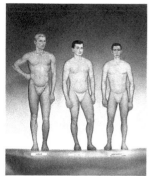

FIGURE 7.6. Racial types Negroid, Mongoloid, Caucasoid displayed in *The Natural History of Man* (1931). Courtesy of American Museum of Natural History Library, #320570.

as the pinnacle of human evolution and the lower animals, in a linear evolutionary sequence titled "Our Face From Fish to Man."[126] To these exhibits were added two cases containing colored life masks to show the "enormous diversity" of "physiognomy" throughout the world. Lodged between the cases of masks was another case filled with nine full-size, two-dimensional, full-color, partially clad figures illustrating the traditional racial triumvirate – Caucasian, Mongoloid, Negroid – each divided into three subtypes[127] (Figure 7.6). The Caucasian races included figures depicting the Nordic, Alpine, and Mediterranean types; Negroid subdivisions included the Negro, Bushman, and Australian types (the last

[126] Rainger, *An Agenda for Antiquity*; David Bindman, *Ape to Apollo: Aesthetics and the Idea of Race in the Eighteenth Century* (Ithaca: Cornell University Press, 2002); Kirk Savage, *Standing Soldiers, Kneeling Slaves: Race, War, and Monument in Nineteenth-Century America* (Princeton: Princeton University Press, 1997); Stephen Jay Gould, *The Mismeasure of Man* (New York: W. W. Norton & Company, 1996); Mary K. Coffey, "The American Adonis: A Natural History of the 'Average American' (Man), 1921–1932," in Susan Currell and Christina Codgell, eds., *Popular Eugenics: National Efficiency and American Mass Culture* (Athens, OH: Ohio University Press, 2006), pp. 185–216. This practice is reminiscent of Josiah Nott and George Gliddon's infamous illustration contrasting the head of Apollo Belvedere with a black man and a chimpanzee, *Types of Mankind* (Philadelphia: Lipponcott, Grambo & Co., 1854). For William King Gregory's account of the science portrayed in the exhibition, see *Our Face from Fish to Man: A Portrait Gallery of Our Ancient Ancestors and Kinfolk Together with a Concise History of Our Best Features* (New York: G. P. Putnam's Sons, 1929). For a discussion of visual hierarchies used to depict evolutionary science, see Constance A. Clark, *God – or Gorilla: Images of Evolution in the Jazz Age* (Baltimore: The Johns Hopkins University Press, 2008).

[127] "Natural History of Man Hall," n.d., Manuscripts, AMNH-HLS.

a controversial designation); and the Mongloid types included the Poly-nesian, Mongolian (usually understood as East Asian), and American Indian. This display remained in place on the third floor of the Museum until at least 1961.[128]

By the time "Human Variation," the final section of the *Biology of Man* hall, was opened in 1974, Harry Shapiro has spent more than forty years crafting presentations of race and racial types for a variety of exhibitions at the American Museum.[129] Over the course of more than twenty years of planning the Human Variation section of the *Biology of Man* hall, Shapiro consistently included racial differentiation as well as somatotypes[130] and the effects of the environment on human bodies.[131] By the 1950s, these components of section 3 were presented through the lens of population genetics as an exposition on "group biology and man's relation to his physical environment."[132] When the hall opened to the public with the first two sections complete in 1961, Shapiro promised museum visitors a third section that would include such "topics as heredity, population genetics, racial as well as other systems of variation, and population dynamics."[133] By 1968, plans for "Group Biology" included efforts to present a history of the race concept showing the gradual replacement of an older typological "blood-theory of heredity" with a populational concept of "human diversity" as "constantly chang-ing." A press release announcing the opening of "Human Variation" in 1974 stressed the "crucial similarities" that "make us all members of the human species," despite "our many physical and cultural differences."[134] But the 1968 plans also called for a huge 40′ × 14′ map of the world,

---

[128] Catalog date on Comparative Anatomy Hall photo showing display, Photographs, Image number 328291, Special Collections, American Museum of Natural History, New York.

[129] *The American Museum of Natural History, 106th Annual Report, July 1974, through June 1975* (The City of New York, 1975), p. 24; American Museum, *Annual Report*, 1931, pp. 46–47; 1961 press release.

[130] The theory of constitutional body types – mesomorph, endomorph, ectomorph – was devised by William Sheldon in the 1940s. Sheldon, *The Varieties of Human Physique: An Introduction to Constitutional Psychology* (New York: Harper, 1940).

[131] "Suggestions," p. 4; "Note on Proposed Installation," p. 2.

[132] "Biology of Man Hall" (handwritten notation reads "Museum Guide 1958"), p. 4, Exhibition Plans – Dept. of Anthropology, AMNH-HLS.

[133] "The Biology of Man at the American Museum of Natural History," *Museum News*, vol. 39, no. 9 (Jun. 1961), typescript, p. 5, AMNH-HLS.

[134] Museum press release announcing the opening of the final section of the hall of the *Biology of Man*, no title, no date (1974), p. 1, AMNH-HLS.

FIGURE 7.7. "Human Variation" hall, *The Biology of Man* exhibition (1974). Courtesy of American Museum of Natural History Library, #65879_25.

"spotted" with seventy life masks "in natural skin tones" showing "the areas of their origin" to be the focal point of the section. Intended to emphasize "the amazing diversity of skin color and appearance across the face of the earth" (utilizing "dramatic lighting" to "bring out the facial characteristics and bone structure of each mask"), the display continued to map variation in terms of skin color and facial physiognomy tied to geography.[135] By the time the section on "Human Variation" opened in 1974, the mural eschewed a map and face masks for a wall-size collage of photographs depicting people from all over the world, reminiscent of Edward Steichen's "Family of Man" exhibition nearly twenty years earlier.[136] A nearby case paired a giant model of a chromosome with a bank of human figures whose heads, torsos, and lower bodies turned independently, mixing and matching a range of physically and culturally diverse individuals in a visual metaphor for the process of genetic mixing that produces human diversity (Figure 7.7). Despite the emphasis in the exhibition on genetics and "environmental factors," when "Human Variation" opened remnants of earlier preoccupations and theoretical associations lingered. "Genes, chromosomes, mutations, karyotypes, skin color, climate, and Watson and Crick," were all

[135] "Group Biology," May 1968, n.p., Box 66, AMNH-HLS.
[136] Edward Steichen, *The Family of Man; The Greatest Photographic Exhibition of All Time* (New York: Published for the Museum of Modern Art by the Maco Magazine Corp., 1955).

"translated into cartoons, films or photographs to tell the Museum-goer why one person's nose is longer than another's and why some Africans grow to be seven feet tall."[137] The discussion of human diversity continued to be framed in terms of skin color and climate, still exemplified through Africa, catering to a persistent EuroAmerican fascination with primitivism and exoticism, now translated through biology. Straining to capture popular interest, the museum touted the exhibition as a set of scientific answers, grounded in universal biological principles, to a set of nagging questions about human variation.

The vacillation between lingering racialization and evolutionary humanism captured the way physical anthropologists in the postwar period struggled with race and human diversity. It was much more difficult to disentangle bodies and cultures than it had seemed in the 1940s when anthropologists started vigorously urging the public to see race and culture as separate realms of knowledge and consequence. By the 1960s, anthropologists found themselves conveying dual, sometimes apparently conflicting messages. On one hand, they continued to offer the public a narrative that implied human variation, exemplified in age-old distinctions such as skin color or African exoticism, remained a topic of legitimate scientific interest. On the other hand, anthropologists strove to contain the topic of human diversity within a rhetoric of unity and "brotherhood," grounded in an evolutionary rubric of almost limitless variation, constant migration and mixing, and unremitting change. Both messages seemed necessary. By the late 1960s, the term "race" was widely shunned among physical – or biological – anthropologists, but interest in those things that once fell under that rubric continued in many quarters. In the wider world, the politics and culture of race and science had changed dramatically since the interwar years. Civil rights activism, post-colonial movements, and challenges to authority – to the authority of scientists to name and classify, and to the sociopolitical establishment to dominate those who had formerly been subjects of racial science – all created an environment in which race and culture continued to be in tension and intertwined, a perpetually unstable but powerful entanglement.

Conclusion

In the 1930s and 1940s, Harry Shapiro, Ruth Benedict, Ashley Montagu, and many other scientists lived in a climate in which it seemed politically,

---

[137] Press release, 1974, p. 1.

scientifically, and socially necessary to assert a humanist framework over and against a racialist, and to many, a racist one. But race and culture proved difficult to disentangle. And racism proved much more resistant to the persistent application of facts than anthropologists and other scientists anticipated.

Confronted by war and divisive racial policies at home and abroad, these anthropologists attempted to grapple with how to present human physical and cultural diversity for public consumption. Their early, sometimes awkward efforts to emphasize culture and brotherhood over divisive somatic taxonomies reveal how hard it was at that time to disentangle bodies and cultures, even when the social and political moment seemed to call for it. Straddling a world in which diverse bodies and lifeways intersected in complex ways across time and space, anthropologists such as Benedict, Weltfish, Montagu, and Shapiro – along with many others in America and Europe – embraced culture but did not fully reject race. The powerful residue of racial essence and type in their texts and images illuminates how deeply intertwined notions of bodies and cultures still were, and how deeply ingrained the "common sense" that human beings should be divisible into natural kinds, even as anthropologists and others strove to dismantle the scientific and social apparatus that had erected such notions.

Traditional notions of race and type were especially difficult to dislodge without a new theory of human development and biology with which to make sense of evident human variation. In the early twentieth century, genetics by itself had not provided such a new framework for thinking about race. If anything, genetics had been used, and abused, to essentialize racial typologies more firmly. Ruth Benedict and other Boasians hoped a thorough appreciation for cultural relativity, combined with a strong sense of social justice and democracy, and a bracing dose of facts about the limits of hereditary differences would overcome racial prejudice. Their hopes may now seem naively humanist and excessively positivist. As the years wore on, in the face of the persistent intermingling of race and culture in essentialist and hereditarian terms, in both folk and scientific realms, most cultural anthropologists and many others abandoned the race concept as Ashley Montagu had urged. Cultural anthropologists left race, or its remnants, to the biological anthropologists and geneticists. No longer would one find cultural anthropologists such as Ruth Benedict asserting the reality and utility of studying human variation, or, like Margaret Mead, setting off to study both the biological and cultural life of a people. Eventually, the holistic racialism of Boas, Benedict, Mead, and others was first minimized and then lost altogether. By the 1950s, Montagu, Mead, Weltfish, and others claimed the mantle of Boasian cultural relativism and the rejection of racism, now understood as the rejection of race, as well. Supporters and defenders of Franz Boas began to reimagine him as the father of a cultural relativism in ways that went well beyond Boas himself in rejecting race for a thoroughgoing cultural relativism.[138]

Physical anthropologists also struggled with how to reimagine their work. Anthropologists such as Harry Shapiro grasped eagerly at the rapidly consolidating field of population genetics for a more promising way to interpret human variation, one many hoped would be more likely to yield real insights about human diversity and less likely to provoke misguided racialism and bloody conflicts. But this new framework, population genetics, was not widely internalized among anthropologists or the general public until at least the 1960s, and probably later for the general public.[139] In the meantime, physical, then biological, anthropologists struggled with the old and the new.

---

[138] Positioning Boasian cultural anthropology as thoroughly relativistic, unrelated to studies of human physical variation, was, in part, also a response to the resurgence in the 1940s of evolutionary anthropology in the work of people such as Leslie White and Julian Steward.

[139] Alice Littlefield, et al., "Redefining Race: The Potential Demise of a Concept in Physical Anthropology [and Comments and Reply]," *Current Anthropology*, vol. 23, no. 6 (Dec. 1982), pp. 641–655.

8

Conclusion

The Persistence of Race

By the 1960s and 1970s, a cursory examination of biology and anthropology would seem to reveal the virtual disappearance of race. But the anxieties, assumptions, and fascinations that had driven racial science and rhetoric for decades had not disappeared. Concerns about human diversity that had once been expressed in the language of racial essentialism and biological determinism were translated into anxiety about world population, interest in human evolution and migration, and arguments about persistent gaps in achievement in works such as the Moynihan Report in the 1960s, and later again in the 1990s, in Richard J. Herrnstein and Charles Murray's *The Bell Curve: Intelligence and Class Structure in American Life.* In human genetics and medical research, the racialized body flourished. In society more broadly, liberal humanist efforts to repudiate race as a concept and racism as an individual failing met with mixed success. Attempts to reorient racial stereotyping into a celebration of ethnic cultural diversity in the sciences as well as in popular culture (e.g., Edward Steichen's *Family of Man* exhibition at the Metropolitan Museum of Art in 1955, Coca-Cola's popular 1971 television advertisement "I'd Like to Buy the World a Coke," the United Colors of Benetton advertisements of the 1980s, and Disney's "Pocahontas" film in the 1990s) often only masked a persistent biological determinism or racially essentializing construct. The waning of race, proclaimed as recently as Elazar Barkan's *Retreat of Scientific Racism,* was once again waxing by the early years of the new century. Books such as *Race: The Reality of Human Differences* (2005), whose authors, anthropologist Vincent Sarich and journalist Frank Miele, decried a Boasian "abyss of deconstructionism," found a wide popular audience. In the late twentieth and early twenty-first century, the very real repercussions of the concept of race in the social, economic, and political lives of the American people continued to coexist uneasily with a "common sense" belief in the biological reality of race that had been promoted, alongside culture, for decades.[1]

Exhibitions of Race and the Demise of the *Races of Mankind*

Harry Shapiro's *Biology of Man* hall, the American Museum's *Races of Man* hall, William K. Gregory's comparative anatomy hall, *The Natural History of Man*, and the Field Museum's *Races of Mankind,* marked the end of an era that had begun more than 100 years earlier with Samuel Morton and the American School of physical anthropology. Opening in 1961, Shapiro's hall ushered in a profoundly biological approach to the "human organism" that sought to divorce anthropology from its racialized praxis. The aging *Natural History of Man* and its Midwest companion, the *Races of Mankind*, were by that time museological and scientific dinosaurs. They were the last time a venue of the size, scope, and stature of the American Museum and the Field Museum would present images of Caucasian, Mongoloid, and Negroid figures to

---

[1] Daniel Patrick Moynihan, "The Negro Family: The Case for National Action," United States Department of Labor, Office of Policy Planning and Research (Washington, DC: U.S. Gov't Printing Office, 1965). Moynihan's (in)famous report drew upon Oscar Lewis, *Five Families: Mexican Case Studies in the Culture of Poverty* (New York: Basic Books, 1959). Richard J. Herrnstein and Charles Murray, *The Bell Curve: Intelligence and Class Structure in American Life* (New York: Free Press, 1994); Edward Steichen, *The Family of Man; The Greatest Photographic Exhibition of All Time* (New York: Published for the Museum of Modern Art by the Maco Magazine Corp., 1955); "The 'Hilltop' Ad: The Story of a Commercial," American Memory, Library of Congress, retrieved from http://memory.loc.gov/ammem/ccmphtml/colaadv.html (accessed July 26, 2013). On race and the advertising of Benetton, see Marion Maguire, *United Colors of Benetton: A Company of Colors and Controversies* (GRIN Verlag, 2007); Leigh H. Edwards, "The United Colors of 'Pocahontas': Synthetic Miscegenation and Disney's Multiculturalism," *Narrative*, vol. 7, no. 2, Multiculturalism and Narrative (May 1999), pp. 147–168; David Roediger, "Guineas, Wiggers, and the Dramas of Racialized Culture," *American Literary History*, vol. 7, no. 4 (Winter, 1995), pp. 654–668; Henry A. Giroux, "Consuming Social Change: The 'United Colors of Benetton,'" *Cultural Critique*, no. 26 (Winter 1993–1994), pp. 5–32; and www.benetton.com. Elazar Barkan, *The Retreat of Scientific Racism: Changing Concepts of Race in Britain and the United States between the World Wars* (Cambridge: Cambridge University Press, 1992); Vincent Sarich and Frank Miele, *Race: The Reality of Human Differences* (Boulder, CO: Westview Press, 2004), pp. 90–91.

the public as science. While the *Biology of Man* was in many respects contemporary, even forward looking, in both content and presentation, the hall as it evolved over the course of some fifteen years reflected the profound changes that physical anthropology had been through in the postwar years. Viewing the history of race through exhibition, the *Races of Mankind*, first at the Field Museum and then as the moniker of a quite different racial exposition at the Cranbrook Institute of Science, the *Natural History of Man*, and finally the *Biology of Man*, capture the complex trajectory of racial science in anthropology and the natural history museum.

By 1967, when the Field Museum's *Races of Mankind* exhibition was dismantled, the premise that Malvina Hoffman's sculptures represented accurate scientific rendering of racial types, as well as moving artistic reflections of humanity, was vigorously contested.[2] By the late 1960s, Hoffman, who had been widely known and praised following publicity surrounding the *Races of Mankind* in the 1930s, was fading into obscurity. Art critics and historians have viewed her effort to accurately register human characteristics in bronze as an "anthropological exercise" that excluded the Field Museum sculptures from the pantheon of serious art.[3] The figures themselves and the exhibit as a whole came under attack as not only inaccurately rendered and categorized, but fundamentally racist in their construction.

Activist Leroi Jones, writing on behalf of the Committee for Unified Newark in 1969, addressed a scathing letter to the museum condemning the "Map of Mankind" (Figure 8.1) based on the *Races of Mankind* figures and approved by the Museum, as full of "consistent and glaring inaccuracies... which can only be the result of ignorant white nationalism and white racism.... We wish to protest the existence of this map and any other product put out by Hammond and Company which incorporates [*sic*] this kind of white racist pseudo anthropology."[4] Malvina Hoffman undertook the "Map of Mankind" project in the 1940s with C. S. Hammond and Company. Initially to be called "The Races of

[2] The exhibition was dismantled in 1967, just after Malvina Hoffman's death, on the pretext that the space was needed for other uses. Races of Mankind, Department of Anthropology Archives, The Field Museum (hereafter FM-ROM); Archives, Department of Anthropology, The Field Museum, Chicago.

[3] Wayne Craven, *Sculpture in America* (New York: Thomas Y. Crowell Company, Inc., 1968), p. 560.

[4] Leroi Jones to E. Leland Webber, Director, and Leon Siroto, Assistant Curator of African Ethnology, August 7, 1969. FM-ROM. Jones later changed his name to Amiri Baraka.

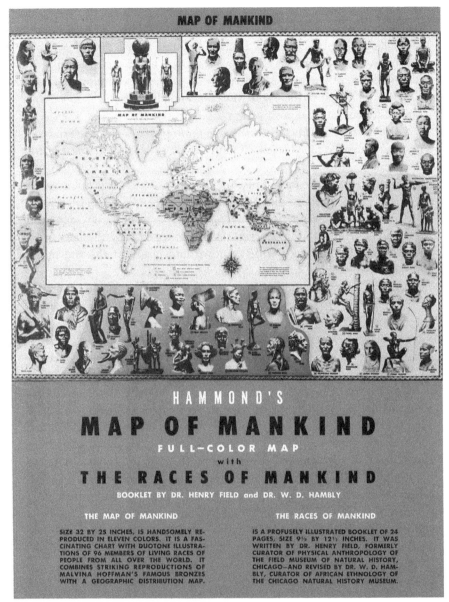

FIGURE 8.1. Malvina Hoffman, Henry Field, C. S. Hammond and Co. "Map of Mankind," cover (1946). *Source: Hammond's Map of Mankind, Full-Color Map with The Races of Mankind*, by Dr. Henry Field and Dr. W. D. Hambly (New York: C. S. Hammond and Co., 1946).

the World and Where They Live," it was a map of the world in color, bordered by photographs of seventy-seven of Hoffman's Field Museum figures, represented with color-coded backgrounds to indicate the part of the world they inhabited. Hoffman had authorized the company to use any text from *Heads and Tales* to enhance the depictions of her bronzes. Although the Field Museum authorized the depictions of Hoffman's sculptures, Curator Paul Martin, who took over after Berthold Laufer's death, was uncomfortable with the project from the start, concerned that the map was anthropologically questionable. After the map had already been printed, with anticipated sales of 25,000 to IBM for distribution to public schools, Martin wrote to the company expressing his concerns. He noted that although "any child" could easily discern where the "various types of human beings" lived, there was no way

of ascertaining to what racial types the various figures belong. The names of the pictures refer to linguistic, social, geographical, cultural and racial classifications, and in a number of cases diverse racial types are grouped on the same background color. The groupings on colored backgrounds imply single racial types and this is incorrect. From this point of view the map is misnamed.[5]

In 1944, the American Council of the Institute of Pacific Relations refused to "push the sale" of the map, asserting that it was "unsuitable for school purposes." Acknowledging that Hoffman's sculptures were among "the great artistic achievements of this generation," the map, however, depicted "small minority groups rarely seen" that would be at odds with teachers' efforts to

avoid any possible association in his students' minds of the great majority of colored peoples of the world with small remnants of aboriginal peoples that dwell somewhere on inaccessible mountain slopes. Moreover, he would stress similarities rather than differences, by singling out representative types that are at least not repulsive. He would prefer . . . some indication of the dignity of the occupations and livelihood of the world's common people to any intended or unintended indication of savagery.[6]

---

[5] Copy of letter from Paul C. Martin to Davis (at Hammond and Company) no date; Malvina Hoffman to (George) Davis, December 29, 1943, Malvina Hoffman Collection, 850042, Special Collections, The Getty Center for the History of Art and the Humanities, Los Angeles, California (hereafter MHC).

[6] Edward C. Carter to (George) Davis, January 5, 1944, MHC. Carter also noted "one or two unfortunate political implications." The recognition of the "Manchukuo in view of the war aims of the United Nations seems a trifle unfortunate," for example. Hoffman characterized Carter's remarks as simply "hit or miss," "hasty" criticisms, noting that it was much easier to criticize than to create. Malvina Hoffman to Paul Martin, February 18, 1944, MHC.

Six months after the museum received Leroi Jones's letter denouncing the "Map" in 1969, the Museum asked C. S. Hammond & Company to discontinue the "Map" and the "Races of Mankind" atlas that accompanied it, stating that "We do not wish the Field Museum's name to be associated with it. . . . 'The Races of Mankind' as it now stands is scientifically indefensible and social [*sic*] objectionable."[7]

In the 1970s, the museum reinstalled the figures outside the exhibition halls as a "Portrait of Man," intentionally placed to "disassociate them from the Field Museum's anthropology exhibits" and instead to present them as random "individuals from here and there about the world" so that "the viewer will be less apt to take them as a statement on race."[8] Still scattered around the museum, the sculptures are presented by the museum without any order or hierarchy and without any anthropological information – they are merely decorative objects, displaying human physical and cultural diversity, but explicitly not typology. For anthropologists, the artistic power of the bronze sculptures was insufficient to justify their exhibition in a museum of science because the racial hierarchy they embodied was no longer acceptable. The very permanence and plasticity of bronze that had appealed to anthropologists in 1933 had eventually made Hoffman's sculptures useless to the museum except as decoration.[9] In a postcolonial era when many viewed race as no more than a contingent mixture of genes resulting in an indivisible range of characteristics and racial and ethnic groups actively protested formerly legitimate race constructions, the Field Museum could no longer display Hoffman's sculptures as artifacts of natural history.

Yet, despite their precautions, museum anthropologists could neither rein in the authority of the science museum setting nor control how visitors understood Hoffman's sculptures. As recently as 1993, Hoffman's sculptures continued to be viewed by some visitors as remarkably evocative and authoritative depictions of humanity, not as monuments to racial hierarchy or despicable totems of institutionalized racism, but as hopeful tributes to human spirit and diversity. A review of the *Races of Mankind* sculptures still displayed in the Museum argued once again that the "statues depict not only racial types but also the inner character of the subjects, capturing the human spirit as it's expressed in each

---

[7] Draft, to Ashley P. Talbot, Senior Editor, C. S. Hammond & Company, January 29, 1970, FM-ROM.

[8] Glen Cole, "Suggested Letter Re: Display of Malvina Hoffman Sculptures," December 1984. Archives, Department of Anthropology, The Field Museum, Chicago.

[9] Ibid.

individual ... balanced in a magically arrested moment that's both immediate and timeless.... Malvina Hoffman's remarkable bronzes ... speak
eloquently of the spirit we share as well as of the characteristics that
distinguish us."[10] Significantly, these viewers saw in the aesthetic appeal
of the sculptures a humanistic counterpoint to scientific claims about
human physical diversity that made the message they took away about
race more, rather than less, coherent and compelling. They argued that
the "subjectivity of art has the potential to enhance scientific inquiry
rather than impede it," and contended that "the disparate unity captured
in these statues, their varied wholeness, is Babel reformed into coherence
by artistic vision."[11] Moreover, these visitors seem to have taken away
an altogether different message about permanence than the one originally
envisioned by Laufer, Field, and Keith. Where the curators understood
Hoffman's sculptures to be preserving for posterity specimens of peoples
who were vanishing from the earth, these visitors understood them as
an emblem of persistence: "Despite their variety, each [sculpture] says
the same thing: not only will we survive, but we will do it in all of our
glorious diversity."[12]

Malvina Hoffman's sculptures have an ambiguous legacy that has
bedeviled the Field Museum since they dismantled the *Races of Mankind*.
The exhibit itself, and the anthropology embedded in it, eventually had
a much less ambiguous fate. The 1930s was probably the last time in
America a museum could have mounted exhibitions such as the *Races
of Mankind*, or the *Races of Man* at the American Museum. A growing critique within anthropology and without, depleted budgets in the
1930s and 1940s, and finally the Second World War and the subsequent
broad repudiation of racial science meant the end of such exhibitions
(although not of race, even in museum display). If we look for a postwar

---

[10] David C. Conrad and Karen J. Tillotson, "The Races of Man," *Reader, Chicago's
Free Weekly*, vol. 22 (1993), no. 28, section 1, pp. 8–9, 30. For similar reactions, see
also, Theodore Wolff, "'Old-Fashioned' Art You Would Love to Hold," *The Christian
Science Monitor*, Wednesday, Apr. 30, 1980, p. 19. The belief that art and science
are compatible representational practices, that art can enhance the ability of scientists
to reveal truths about the natural world, harkens back to the eighteenth century and
earlier representations of the body. See, for example, Lorraine Daston and Peter Galison,
"The Image of Objectivity," *Representations* 40 (Fall 1992), pp. 81–128; Barbara Maria
Stafford, *Body Criticism: Imaging the Unseen in Enlightenment Art and Medicine* (New
York: Routledge, 1991); Pamela H. Smith and Paula Findlen, eds., *Merchants and
Marvels: Commerce, Science and Art in Early Modern Europe* (New York: Routledge,
2002).

[11] Conrad and Tillotson, "The Races of Man," pp. 8–9.

[12] Ibid.

analogue to the *Races of Mankind*, we find it not in the halls of science but in *The Family of Man*, a massive photographic exhibition mounted at the Metropolitan Museum of Art in New York in 1951.[13] It is the modernist counterpart to the Field Museum's interwar brew of racialism and humanism, highbrow art and mass appeal. In broad terms, they have a similar sweeping, global reach; a similar concern to present a reassuring message in a time of social unease; the same preoccupation with human categories and geographic representation; and a provocative mix of race and culture. But instead of bronze, the Met deployed photography; instead of natural history, it was art; instead of displaying race, it celebrated nationality and ethnicity; and instead of the authority of science, the Met proffered human nature and kinship. By 1974, when the American Museum finally opened the last section of the *Biology of Man* hall, it, too, employed a photographic collage of human diversity to speak to the American public about human variation.[14]

The Dilemma of Race: Are We So Different?

Harry Shapiro spanned a period of remarkable change in the science and public perceptions of race. Trained in physical anthropology in the

[13] Edward Steichen, *The Family of Man*. Greg Foster-Rice argues that traditional artistic and representational styles, and antimodern aesthetics, were allied in the twentieth century to essentialized racial categories and racial hierarchies that promoted the kind of fixed racial types and notions of superiority seen in classificatory schemes such as Aleš Hrdlička's "Old Americans." A more modern aesthetic was associated with Boasian cultural relativism and a rejection of primitivism as seen in the modernism of the New Negro movement's representations of blackness. A similar contrast can be seen between the Field Museum's *Races of Mankind* hall and the WWII-era and postwar exhibitions discussed in this book. See Foster-Rice "The Visuality of Race: 'The Old Americans,' 'The New Negro' and American Art, c. 1925" (Ph.D diss., Northwestern University, 2003).

[14] The American Museum completed the transformation of physical anthropology's museological shift in the twentieth century from racial typology to physiology and evolutionary biology when it demolished the Biology of Man hall in 1988 to make way for the Hall of Human Biology and Evolution that opened in 1993. This new hall, funded in part with a $500,000 grant from the National Science Foundation, showcased genetics, physiology, and especially human evolutionary theory with fleshed-out reconstructions of human ancestors in recreated prehistoric sites, including an *Australopithecus afarensis* couple (based on the fossil skeleton called "Lucy") strolling through the African savannah and a portion of the Lascaux caves in France. The new hall retained one vestige of the old *Biology of Man* hall: a hologram of the "Transparent Woman." *The American Museum of Natural History, 120th Annual Report 1988/89* (The City of New York, 1989), pp. 1, 9, 45. *The American Museum of Natural History, 122nd Annual Report 1990/91* (The City of New York, 1991), pp. 1, 10, 46. *The American Museum of Natural History, 123rd Annual Report 1991/92* (The City of New York, 1992), pp. 13, 63, 67. *The American Museum of Natural History, 124th Annual Report 1992/93* (The City of New York, 1993), pp. 9, 14, 71, 75, 77.

1920s, Shapiro continued actively working into the 1980s.[15] Starting out as a Boasian anthropometrist, compiling thousands of measurements and indices to test hypotheses about racial mixing and the impact of the environment on the human constitution, in the postwar phase of his career Shapiro transformed his own early interests in race toward projects more in line with the evolving biological and paleontological emphasis in his field, pursuing an interest in Peking Man, planning and mounting a hall devoted to human biology, and developing the field of forensic anthropology, including the racial identification of human remains. The transformation of his career mirrored the trajectory of race in American anthropology.[16] Race had been transformed from an organizing concept in anthropology, especially physical anthropology, into an increasingly marginalized and discredited notion. Some physical anthropologists such as Carleton Coon and Stanley Garn committed to a typological rubric they had adopted decades earlier, continued to publish racial classifications, but they were greeted with increasingly hostile and dismissive reviews.[17] Committed racialists and hereditarians who rejected the cultural turn in anthropology, they were largely exiled to the margins, funded by a handful of like-minded private foundations such as the Pioneer Fund.[18] But the race concept hung on in medical, public health, and forensic work, where the intermingling of bodies and cultures persisted. Race also migrated into international population studies and human genetics. In these fields, although the term "race" was often avoided, the consequences of decades of racialization among the populations they studied, and the racial assumptions embedded in the way problems were framed, meant these were fields where race persisted.

[15] Shapiro retired from the museum in 1970 but continued to work and publish throughout the 1970s and into the 1980s. Shapiro spent his entire professional career, apart from adjunct teaching at Columbia University, at the museum. He continued to come into his office in the museum until shortly before his death on January 7, 1990, at eighty-seven. Frank Spencer, "Harry Lionel Shapiro, March 19, 1902–January 7, 1990," *Biographical Memoirs*, vol. 70 (1996), National Academy of Sciences, pp. 369–387.

[16] His career also mirrored the peak and decline of museums as a locus for anthropological research and discipline building. By the 1940s, the trend in American anthropology was clearly away from museum-based anthropology and toward universities, as funding sources shifted from private philanthropy and individual wealthy donors to large foundations and government support. George W. Stocking, Jr., "Ideas and Institutions in American Anthropology," *The Ethnographer's Magic* (Madison: University of Wisconsin Press, 1992).

[17] Carleton Coon, *The Origin of Races* (New York: Knopf, 1962); Carleton Coon, *The Living Races of Man* (New York: Knopf, 1965); Stanley M. Garn, *Human Races* (Springfield, Ill: Thomas, 1960).

[18] William H. Tucker, *The Funding of Scientific Racism: Wickliffe Draper and the Pioneer Fund* (Urbana and Chicago: University of Illinois Press, 2002).

Franz Boas, Ruth Benedict, and Harry Shapiro faced a dilemma that still haunts efforts to speak about race, without further reifying it. While Boas, Benedict, and Shapiro believed in the reality of race in ways few modern anthropologists do, they faced a similar skepticism from members of the public who felt their everyday lives confirmed the reality of race and justified their prejudices. For Benedict and the rest, their strategy was to grant the reality of race; disseminate accurate information gathered by leading anthropologists, psychologists, and other scientists that limited the purview of racial analysis to hereditary physical features of negligible import; and then move on to debate racist claims in cultural, social, and political terms. Strategically, culture was likely to be a more fruitful ground for debate, more likely to yield changed perceptions and behavior. Unfortunately, while they put aside race in the short term by naturalizing supposedly innocuous differences as irrelevant to the cultural problem of racism, in the long run, setting aside race as a purely natural category has lent the authority of science to divisive hereditarian claims ever since.[19]

In 2007, anthropologists, for the first time in decades, actively injected themselves back into debates about race with the traveling exhibition *Race: Are We So Different?*[20] The exhibition title echoes the central message in *The Brotherhood of Man*, the animated version of Ruth Benedict and Gene Weltfish's *Races of Mankind* wartime pamphlet, released sixty years earlier. When Henry eagerly heads out into his backyard to meet his new neighbors from around the world, his "Green devil" restrains him (see Figure 6.9):

> Green [devil]: It'll never work! You can't get along with those people – they're too different!
> Henry: We'll get along – we've got to! The future of our civilization depends on brotherhood!
> [a fight, spurred by each neighbor's Green devil, ensues]
> Narrator: Wait a minute – what about this business of brotherhood?
> Henry: But we're all different.
> Narrator: Are you? Let's take a look at the facts."[21]

[19] See John P. Jackson, Jr., "'In Ways Unacademical': The Reception of Carleton S. Coon's *The Origin of Races*," *Journal of the History of Biology* 34 (2001), pp. 247–285; Tucker, *The Funding of Scientific Racism*.

[20] For further information about the *Race: Are We So Different?* exhibition, including the associated educational materials and scholarship, see http://www.understandingrace.org/home.html.

[21] Ring Lardner, Jr., Maurice Rapf, John Hubley, and Phil Eastman, "'Brotherhood of Man': A Script," *Hollywood Quarterly*, vol. 1, no. 4 (Jul. 1946), pp. 353–359 (354–355).

Alarmed by the resurgence of race or its proxies in medicine, pharmacology, and especially human genetics in the 1990s, many anthropologists perceived a need to craft an alternative expert, scientific discourse to counter persistent essentialized, deterministic, and racialized frameworks for understanding human variation.[22] Their response, even to the level of their rhetorical claims and catch phrases, was strikingly reminiscent of wartime anthropology. After decades of distancing themselves from race and racial debates, *Race: Are We So Different?*, and the disciplinary discussion within anthropology that preceded it, signaled their reentry into the fray through a highly visible repudiation and rebuttal of the sorts of racial science and concepts of race once closely associated with anthropology.

Designed and originally mounted at the Science Museum of Minnesota, the location in a museum of science reinforced the perception that the AAA was countering the claims of physicians, biologists, and geneticists – and the folk conception of race as biologically, scientifically grounded – with equally scientific arguments and evidence (Figure 8.2). The exhibition approached race historically, culturally, and scientifically, conveying two broad messages. First, race, as historically and culturally constructed, is not a conceptual category that accurately describes or explains naturally occurring human variation: Race is not real. Second, as a social construct that shapes how people live, race is very real. These apparently contradictory messages were intended to address two key features of popular notions of race: that it is rooted in essential biological features and processes, and that everyday experiences provide evidence of this. Or as Vivian Ota Wang, of the National Human Genome Research Institute, put it, "You may tell people that race isn't real and doesn't matter, but they can't catch a cab. So unless we take that into account it makes us sound crazy."[23]

The exhibit asks, "Are We So Different?" Of course, the answer is supposed to be, "No." Echoing Edward Steichen's "Family of Man," the answer to the AAA's overarching question is that we are really all the same. Except this time, instead of invoking a neo-humanist common "human nature," anthropologists invoked genetics to argue that all human beings are the product of evolution and continuous mixing, a

---

[22] C. C. Mukhopadhyay and Yolanda T. Moses, "Reestablishing 'Race' in Anthropological Discourse," *American Anthropologist*, 99(3), 1997, pp. 517–533; Harrison, "Introduction: Expanding the Discourse on 'Race,'" *American Anthropologist*, 100(3), 1999, pp. 609–631.

[23] R. Weiss, "Scientists Find a DNA Change That Accounts for White Skin," *Washington Post*, Dec. 16, 2005, p. A1.

FIGURE 8.2. Multiracial logo image from *Race: Are We So Different?* (2007). Courtesy of American Anthropological Association.

commonality reflected in a genome that is overwhelmingly similar wherever one looks at living people. Presented convincingly, this is a powerful argument against a key element of American racial logic. But there is a danger in this humanism, in elevating sameness at the expense of diversity. Asking "Are We So Different?" can be interpreted as denigrating difference, and particularly blackness, and taken to an extreme, might suggest that the best way to eliminate racism and inequality is mixing to the point that all differences of appearance and behavior disappear. The more common trend of late is not the invocation of humanism and sameness, but the embrace of multiculturalism and diversity, including the rejection of rigid categories and dichotomies in favor of a proliferation of identities. For people who reject race as a biological category but embrace diversity, invoking sameness or its cousin, color-blindness, risks erasing differences and identities that provide meaning and enable resistance. The AAA exhibition must attempt to convince visitors that humans are fundamentally the same without denigrating diversity, and promote diversity without reinforcing essentialist racialism.[24]

[24] Harrison, pp. 618–619.

There is another side to the question "Are We So Different?" and the unstated reply, which also links this modern anthropological response to racism with mid-century arguments that also marshaled genetics on all sides of the debate. Like the current exhibition, Ruth Benedict, Harry Shapiro, Ashley Montagu, and a host of other anthropologists and geneticists argued strenuously that although hereditary, genetic differences existed among people, their manifestations were minor compared to the overwhelming similarities among human beings. They argued repeatedly on the basis of both scientific evidence and human experience that apparent human variation was a matter of diversity within a much deeper unity, that genetically and culturally human beings were more alike than different. But the question of just what sorts of differences there were, where they came from, and what import they might have for society would not go away, and threatened to undermine antiracist formulations that relied upon the assertion that human beings were either entirely the same or so similar that any differences were small and innocuous. In the 1950s and 1960s, physical anthropologists and geneticists, as well as a number of psychologists, argued strenuously with cultural anthropologists and other social scientists about the basis for such claims, and their consequences. What if the answer to "Are We So Different?" is "Yes"? In the postwar period, many of those who continued to study human variation argued vociferously that there was little hard evidence to assert that human beings were virtually identical (of course, they could not be perfectly identical, if they were products of evolution, as all agreed they were), nor evidence to assert with confidence that what differences there were or that might be more conclusively demonstrated were merely innocuous "variations on a theme." What happens to arguments for equality and justice if they are founded upon the requirement of "equal endowments"? This, of course, is not a new debate in American history. Americans in the nineteenth century were consumed with this very question, most often answering that unequal endowments justified unequal treatment and conditions.

Worried that the answer to the question of whether more notable heritable differences existed might ultimately be "Yes," and wary of offering racists an opening to again marshal evidence of hereditary differences to support social, economic, and political inequity, anthropologists in the mid-twentieth century asserted not only that such differences as may exist were harmless, but also that differences of any kind or degree were irrelevant. Mid-century anthropologists did indeed reify race further by insisting that race was real, demarcating it as a legitimate subject alongside culture. But they also made the argument that differences, even possibly large differences (which most argued were quite unlikely, and even where they existed were subject to modification through cultural interventions), were irrelevant to justice in a democratic society. This argument is different, and more powerful, than simply insisting that differences are small and harmless, that they don't differentiate people very much, nor impact life in any meaningful way. An argument about the utter irrelevance of hereditary differences, whatever their magnitude, to the determination of values, rights, security, or justice removes the reality of race, whether biological or cultural, from the ethical question of equality in way that trying to limit race, as bodies, to a privileged domain of untainted science cannot. Irrelevance is a more powerful strategy, one many also happen to believe is true.

The problem, of course, is that many scientists and much of the American public are not at all convinced that hereditary differences are actually irrelevant, and this is an argument that cannot be conclusively resolved with scientific evidence alone. Anthropologists and geneticists have been presenting the "facts" for more than sixty years now, and yet we continue to have the same debates. Racial science as it was once practiced may have been discredited, but many people continue to believe its most fundamental tenets – that the physical differences people see all around are evidence of some basic biological difference among people that can be understood in terms of race, and that these differences may well be related in some significant way to a wide variety of social issues that impinge on many people's lives – although many people are reluctant to admit they hold such views. As a society and frequently as individuals, Americans exist with a contradictory double consciousness and confusion about race – race does not exist, and race is real. On one hand, current theories of evolution and population genetics dictate that human variation cannot be grouped into biologically meaningful kinds, that each human being carries a genome that varies in some fashion, however little, from all others. Thus, there is no such thing as discretely bounded racial types, and no individual can be said to be racially "typical." The patent variation in human appearance, upon which theories of race have been built in the Western scientific tradition, can only be studied in terms of distributions of variations across time and space, and what is more, many of the characteristics, such as skin color, hair texture, or nose shape are of dubious biological, physiological, medical, or evolutionary significance. This populational view of human variation is compatible with another literature, which asserts that race is entirely a social construct. This scholarship argues that race has been constructed as means of naturalizing human difference, for a mix of epistemological, political, social, and economic reasons. As a construct, instead of an immutable foundation of human existence, race becomes a political and social problem, rather than a biological one. Of course, in American life, the idea of race has very real repercussions in people's identities, on their bodies, and in the political and social structure. Many scholars and evolutionary biologists may regard race as a conceptual fiction, but most Americans find it very real, indeed. This common-sense feeling that race exists, combined with the fact that scientific efforts to unravel the origins and significance of human diversity persist, as does racial typing (in forensic science and medicine, to name two areas), continues to create heated controversies in a society that continues to struggle with racism and the economic, social, and political disparities associated with it.[25]

The history presented in *Constructing Race* demonstrates the difficulty of treading the line Franz Boas tried to define. His attempts to explore physical differences but divorce that study from unwarranted and dangerous associations, speculation, or extension into inappropriate realms met with only limited success among his colleagues and in the wider world. The history and power of the way race has been understood and practiced for more than 400 years do not offer those hoping to maintain a firm barrier between a delimited conception of human variation and the culture concept much reason for optimism. Those arguing that race is in fact a social construction, a tool of oppression that exploited human differences without being limited to any particular constellation of them, will likely find little solace in the equivocal history of cultural relativism and racial science presented here. The racial and cultural theories explored in *Constructing Race* formed the foundation of both the shift toward cultural explanations of human diversity as well as the persistence of race thinking that are reflected in the continuing tension between biological determinism and cultural theory. But as Jenny Reardon has so persuasively argued, and as I hope this study of mid-century racial science has demonstrated, the creation of both race and culture in anthropology were, and are, products of a process of knowledge formation and dissemination

---

[25] Henry Louis Gates, Jr., has discussed the dangers of "cultural nominalism" in which arguments that race is not a biologically meaningful, or real, object are extended to deny the social reality of race as well. Gates argues that rejecting the idea that racial categories reveal something essential about individuals or groups should not be used to deny the very real social experience of race. Socially constructed is not equivalent to unreal. "Writing 'Race' and the Difference it Makes," in Gates, ed., *Race, Writing and Difference* (Chicago: University of Chicago Press, 1986), pp. 1–20.

that is constrained not so much by what is true about human diversity (as that question has been posed by scientists and their publics), as by how dilemmas posed by human variation have been defined, by the conceptual tools available to craft answers to those problems, and, perhaps most critically, by the sociopolitical environments in which problems of diversity are framed and solutions devised.

The historical question is not so much the success or failure of race and the science of race, but rather how, when, and in what incarnations did they prosper or languish. Or as Reardon put it, we ask not, "Is race real?" but, "How did race happen?"[26] In that question lies some hope of understanding the complexities of race and culture in America, past and present.

[26] Jenny Reardon, Brady Dunklee, and Kara Wentworth, "Race and Crisis," Social Science Research Council Forum on "Is 'Race' Real?," Social Science Research Council Forum on "Is 'Race' Real?," retrieved from http://raceandgenomics.ssrc.org/Reardon/, 2006. Also, Jenny Reardon, *Race to the Finish: Identity and Governance in an Age of Genomics* (Princeton: Princeton University, 2005).

Bibliography

Archives

AMNH-ANTH Anthropology Department Papers, Special Collections, Library, American Museum of Natural History, New York.

AMNH-HLS Harry L. Shapiro Papers, Special Collections, Library, American Museum of Natural History, New York.

RBP-VC Ruth Fulton Benedict Papers, Vassar College Library Archives and Special Collections Library, Vassar College, Poughkeepsie, New York.

Cranbrook-ROM Races of Mankind Collection, Archives, Cranbrook Center for Collections and Research, Bloomfield Hills, Michigan.

FM-BL Berthold Laufer Papers, Department of Anthropology, Museum Archives, The Field Museum.

FM-HF Henry Field Papers, Races of Mankind Correspondence, vol. 12, 1920–1950, Museum Archives, The Field Museum.

FM-ROM Malvina Hoffman, Races of Mankind, Department of Anthropology, Museum Archives, The Field Museum.

MHC Malvina Hoffman Collection, 850042-1, Special Collections, The Getty Center for the History of Art and the Humanities, Los Angeles, California.

Race Relations Department Archives, United Church Board for Homeland Ministries, 1942–1976, Amistad Research Center, Tulane University, New Orleans, Louisiana.

Primary and Secondary Sources

Adams, Richard E. W. *Prehistoric Mesoamerica*. Norman, OK: University of Oklahoma Press, revised edition, 1996.

Allen, Garland E. "The Misuse of Biological Hierarchies: The American Eugenics Movement, 1900–1940." *History and Philosophy of the Life Sciences*, 5 (1984): 105–128.

Allen, John S. "Franz Boas's Physical Anthropology: The Critique of Racial Formalism Revisited." *Current Anthropology*, vol. 30, no. 1 (February 1989): 79–84.

The American Museum of Natural History. *Fifty-Fourth Annual Report for the Year 1922.* The City of New York, 1923.

The American Museum of Natural History. *Fifty-Fifth Annual Report for the Year 1923.* The City of New York, 1924.

The American Museum of Natural History. *Sixty-Third Annual Report of the Trustees for the Year 1931.* The City of New York, 1931.

The American Museum of Natural History. *Sixty-Seventh Annual Report for the Year 1935.* The City of New York, 1936.

The American Museum of Natural History. *Sixty-Eighth Annual Report for the Year 1936.* The City of New York, 1937.

The American Museum of Natural History. *Sixty-Ninth Annual Report for the Year 1937.* The City of New York, 1938.

The American Museum of Natural History. *Seventieth Annual Report for the Year 1938.* The City of New York, 1939.

The American Museum of Natural History. *Seventy-Third Annual Report for the Year 1941.* The City of New York, 1942.

The American Museum of Natural History. *Seventy-Fourth Annual Report, For the Year 1942.* The City of New York, May 1, 1943.

The American Museum of Natural History. *Ninety-Second Annual Report, July, 1960 Through June, 1961.* The City of New York, 1961.

The American Museum of Natural History. *106th Annual Report, July 1974, Through June 1975.* The City of New York, 1975.

Anastasi, Anne. *Differential Psychology: Individual and Group Differences in Behavior.* New York: The Macmillan Company, 1937.

Anderson, Warwick. *Colonial Pathologies: American Tropical Medicine, Race, and Hygiene in the Philippines.* Durham, NC: Duke University Press, 2006.

———. "Hybridity, Race, and Science: The Voyage of the *Zaca*, 1934–35." *Isis*, vol. 103, no. 2 (June 2012): 229–253.

Annual Report of the Director to the Board of Trustees, 1931, vol. VIII, no. 1. Chicago: Field Museum of Natural History, 1932.

Annual Report of the Director to the Board of Trustees, 1933, vol. X, no. 1, publication 328. Chicago: Field Museum of Natural History, January 1934.

Annual Report of the Director to the Board of Trustees, 1935. Chicago: Field Museum of Natural History, 1936.

Annual Report of the Director to the Board of Trustees, 1964. Chicago: Chicago Museum of Natural History, 1965.

Armstrong, Elizabeth. "Interview with Kara Walker." *no place (like home)*, Richard Flood et al., eds. Minneapolis: Walker Art Center, 1997.

"Army Drops Race Equality Book; Denies May's Stand Was Reason." *New York Times*, March 6, 1944: 1.

Arnoldi, Mary Jo. "Herbert Ward's Ethnographic Sculptures of Africans." Amy Henderson and Adrienne L. Kaeppler, eds. *Exhibiting Dilemmas, Issues of Representation at the Smithsonian.* Washington, DC: Smithsonian Institution Press, 1997: 70–91.

Ayer, Edward Everett. "In Re; Founding of the Field Museum." Stephen E. Nash and Gary M. Feinman, eds. *Curators, Collections and Contexts: Anthropology at the Field Museum, 1893–2002.* Fieldiana: Anthropology, New Series, no. 36. Chicago: Field Museum of Natural History, 2003: 49–52.

Baker, Lee D. *From Savage to Negro: Anthropology and the Construction of Race, 1896–1954.* Berkeley and Los Angeles: University of California Press, 1998.

Bamshad, Michael J., et al. "Human Population Genetic Structure and Inference of Group Membership." *American Journal of Human Genetics*, vol. 72, no. 3 (March 2003): 578–589.

Bamshad, Michael J. and Steve E. Olson. "Does Race Exist?" *Scientific American* (December 2003): 78–85.

Banner, Lois W. *Intertwined Lives: Margaret Mead, Ruth Benedict, and Their Circle.* New York: Alfred A. Knopf, 2003.

Banta, Melissa and Curtis Hinsley, eds. *From Site to Sight: Anthropology, Photography, and the Power of Imagery.* Cambridge, MA: Peabody Museum Press, 1986.

Barkan, Elazar. "Mobilizing Scientists against Nazi Racism, 1933–1939." George W. Stocking, Jr., ed. *Bones, Bodies, Behavior: Essays on Biological Anthropology*, History of Anthropology, vol. 5. Madison: University of Wisconsin Press, 1988: 180–205.

———. *The Retreat of Scientific Racism: Changing Concepts of Race in Britain and the United States between the World Wars.* Cambridge: Cambridge University Press, 1992.

Barkan, Elazar and Ronald Bush, eds. *Prehistories of the Future: The Primitivist Project and the Culture of Modernism.* Stanford, CA: Stanford University Press, 1995.

Barzun, Jacques. *Race: A Study in Modern Superstition.* New York: Harcourt, Brace & Co., 1937.

Baulch, Vivian M. and Patrical Zacharias. "The 1943 Detroit Race Riots." *The Detroit News*, Feb. 11, 1999.

Baum, Bruce. *The Rise and Fall of the Caucasian Race: A Political History of Racial Identity.* New York: New York University Press, 2006.

Becker, Stephen. *Marshall Field III: A Biography.* New York: Simon and Schuster, 1964.

Bederman, Gail. *Manliness and Civilization: A Cultural History of Gender and Race in the United States.* Chicago: University of Chicago Press, 1995.

Benedict, Burton. *The Anthropology of World's Fairs: San Francisco's Panama Pacific International Exposition of 1915.* Berkeley, CA: Lowie Museum of Anthropology; London and Berkeley: Scolar Press, 1983.

Benedict, Ruth. *Patterns of Culture.* New York: The New American Library of World Literature, Inc., 1934.

———. *Race: Science and Politics*, with a foreword by Margaret Mead. New York: The Viking Press, 1940.

———. *Race: Science and Politics*, with a foreword by Margaret Mead. Revised edition. New York: The Viking Press, 1943.

———. *Race: Science and Politics*, with a foreword by Margaret Mead. New edition, including *The Races of Mankind.* New York: The Viking Press, 1945.

_____. *The Chrysanthemum and the Sword: Patterns of Japanese Behavior.* Boston: Houghton Mifflin, 1946.

_____. *Patterns of Culture.* New American Library of World Literature, Mentor Book, 13th edition, 1956.

_____. *Race: Science and Politics*, with a foreword by Margaret Mead. New York: The Viking Press, 1959.

_____. *Patterns of Culture*, with a new preface by Margaret Mead. Boston: Houghton Mifflin Company, 1961.

_____. *Patterns of Culture*, preface by Margaret Mead, foreword by Louise Lamphere. Boston: Houghton Mifflin Harcourt, 2005.

Benedict, Ruth and Gene Weltfish. *The Races of Mankind*, Public Affairs Pamphlet No. 85. New York: Public Affairs Committee, Incorporated, 1943.

_____. *In Henry's Backyard. The Races of Mankind.* New York: Henry Schuman, Inc., 1948.

Benn Michaels, Walter. *The Trouble with Diversity: How We Learned to Love Identity and Ignore Inequality.* New York: Metropolitan Books, Henry Holt and Company, 2006.

Bennett, Tony. *The Birth of the Museum: History, Theory, Politics.* London and New York: Routledge, 1995.

Bilbo, Theodore. *Take Your Choice: Separation or Mongrelization.* Poplarville, MS: Dream House Publishing Co., 1947.

Bindman, David. *Ape to Apollo: Aesthetics and the Idea of Race in the Eighteenth Century.* Ithaca: Cornell University Press, 2002.

Blackie, Walter Graham. *The Comprehensive Atlas & Geography of the World; Comprising an Extensive Series of Maps, a Description, Physical and Political, of All the Countries of the Earth; A Pronouncing Vocabulary of Geographical Names, and a Copious Index of Geographical Positions: Also Numerous Illustrations Printed in the Text, and a Series of Coloured Engravings Representing the Principal Races of Mankind.* London: Blackie & Son, 1882.

Blakey, Michael L. "Skull Doctors: Intrinsic Social and Political Bias in the History of American Physical Anthropology, with Special Reference to the Work of Aleš Hrdlička." *Critique of Anthropology*, vol. 7, no. 2 (1987): 7–35.

Blanchard, Pascal, et al., *Human Zoos: Science and Spectacle in the Age of Colonial Empires.* Liverpool University Press, 2008.

William Bligh, and Edward Christian. *The Bounty Mutiny.* New York: Penguin, 2001.

Blumenbach, Johann Friedrich. *On the Natural Varieties of Mankind. De Generis Humani Varietate Nativa.* Translated and edited from the Latin, German, and French originals by Thomas Bendyshe. New York: Bergman Publishers, 1969.

Boas, Franz. "The History of Anthropology." Address at the International Congress of Arts and Sciences, St. Louis, September 1904. Reprinted in H. J. Rogers, ed. *Congress of Arts and Science*, vol. 5. Boston: Houghton Mifflin, 1906: 468–482.

_____. Review of *The Racial History of Man* by Roland B. Dixon. *Science*, New Series, vol. 57, no. 1481 (May 18, 1923): 587–590.

_____. "The Tempo of Growth of Fraternities." *Proceedings of the National Academy of Sciences*, vol. 21, no. 7 (July 15, 1935): 413–418.

———. "Heredity and Environment." *Jewish Social Studies*, vol. 1, no. 1 (Jan. 1939): 5–14.

———. "The Relation between Physical and Mental Development." *Science*, new series, vol. 93, issue 2415 (Apr. 11, 1941): 339–342.

———. "The Anthropology of the North American Indian." *Memoirs of the International Congress of Anthropology*. Chicago: Schulte, 1894: 37–49. Reprinted in Franz Boas, *Race, Language, and Culture*. New York: Macmillan Company, 1948: 191–201.

———. "Changes in Bodily Form of Descendants of Immigrants." *American Anthropologist*, new series, vol. 14 (1912): 530–562. Reprinted in Franz Boas, *Race, Language, and Culture*. New York: Macmillan Company, 1948: 60–75.

———. "The Half-Blood Indian." *Popular Science Monthly* (October 1894). Reprinted in Franz Boas, *Race, Language, and Culture*. New York: Macmillan Company, 1948: 138–148.

———. "History and Science in Anthropology: A Reply." *American Anthropologist*, new series, vol. 38 (1936): 137–141. Reprinted in Franz Boas, *Race, Language, and Culture*. New York: Macmillan Company, 1948: 309–310.

———. "The Measurement of Differences between Variable Quantities." *Quarterly Publication of the American Statistical Association* (December 1922): 425–445. Reprinted in Franz Boas, *Race, Language, and Culture*. New York: Macmillan Company, 1948: 189–190.

———. *The Mind of Primitive Man*. Revised edition. New York: Macmillan Company, 1948.

———. "Modern Populations of America." *Proceedings of the 19th International Congress of Americanists, Washington, December, 1915*, Washington, DC, 1917: 569–575. Reprinted in Franz Boas, *Race, Language, and Culture*. New York: Macmillan Company, 1948: 18.

———. *Race, Language, and Culture*. New York: Macmillan Company, 1948.

———. "The Relations between Physical and Social Anthropology." *Essays in Anthropology in Honor of Alfred Louis Kroeber*. Berkeley: University of California Press, 1936. Reprinted in Franz Boas, *Race, Language, and Culture*. New York: Macmillan Company, 1948: 172–175.

———. "Report of an Anthropometric Investigation of the Population of the United States." *Journal of the American Statistical Association*, vol. 18 (June, 1922): 181–209. Reprinted in Franz Boas, *Race, Language, and Culture*. New York: Macmillan Company, 1948: 28–59.

———. "Review of William Z. Ripley, 'The Races of Europe.'" *Science*, new series, vol. 10 (September 1, 1899). Reprinted in Franz Boas, *Race, Language, and Culture*. New York: Macmillan Company, 1948: 155–159.

———. "Some Recent Criticisms of Physical Anthropology." *American Anthropologist*, new series, vol. 1 (January, 1899). Reprinted in Franz Boas, *Race, Language, and Culture*. New York: Macmillan Company, 1948: 165–171.

———. "Changes in Immigrant Body Form," letter from Franz Boas to J. W. Jenks, March 23, 1908. Reprinted in George W. Stocking, Jr., ed. *The Shaping of American Anthropology, 1883–1911: A Franz Boas Reader*, Midway Reprint. Chicago and London: The University of Chicago Press, 1989): 202.

———. "The History of Anthropology." *Science*, vol. 20 (1904): 513–524. Reprinted in George W. Stocking, Jr., ed. *The Shaping of American Anthropology, 1883–1911: A Franz Boas Reader*, Midway Reprint. Chicago and London: The University of Chicago Press, 1989): 23–36.

———. "Human Faculty as Determined by Race." American Associations for the Advancement of Science, *Proceedings* 43 (1894): 301–327. Reprinted in George W. Stocking, Jr., ed. *The Shaping of American Anthropology, 1883–1911: A Franz Boas Reader*, Midway Reprint. Chicago and London: The University of Chicago Press, 1989: 221–242.

Boittin, Jennifer Ann. *Colonial Metropolis: The Urban Grounds of Anti-Imperialism and Feminism in Interwar Paris*. Lincoln: University of Nebraska Press, 2010.

Bokovoy, Matthew F. *The San Diego World's Fairs and Southwestern Memory, 1880–1940*. Albuquerque: University of New Mexico Press, 2005.

Boyd, Valerie. *Wrapped in Rainbows: The Life of Zora Neale Hurston*. New York: Scribner, 2003.

Boyle, Kevin. *Arc of Justice: A Saga of Race, Civil Rights, and Murder in the Jazz Age*. New York: Macmillan, 2007.

Brace, C. Loring. "Race in American Physical Anthropology." Frank Spencer, ed. *The History of American Physical Anthropology 1930–1980*. New York: Academic Press, 1982: 11–30.

Brantlinger, Patrick. *Dark Vanishings: Discourse on the Extinction of Primitive Races, 1800–1930*. Ithaca: Cornell University Press, 2003.

Brattain, Michelle. "Race, Racism, and Antiracism: UNESCO and the Politics of Presenting Science to the Postwar Public." *The American Historical Review*, vol. 112, no. 5 (December 2007): 1386–1413.

Broadluck, Cephas. *Races of Mankind; with Travels in Grubland*. Cincinnati: Longley, 1856.

Brown, Robert. *The Races of Mankind, Being a Popular Description of the Characteristics, Manners and Customs of the Principal Varieties of the Human Family*. London and New York: Cassell, Peter & Galpin (1873–1876).

Buck, Peter H. (Ti Rangi Hīroa). *Vikings of the Sunrise*. New York: Frederick A. Stokes Company, 1938.

Bulmer, M. G. *Francis Galton: Pioneer of Heredity and Biometry*. Baltimore: The Johns Hopkins University Press, 2003.

Bunzl, Matti. "Franz Boas and the Humboldtian Tradition: From Volkgeist and Nationalcharakter to an Anthropological Concept of Culture." George W. Stocking, Jr., ed. *Volkgeist as Method and Ethic, Essays on Boasian Ethnography and the German Anthropological Tradition*, History of Anthropology, vol. 8. Madison: University of Wisconsin Press, 1996: 17–78.

Burba, Juliet. "'Whence Came the American Indians?': American Anthropologists and the Origins Question, 1880–1935." PhD dissertation, University of Minnesota, 2006.

Burkholder, Zoë. "'With Science as His Shield': Teaching Race and Culture in American Public Schools, 1900–1954." PhD dissertation, New York University, 2008.

_____. *Color in the Classroom: How American Schools Taught Race, 1900–1954*. Oxford: Oxford University Press, 2011.

Burton, J. W. "Shadows at Twilight: A Note on History and the Ethnographic Present." *Proceedings of the American Philosophical Society*, 1988, vol. 132: 420–433.

Buxton, L. H. Dudley. *The Peoples of Asia*. New York: Alfred Knopf, 1925.

_____. "The Essential Craniological Technique." *Journal of the Royal Anthropological Institute of Great Britain and Ireland*, vol. 63 (Jan.–Jun. 1933): 19–47.

Caffrey, Margaret M. *Ruth Benedict: Stranger in This Land*. Austin: University of Texas Press, 1989.

Carter, Julian B. *The Heart of Whiteness: Normal Sexuality and Race in America, 1890–1940*. Durham and London: University of North Carolina Press, 2007.

Castañeda, Quetzil E. "The Carnegie Mission and Vision of Science: Institutional Contexts of Maya Archaeology and Espionage." *Histories of Anthropology Annual*, vol. 1, 2005.

Castle, W. E. *Genetic and Eugenics*. Cambridge: Harvard University Press, 1930.

_____. "Race Mixture and Physical Disharmonies." *Science*, New Series, vol. 71, no. 1850 (Jun. 13, 1930): 603–606.

Chamberlain, Alexander F. Review: Albert Earnest Jenks, *The Bontoc Igorot*, Department of the Interior, Ethnological Survey Publications, vol. 1, Manila: Bureau of Public Printing, 1905. *American Anthropologist*, New Series, vol. 7, no. 4 (Oct.–Dec., 1905): 696–701.

"Chicago Quickens World's Fair Work." *New York Times* (April 13, 1931): 5.

Clark, Constance Areson. *God – or Gorilla: Images of Evolution in the Jazz Age*. Baltimore, MD: The Johns Hopkins University Press, 2008.

Clifford, James. "On Ethnographic Self-Fashioning: Conrad and Malinowski." *The Predicament of Culture: Twentieth-Century Ethnography, Literature, and Art*. Boston: Harvard University Press, 1988: 92–116.

_____. *The Predicament of Culture: Twentieth-Century Ethnography, Literature, and Art*. Boston: Harvard University Press, 1988.

Clifford, James and George E. Marcus, eds. *Writing Culture: The Poetics and Politics of Ethnography*. Berkeley: University of California Press, 1986.

Coffey, Mary K. "The American Adonis: A Natural History of the 'Average American' (Man), 1921–1932." Susan Currell and Christina Codgell, eds. *Popular Eugenics: National Efficiency and American Mass Culture*. Athens, OH: Ohio University Press, 2006: 185–216.

Cole, Douglas. *Franz Boas: The Early Years, 1859–1906*. Vancouver, BC, Canada: Douglas & McIntyre; Seattle: University of Washington Press, 1999.

Conklin, Alice. *A Mission to Civilize: The Republican Idea of Empire in France and West Africa, 1895–1930*. Stanford, CA: Stanford University Press, 1997.

_____. *In the Museum of Man: Race, Anthropology, and Empire in France, 1850–1950*. Ithaca: Cornell University Press, 2013.

Conn, Steven. *Museums and American Intellectual Life, 1876–1926*. Chicago: University of Chicago Press, 1998.

Conrad, David C. and Karen J. Tillotson. "The Races of Man." *Reader, Chicago's Free Weekly*, vol. 22, no. 28, section 1 (1993): 8–9, 30.

Coon, Carleton. *The Races of Europe*. New York: The Macmillan Company, 1939.

_____. *The Origin of Races*. New York: Knopf, 1962.

_____. *The Living Races of Man*. New York: Knopf, 1965.

Cowlishaw, Gillian. "Censoring Race in 'Post-Colonial' Anthropology." *Critique of Anthropology*, vol. 20, no. 2 (2000): 101–123.

Cox, Earnest Sevier. *White America*. Richmond, VA: Mitchell and Hotchkiss, 1923.

_____. *The Races of Mankind: A Review*, foreword by Arthur Daugherty. Jellico, TN: Arthur Daugherty, 1951.

Crais, Clifton and Pamela Scully. *Sara Baartman and the Hottentot Venus: A Ghost Story and a Biography*. Princeton, NJ: Princeton University Press, 2008.

Cravens, Hamilton. *The Triumph of Evolution: American Scientists and the Heredity-Environment Controversy 1900–1941*. Philadelphia: University of Pennsylvania Press, 1978.

Craven, Wayne. *Sculpture in America*. New York: Thomas Y. Crowell Company, Inc., 1968.

Currell, Susan and Christina Codgell, eds. *Popular Eugenics: National Efficiency and American Mass Culture in the 1930s*. Columbus: Ohio State University Press, 2006.

Dalton, Karen C. C. and Henry Louis Gates, Jr., "Josephine Baker and Paul Colin: African American Dance Seen Through Parisian Eyes," *Critical Inquiry*, vol. 24, no. 4 (Summer 1998): 903–934.

Darnell, Regna. *And Along Came Boas: Continuity and Revolution in Americanist Anthropology*. Amsterdam and Philadelphia, PA: J. Benjamins, 1998.

_____. "Review: Reenvisioning Boas and Boasian Anthropology." *American Anthropologist*, New Series, vol. 102, no. 4 (Dec., 2000): 896–899.

_____. *Invisible Genealogies: A History of Americanist Anthropology*, Critical Studies in the History of Anthropology. Lincoln and London: University of Nebraska Press, 2001.

Daston, Lorraine. "Type Specimens and Scientific Memory." *Critical Inquiry* 31 (Autumn 2004): 153–182.

Daston, Lorraine, and Peter Galison. "The Image of Objectivity." *Representations* 40 (Fall 1992): 81–128.

Davenport, Charles B. and Morris Steggerda. *Race Crossing in Jamaica*. Washington, DC: Carnegie Institution of Washington, 1929.

Davis, Lucille. "Sarah Baartman, at Rest at Last." SouthAfrica.info. Retrieved from http://www.southafrica.info/about/history/saartjie.htm#.Ue2ytWTuU5s (accessed July 22, 2013).

"The Death of Dr. Laufer, Curator of Anthropology." *Field Museum News* (October 1934): 2.

Decoteau, P. H. "Malvina Hoffman and the 'Races of Mankind.'" *Art Journal* (Fall 1989/Winter 1990): 7–12.

Deniker, Joseph. *The Races of Mankind: An Outline of Anthropology and Ethnography*. London: Walter Scott Publishing Co., Ltd.; New York: Charles Scribner's Sons, 1913.

Detroit African American History Project, Wayne State University. Retrieved from http://www.daahp.wayne.edu/biographiesDisplay.php?id=7 (accessed July 26, 2013).

di Leonardo, Michaela. *Exotics at Home: Anthropologies, Others, American Modernity*. Chicago: University of Chicago Press, 1998.

Dias, Nelia. "The Visibility of Difference: Nineteenth Century French Anthropological Collections." Sharon Macdonald, ed. *The Politics of Display: Museums, Science, Culture*. New York and London: Routledge, 1998: 36–52.

"Display Aids Tolerance. 'Races of Mankind' Exhibition Opened at Public Library." *New York Times*, February 2, 1945: 17.

"Disputes Keith View as to Race Prejudice." *The New York Times* (June 8, 1931): 13.

Dixon, Roland B. "A New Theory of Polynesian Origins." *Proceedings of the American Philosophical Society*, vol. 59, no. 4 (1920): 261–267.

_____. *The Racial History of Man*. New York: Charles Scribner's Sons, 1923.

Dobzhansky, Theodosius. *Genetics and the Origin of Species*. New York: Columbia University Press, 1937.

Dominguez, Virginia. "A Taste for 'the Other': Intellectual Complicity in Racializing Practices." *Current Anthropology*, vol. 35, no. 4 (1994): 333–348.

Dorr, Gregory Michael. *Segregation's Science: Eugenics and Society in Virginia*. Charlottesville, VA: University of Virginia Press, 2008.

Douglas, Bronwen. "Foreign Bodies in Oceania." *Foreign Bodies: Oceania and the Science of Race 1750–1940*, Bronwen Douglas and Chris Ballard, eds. ANU E Press: The Australian National University, 2008. Retrieved from http://epress.anu.edu.au/foreign_bodies/html/frames.php.

Drake, St. Clair and Horace R. Cayton. *Black Metropolis: A Study of Negro Life in a Northern City*. Chicago: University of Chicago Press, 1993.

Duffus, R. L. "The Fair: A World of Tomorrow." *New York Times* (May 28, 1933): SM1, 3 pgs.

Dunn, L. C. "An Anthropological Study of Hawaiians of Pure and Mixed Blood." *Papers of the Peabody Museum*, vol. 11 (1928): 90–211.

Dunn, L. C. and Theodosius Dobzhansky. *Heredity, Race and Society*. New York: New American Library of World Literature, 1946.

Dyer, Richard. *White*. New York and London: Routledge, 1997.

East, Edward M. *Mankind at the Crossroads*. New York: Scribners, 1923.

_____. *Heredity and Human Affairs*. New York: Scribners, 1927.

_____. *Biology in Human Affairs*. New York: McGraw-Hill, 1931.

East, Edward M. and Donald F. Jones. *Inbreeding and Outbreeding: Their Genetic and Sociological Significance*. Philadelphia: Lippincott, 1919.

"Education: Race Question." *TIME*, Monday, Jan. 31, 1944.

"Education Notes." *New York Times*, May 13, 1945: E9.

Edwards, Leigh H. "The United Colors of 'Pocahontas': Synthetic Miscegenation and Disney's Multiculturalism." *Narrative*, Multiculturalism and Narrative, vol. 7, no. 2 (May 1999): 147–168.

Edwards, Violet. "Note on *The Races of Mankind.*" Ruth Benedict. *Race: Science and Politics*, with a foreword by Margaret Mead. New York: The Viking Press, 1959: 167–168.

"Eight New Sculptures of Racial Types Added to Chauncey Keep Memorial Hall." *Field Museum News*, vol. 5, no. 11 (November 1934): 3.

Ellen, R. F., et al. *Malinowski between Two Worlds: The Polish Roots of an Anthropological Tradition.* Cambridge: Cambridge University Press, 1988.

Evans, Andrew. *Anthropology at War: World War I and the Science of Race in Germany.* Chicago: University of Chicago Press, 2010.

Evans-Pritchard, Edward E. *The Nuer: A Description of the Modes of Livelihood and Political Institutions of a Nilotic People.* Oxford: Clarendon Press, 1940.

"Expedition at Kish Resumes Operations." *Field Museum News*, Chicago: Field Museum of Natural History (December 1931): 3.

Farber, Paul. "Race-Mixing and Science in the United States." *Endeavor*, vol. 27, no. 4 (2003): 166–170.

Fausto-Sterling, Anne. "Gender, Race, and Nation: The Comparative Anatomy of 'Hottentot' Women in Europe, 1815–1817." Jennifer Terry and Jacqueline Urla, eds. *Deviant Bodies: Critical Perspectives on Difference in Science and Popular Culture.* Bloomington: Indiana University Press, 1995: 19–48.

Field, Henry. *The Races of Mankind.* Popular series, Anthropology Leaflet 30, Chicago: Field Museum of Natural History, 1933.

―――. *The Races of Mankind.* Popular series, Anthropology Leaflet 30, Second Edition, Chicago: Field Museum of Natural History, 1934.

―――. "The Story of Man." Science Service Radio Talks Presented Over the Columbia Broadcasting System. *The Scientific Monthly* (July 1935): 61–65.

―――. *The Anthropology of Iraq, the Upper Euphrates.* Anthropological Series, Publication 469, Field Museum of Natural History, vol. 30, part I, no. 1, May 31, 1940.

―――. *Folklore and Customs of Southwestern Asia.* Anthropological series, Publication 484, Field Museum of Natural History, vol. 33, no. 1, Dec. 30, 1940.

―――. *The Races of Mankind.* Popular series, Anthropology Leaflet 30, Fourth Edition, Chicago: Field Museum of Natural History, 1942.

―――. *The Track of Man: Adventures of an Anthropologist.* New York: Greenwood Press, Publishers, reprinted with permission of Doubleday & Co., 1969.

Field, Henry and W. D. Hambly. *Hammond's Map of Mankind, Full-Color Map with The Races of Mankind.* New York: C. S. Hammond and Co., 1946.

Field Museum News. Chicago: Field Museum of Natural History (November 1941): 6.

Field Museum Bulletin, Chicago: Field Museum of Natural History, vol. 54 (Feb. 1983): 7.

Fields, Barbara. "Ideology and Race in American History." J. Morgan Kousser and James M. McPherson, eds. *Region, Race, and Reconstruction: Essays in Honor of C. Vann Woodward.* New York: Oxford University Press, 1982): 143–177.

Fischer, Eugen. *Die Rehobother Bastards und das Bastardierungsproblem beim Menschen.* Jena: Gustav Fischer, 1913.

Fleure, H[enry] J[ohn]. *The Races of Mankind.* London: E. Benn, and Garden City, NY: Doubleday, Doran: 1928.

Forbes, John Ripley. "Valuable Racial Exhibit." *Hobbies: The Magazine for Collectors*, vol. 49 (March 1944): 22.

Foster-Rice, Gregory. "The Visuality of Race: 'The Old Americans,' 'The New Negro' and American Art, c. 1925." PhD Dissertation, Northwestern University, 2003.

Freeman, Frank S. *Individual Differences: The Nature and Causes of Variations in Intelligence and Special Abilities.* New York: H. Holt and Company, 1934.

Fuchs, Eckhardt. "The Politics of the Republic of Learning: International Scientific Congresses in Europe, the Pacific Rim, and Latin America." Eckhardt Fuchs and Benedikt Stuchtey, eds. *Across Cultural Borders: Historiography in Global Perspective.* Lanham, MD: Rowman & Littlefield Publishers, Inc., 2002: 205–246.

Fuller, Steve. *Thomas Kuhn: A Philosophical History for Our Times.* Chicago: University of Chicago Press, 2000.

Galton, Francis. "Composite Portraits, Made by Combining Those of Many Different Persons into a Single Resultant Figure." *The Journal of the Anthropological Institute of Great Britain and Ireland*, Vol. 8 (1879): 132–144.

Garn, Stanley. *Human Races.* Springfield, IL: Thomas, 1961.

Garn, Stanley Marion and Carleton S. Coon. *On the Number of Races of Mankind.* Fels Research Institute, *Publications*, no. 3. Yellow Springs, OH: Antioch College, 1955: 100–105.

Garn, Stanley and Eugene Giles. "Earnest Albert Hooton, November 20, 1887–May 3, 1954." National Academy of Sciences, *Biographical Memoirs* 68 (1995): 167–179.

Garth, T. R. *Race Psychology: A Study of Racial Mental Differences.* New York: Whittlesey House, McGraw-Hill Book Company, Inc. 1931.

Gates, Henry Louis, Jr. "Writing 'Race' and the Difference It Makes." Henry Louis Gates, Jr., ed. *Race, Writing and Difference.* Chicago: University of Chicago Press, 1986): 1–20.

———. *Finding Oprah's Roots, Finding Your Own.* New York: Random House, Crown Publishing Group, 2007.

Gates, Reginald Ruggles. "A Pedigree Study of Amerindian Crosses in Canada." *Journal of the Royal Anthropological Institute*, vol. 58 (1928): 511–532.

Geertz, Clifford. "Us/Not-Us: Benedict's Travels." *Works and Lives: The Anthropologist as Author.* Stanford, CA: Stanford University Press, 1988: 102–128.

———. *Works and Lives: The Anthropologist as Author.* Stanford, CA: Stanford University Press, 1988.

Gerstle, Gary. *American Crucible: Race and Nation in the Twentieth Century.* Princeton, NJ: Princeton University Press, 2001.

Gilkeson, John S., Jr. "The Domestication of 'Culture' in Interwar America, 1919–1941." *The Estate of Social Knowledge*, JoAnne Brown and David K. van Keuren, eds. Baltimore and London: The Johns Hopkins University Press, 1991): 153–174.

Gilman, Sander L. "Black Bodies, White Bodies: Toward an Iconography of Female Sexuality in Late Nineteenth-Century Art, Medicine, and Literature." *Critical Inquiry*, vol. 12 (Autumn 1985): 212–223.

Giroux, Henry A. "Consuming Social Change: The 'United Colors of Benetton.'" *Cultural Critique*, no. 26 (Winter 1993–1994): 5–32.

Goodman, Alan and Evelynn Hammonds. "Reconciling Race and Human Adaptability: Carleton Coon and the Persistence of Race in Scientific Discourse." Jonathan Marks, ed. *Racial Anthropology: Retrospective on Carleton Coon's The Origin of Races* (1962), Kroeber Anthropological Society papers, no. 84. Berkeley, CA: Kroeber Anthropological Society, 2000: 28–55.

Goodman, Alan, Deborah Heath, and M. Susan Lindee, eds. *Genetic Nature/Culture: Anthropology and Science beyond the Two-Culture Divide*. Berkeley: University of California Press, 2003.

Goodspeed, Thomas Wakefield. "Marshall Field." *The University of Chicago Biographical Sketches*, vol. 1. Chicago: University of Chicago Press, 1922: 1–34.

Gossett, Thomas F. *Race: The History of an Idea in America*. New York: Shocken Books, 1963.

Gould, Stephen Jay. *The Mismeasure of Man*. New York: Norton, 1981.

Graham, Clarence H. "Robert Sessions Woodworth, 1869–1962, a Biographical Memoir." Washington, DC: National Academy of Sciences, 1967: 541–572.

Gravelee, Clarence C., H. Russell Bernard, and William R. Leonard, "Boas's *Changes in Bodily Form*: The Immigrant Study, Cranial Plasticity, and Boas's Physical Anthropology." *American Anthropologist*, vol. 105, no. 2 (2003): 326–332.

———. "Heredity, Environment, and Cranial Form: A Re-Analysis of Boas's Immigrant Data." *American Anthropologist*, vol. 105, no. 1 (2003): 123–136.

Gregory, William King. *Our Face from Fish to Man: A Portrait Gallery of Our Ancient Ancestors and Kinfolk Together with a Concise History of Our Best Features*. New York: G. P. Putnam's Sons, 1929.

Griffiths, Alison. *Wondrous Difference: Cinema, Anthropology & Turn-of-the-Century Visual Culture*. New York: Columbia University Press, 2002.

Gruenberg, H. "Men and Mice at Edinburgh: Reports from the Genetics Congress." *The Journal of Heredity*, vol. 30, no. 9 (September 1939): 371–374.

Guglielmo, Thomas A. *White on Arrival: Italians, Race, Color and Power in Chicago, 1890–1945*. New York: Oxford University Press, 2003.

Guha, Biraja Sankar, Ranjit Kumar Bhattacharya, and Jayanta Sarkar. *Anthropology of B. S. Guha: A Centenary Tribute*. Anthropological Survey of India, Ministry of Human Resource Development, Dept. of Culture, Govt. of India, 1996.

Guterl, Matthew Pratt. *The Color of Race in America, 1900–1940*. Harvard University Press, 2001.

———. "The Importance of Place in Post-Everything American Studies." *American Quarterly*, vol. 61, no. 4 (December 2009): 931–941.

Haardt, Georges-Marie and Louis Audouin-Dubreuil. *The Black Journey; Across Central Africa with the Citroën Expedition.* New York: Cosmopolitan Book Corporation, 1927.

Haddon, A[lfred]. C[ort]. "The Urgent Need for Anthropological Investigation." *Nature*, no. 2329, vol. 93 (June 18, 1914): 407–408.

Hall, Jacquelyn Dowd. "The Long Civil Rights Movement and the Political Uses of the Past." *The Journal of American History*, vol. 91, no. 4 (Mar. 2005): 1233–1263.

Haller, John S. *Outcasts from Evolution: Scientific Attitudes of Racial Inferiority, 1859–1900.* Carbondale: Southern Illinois University, 1995.

Hallowell, A. Irving. "The History of Anthropology as an Anthropological Problem." *Journal of the History of the Behavioral Sciences*, vol. 1 (1965): 24–38.

Hammond, Michael. "The Shadow Man Paradigm in Paleoanthropology, 1911–1945." George W., Stocking, Jr., ed. *Bones, Bodies, Behavior: Essays on Biological Anthropology*, History of Anthropology, vol. 5. Madison: University of Wisconsin Press, 1988: 117–137.

Hammonds, Evelynn. *The Nature of Difference: Sciences of Race from Jefferson to Genomics in the United States* (Cambridge: MIT Press, 2008).

Handler, Richard. "Vigorous Male and Aspiring Female: Poetry, Personality, and Culture in Edward Sapir and Ruth Benedict." George W. Stocking, Jr., ed. *Malinowski, Rivers, Benedict and Others: Essays on Culture and Personality*, History of Anthropology, vol. 4. Madison: University of Wisconsin Press, 1986: 127–155.

Haraway, Donna. "Remodeling the Human Way of Life: Sherwood Washburn and the New Physical Anthropology, 1950–1980." George W. Stocking, Jr. *Bones, Bodies, Behavior: Essays on Biological Anthropology*, History of Anthropology, vol. 5. Madison: University of Wisconsin Press, 1988: 217–218.

_____. *Primate Visions. Gender, Race, and Nature in the World of Modern Science.* New York and London: Routledge, 1989.

_____. "Teddy Bear Patriarchy: Taxidermy in the Garden of Eden, New York City, 1908–36." *Primate Visions. Gender, Race, and Nature in the World of Modern Science.* New York and London: Routledge, 1989: 26–58.

_____. *Modest_Witness@Second_Millennium.FemaleMan©_Meets_Onco Mouse™: Feminism and Technoscience.* New York: Routledge, 1997.

_____. "Universal Donors in a Vampire Culture: It's All in the Family – Biological Kinship Categories in the Twentieth-Century United States." *Modest_Witness@Second_Millennium.FemaleMan©_Meets_OncoMouse™: Feminism and Technoscience.* New York: Routledge, 1997: 213–266.

Harris, Marvin. *The Rise of Anthropological Theory.* New York: Thomas Crowell, 1968.

Harrison, Faye. "The Persistent Power of 'Race' in the Cultural and Political Economy of Racism." *Annual Review of Anthropology*, vol. 24 (1995): 48.

_____. "Introduction: Expanding the Discourse on 'Race.'" *American Anthropologist*, 100(3), 1999: 609–631.

Hart, Mitchell B. "Jews and Race: An Introductory Essay," Mitchel B. Hart, ed., *Jews and Race: Writings on Identity and Difference, 1880–1940*. Waltham, MA: Brandeis University Press, 2011: xiii–xxxix.

Harvey, Penelope. *Hybrids of Modernity: Anthropology, the Nation State and the Universal Exhibition*. New York: Routledge, 1996.

Haskin, Warren. *Anthropology at the Field Museum, 1894–2000*. Unpublished manuscript. Special Collections, Library, Chicago: The Field Museum.

Hatt, Robert T. "The Races of Mankind, Announcing an Exhibition." *Cranbrook Institute of Science News Letter [sic]*, Bloomfield Hills, MI, vol. 13, no. 5 (Jan. 1944): 1–3.

———. "Report of the Director." *Fourteenth Annual Report, July 1, 1943–June 30, 1944*, Cranbrook Institute of Science, Bloomfield Hills, MI: 12–21.

Hazard, Anthony Q., Jr. "Postwar Anti-Racism: The United States, UNESCO and 'Race', 1945–1968." PhD Dissertation, Temple University, 2008.

Hedlin, Ethel Wolfskill. "Earnest Cox and Colonization: A White Racist's Response to Black Repatriation, 1923–1966." PhD Dissertation, Duke University, 1974.

Herrnstein, Richard J. and Charles Murray, *The Bell Curve: Intelligence and Class Structure in American Life*. New York: Free Press, 1994.

Herskovits, Melville J. "Extremes and Means in Racial Interpretation." *The Journal of Social Forces*, vol. 2, no. 4 (May 1924): 550–551.

———. *The American Negro: A Study in Racial Crossing*. New York: Columbia University Press, 1928.

———. *The Myth of the Negro Past*. New York: Harper, 1941.

———. *Franz Boas: The Science of Man in the Making*. New York: Charles Scribner's Sons, 1953.

Hess, Thomas B. "The Art Comics of Ad Reinhardt." *Artforum*, vol. 12, no. 8 (April 1974): 46–51.

Higham, John. *Strangers in the Land: Patterns of American Nativism, 1860–1925*. New Brunswick, NJ: Rutgers University Press, 1955.

"The 'Hilltop' Ad: The Story of a Commercial." American Memory, Library of Congress. Retrieved from http://memory.loc.gov/ammem/ccmphtml/colaadv.html (accessed July 26, 2013).

Hīroa, Te Rangi (Peter H. Buck). *An Introduction to Polynesian Anthropology*. Bernice P. Bishop Museum, Bulletin 187, Honolulu: Bernice P. Bishop Museum, 1945, reprinted in New York: Kraus, 1971.

Hirsch, Arnold R. *Making the Second Ghetto: Race and Housing in Chicago, 1940–1960. Historical Studies of Urban America*. Chicago: University of Chicago Press, 1998.

"Hits 'Races of Mankind.' House Group Says Book Army Used Has Misstatements." *New York Times*, April 28, 1944: 7.

Hoffman, Joann. "A. C. Haddon's Original Vision: An Ethnography of Resistance in a Colonial Archive." PhD Dissertation, Fielding Graduate University, 2008.

Hoffman, Malvina. *Heads and Tales*. New York: Charles Scribner and Sons, 1936.

———. *Sculpture Inside and Out*. New York: Bonanza Books, 1939.

_____. *Yesterday Is Tomorrow: A Personal History.* New York: Crown Publishers, 1965.

Holloway, R. L. "Head to Head with Boas: Did He Err on the Plasticity of Head Form?" *Proceedings of the National Academy of Sciences,* vol. 99 (2002): 14622–14623.

Holt, Tom. *The Problem of Race in the Twenty-First Century.* Cambridge, MA: Harvard University Press, 2000.

Hooton, E[arnest]. A[lbert]. "Louis Robert Sullivan." *American Anthropologist,* New Series, vol. 27, no. 2 (Apr., 1925): 357–358.

Hooton, Earnest A. "Development and Correlation of Research in Physical Anthropology at Harvard University." *American Philosophical Society, Philadelphia. Proceedings,* vol. LXXV, Philadelphia, 1935.

"Hooton, Earnest Albert." *Current Biography,* 1940: 397–400.

Horowitz, Helen Lefkowitz. *Culture and the City: Cultural Philanthropy in Chicago from the 1880s to 1917.* Chicago and London: University of Chicago Press, 1989.

Horsman, Reginald. *Race and Manifest Destiny: The Origins of American Racial Anglo-Saxonism.* Cambridge: Harvard University Press, 1981.

Hough, Walter. "Dr. Berthold Laufer: An Appreciation." *Scientific Monthly,* vol. 39, no. 5 (Nov. 1934): 478–480.

Hrdlička, Aleš. "*The Racial History of Man* by Roland B. Dixon." *The American Historical Review,* vol. 28, no. 4 (Jul., 1923): 723–726.

_____. *The Old Americans.* Baltimore: Williams and Wilkins, 1925.

Hulse, Frederick. "Exogamie et Heterosis," *Archives Suisses d'Anthropologie General,* vol. 22 (1957): 103–125.

Hume, Brad. "The Naturalization of Humanity in America, 1776–1865." PhD Dissertation, Indiana University, 2000.

_____. "Quantifying Characters: Polygenist Anthropologists and the Hardening of Heredity." *Journal of the History of Biology,* vol. 41, 1 (March, 2008): 119–158.

Hummel, Arthur W. "Berthold Laufer: 1874–1934." *American Anthropologist,* N. S., 38, 1936: 101–111.

Hutchinson, H[enry] N[eville]. *Living Races of Mankind: A Popular Illustrated Account of the Customs, Habits, Pursuits, Feasts, and Ceremonies of the Races of Mankind Throughout the World.* London: Hutchinson, 1901.

Hutchinson, Jane Campbell. *Albrecht Dürer: A Guide to Research.* Taylor & Francis, 2000.

Huxley, Julian. *We Europeans: A Survey of 'Racial' Problems.* New York and London: Harper, 1936.

_____. *Evolution: The Modern Synthesis.* London: G. Allen & Unwin Ltd, 1942.

Hyatt, Marshall. *Franz Boas, Social Activist: The Dynamics of Ethnicity,* Contributions to the Study of Anthropology, no. 6. New York and Westport, CT: Greenwood Press, 1990.

Ignatiev, Noel. *How the Irish Became White.* New York and London: Routledge, 1996.

"The Inter-Cultural Exhibition at the Cranbrook Institute of Science Draws Large Number of Interested Visitors." *The Birmingham Eccentric*, Birmingham, MI, vol. 66, no. 45 (Feb. 3, 1944): pt. 2, p. 4.

Jacknis, Ira. "Franz Boas and Exhibits: On the Limitations of the Museum Method of Anthropology." George W. Stocking, Jr., ed. *Objects and Others: Essays on Museums and Material Culture*, History of Anthropology, vol. 3. Madison: University of Wisconsin Press, 1985: 75–111.

Jackson, John P., Jr. "'In Ways Unacademical': The Reception of Carleton S. Coon's *The Origin of Races*." *Journal of the History of Biology*, vol. 34 (2001): 247–285.

_____. *Social Scientists for Social Justice: Making the Case against Segregation*. New York: New York University Press, 2001.

Jackson, John P., Jr. and Nadine M. Weidman, eds. *Race, Racism, and Science: Social Impact and Interaction*, Science and Society Series. New Brunswick: Rutgers University Press, 2006.

Jacobson, Matthew Frye. *Whiteness of a Different Color: European Immigrants and the Alchemy of Race*. Cambridge, MA: Harvard University Press, 1998.

Janiewski, Dolores and Lois Banner, eds. *Reading Benedict/Reading Mead: Feminism, Race, and Imperial Visions*. Baltimore: The Johns Hopkins University Press, 2004.

Jenks, Albert Earnest. "Report on the Science of Anthropology in the Western Hemisphere and the Pacific Islands." W. H. R. Rivers, A. E. Jenks, S. G. Morley, "Reports upon the Present Condition and Future Needs of the Science of Anthropology." Washington, DC: Carnegie Institution of Washington, 1914.

Johnston, Harry Hamilton. *The Backward Peoples and Our Relations with Them*. New York and London: Oxford University Press, 1920.

Jones, Jeannette. *In Search of Brightest Africa: Reimagining the Dark Continent in American Culture, 1884–1936*. Athens, GA: University of Georgia Press, 2010.

Kahn, Jonathan. *Race in a Bottle: The Story of BiDil and Racialized Medicine in a Post-Genomic Age*. New York: Columbia University Press, 2012.

Keane, A[ugustus] H[enry]. *The World's Peoples: The World's Peoples; A Popular Account of Their Bodily & Mental Characters, Beliefs, Traditions, Political and Social Institutions*. New York: G. P. Putnam's sons, 1908.

Keith, Arthur. "Science Ponders Man's Future." *The New York Times* (May 2, 1926): SM1.

_____. "The Evidence for Darwin Is Summed Up," *The New York Times* (September 14, 1927): XX1.

_____. *The Antiquity of Man*. London: Williams and Norgate, Ltd., Seventh impression, January 1929.

_____. "Creating a New American Race." *The New York Times* (June 2, 1929): SM1.

_____. "Whence Came the White Race?" *The New York Times* (October 12, 1930): SM1.

_____. "The Greatest Test for Mankind." *The New York Times* (February 8, 1931): 77.

———. "America: The Greatest Race Experiment." *The New York Times* (January 24, 1932): SM6.

———. "Races of the World: A Gallery in Bronze." *New York Times* (May 21, 1933): SM10, 2 pages.

———. "Introduction." Henry Field, *The Races of Mankind*, Popular series, Anthropology Leaflet 30, Fourth ed. Chicago: Field Museum of Natural History, 1942: 10–11.

———. *The Place of Prejudice in Modern Civilization*. Wheaton, IL: QUEST, 1973.

Kennedy, David M. *Freedom From Fear: The American People in Depression and War, 1929–1945*. Oxford: Oxford University Press, 1999.

Kevles, Daniel. *In the Name of Eugenics. Genetics and the Uses of Human Heredity*. Cambridge, MA: Harvard University Press, 1995.

Kim, Linda. "Malvina Hoffman's *Races of Mankind* and the Materiality of Race in Early Twentieth-Century Sculpture and Photography." PhD dissertation, University of California, Berkeley, 2006.

King, Richard H. *Race, Culture, and the Intellectuals, 1940–1970*. Baltimore and London: The Johns Hopkins University Press, 2004.

Kinkel, Marianne. *Races of Mankind: The Sculptures of Malvina Hoffman*. Urbana, IL: University of Illinois Press, 2011.

Kirchhoff, H[einrich] Leutemann, A[dolf]. *Graphic Pictures of Native Life in Distant Lands: Illustrating the Typical Races of Mankind*. London: George Philip & Son, 1888.

Klimberger, Joseph. "The Races of Mankind Exhibitions at the Detroit Public Library." *Library Journal*, vol. 69 (Nov 1, 1944): 919–921.

Kline, Wendy. *Building a Better Race: Gender, Sexuality, and Eugenics from the Turn of the Century to the Baby Boom*. Berkeley: University of California Press, 2001.

Klineberg, Otto. *Race Differences*. New York: Harper, 1935.

Kramer, Paul. *The Blood of Government: Race, Empire, the United States, and the Philippines*. Chapel Hill, NC: University of North Carolina Press, 2006.

Krauss, Wilhelm W. "Race Biological Impressions in Hawaii." *Mid-Pacific Magazine* (October–December 1935): 301–303.

Kroeber, Alfred. "The Place of Boas in Anthropology." *American Anthropologist*, vol. 58 (1956): 151–159.

———. "A History of the Personality of Anthropology." *American Anthropologist*, vol. 61 (1959): 398–404.

Kroeber, Alfred and Clyde Kluckhohn, "Culture: A Critical Review of Concepts and Definitions." *Papers of the Peabody Museum of American Archeology and Ethnology*, Harvard University, vol. 47, no. 1, 1952.

Krogman, W. M. Review of *Man's Most Dangerous Myth: The Fallacy of Race* by M. F. Ashley Montagu. *American Anthropologist*, New Series, vol. 45, no. 2 (Apr.–Jun., 1943): 292–293.

Kuhn, Thomas S. *The Structure of Scientific Revolutions*, second enlarged edition. Chicago: University of Chicago Press, 1970.

Kuklick, Henrika. *The Savage Within: The Social History of British Anthropology, 1885–1945*. Cambridge: Cambridge University Press, 1991.

_____. "The British Tradition." Henrika Kuklick, ed. *A New History of Anthropology.* Oxford: Blackwell Publishing, 2008: 60–63.

_____. ed., *A New History of Anthropology.* Oxford: Blackwell Publishing, 2008.

Kuznar, Lawrence A. *Reclaiming a Scientific Anthropology.* Lanham, MD: AltaMira Press, 2008.

Lardner, Ring, Jr., Maurice Rapf, John Hubley, and Phil Eastman. "'Brotherhood of Man': A Script." *Hollywood Quarterly,* vol. 1, no. 4 (Jul., 1946): 353–359.

Laufer, Berthold. "The Projected Hall of the Races of Mankind (Chauncey Keep Memorial Hall)." *Field Museum News,* vol. 2, no. 12 (December 1931).

_____. "Sino-American Points of Contact." *The Scientific Monthly,* vol. 34 (March 1932): 243–246.

_____. "East and West," *The Open Court,* vol. 47, no. 8, Number 927 (December 1933): 473–478.

_____. "Hall of the Races of Mankind (Chauncey Keep Memorial Hall) Opens June 6." *Field Museum News,* vol. 4, no. 6 (June 1933).

_____. "Preface." Henry Field, *The Races of Mankind,* Popular Series, Anthropology Leaflet 30, Second Edition, Chicago: Field Museum of Natural History, 1934: 3–6.

Lehman, Christopher P. *The Colored Cartoon: Black Representation in American Animated Short Films, 1907–1954.* Amherst: University of Massachusetts Press, 2007.

Leighton, Isabel. *The Aspirin Age, 1919–1941.* New York: Simon and Schuster, 1949.

Levine, Alison Murray. "Film and Colonial Memory: La Croisière Noire, 1924–2004." Alec G. Hargreaves, ed. *Memory, Empire and Postcolonialism: Legacies of French Colonialism.* Lanham, MD: Lexington Books, 2005: 81–97.

Lewis, Oscar. *Five Families: Mexican Case Studies in the Culture of Poverty.* New York: Basic Books, 1959.

Linné, Carl von. *Systema Naturae, 1735.* Nieuwkoop: B. de Graff, 1964.

Liss, Julia. "The Cosmopolitan Imagination: Franz Boas and the Development of American Anthropology." PhD Dissertation, University of California, Berkeley, 1990.

_____. "German Culture and German Science in the Bildung of Franz Boas." George W. Stocking, Jr., ed. *Volkgeist as Method and Ethic, Essays on Boasian Ethnography and the German Anthropological Tradition,* History of Anthropology, vol. 8. Madison: University of Wisconsin Press, 1996: 155–184.

_____. "Diasporic Identities: The Science and Politics of Race in the Work of Franz Boas and W. E. B. Du Bois, 1894–1919." *Cultural Anthropology,* vol. 13, no. 2 (May 1998): 127–166.

Little, Michael A. "The Development of Ideas on Human Ecology and Adaptation." Frank Spencer, ed., *A History of American Physical Anthropology, 1930–1980.* New York: Academic Press, 1982: 405–434.

Littlefield, Alice, et al. "Redefining Race: The Potential Demise of a Concept in Physical Anthropology (and Comments and Reply)." *Current Anthropology,* vol. 23, no. 6 (Dec., 1982): 641–655.

Locke, Alain, et al. *Diversity within National Unity: A Symposium.* Washington, DC: The National Council for the Social Sciences, February 1945.

Lopez, Ian Haney. "Race and Colorblindness after *Hernandez* and *Brown.*" Paper presented at *Race and Human Variation: Setting an Agenda for Future Research and Education,* American Anthropological Association, September 12–14, pp. 1–14.

Lorimer, Frank and Frederick Osborn. *Dynamics of Population.* New York: Macmillan, 1934.

Lowie, Robert. "Boas Once More." *American Anthropologist,* vol. 58 (1956): 159–164.

———. "Reminiscences of Anthropological Currents in America Half a Century Ago." *American Anthropologist,* vol. 58 (1956): 995–1016.

Ludmerer, Kenneth. *Genetics and American Society.* Baltimore: Johns Hopkins University, 1972.

Lurie, Edward. *Louis Agassiz: A Life in Science.* Chicago: University of Chicago Press, 1960.

Lyon, Gabrielle H. "The Forgotten Files of a Marginal Man: Henry Field, Anthropology and Franklin D. Roosevelt's 'M' Project for Migration and Settlement." Master's thesis, University of Chicago, 1994.

Lyons, Andrew P. "The Neotenic Career of M. F. Ashley Montagu." Larry T. Reynolds and Leonard Lieberman, eds. *Race and Other Misadventures: Essays in Honor of Ashley Montagu in His Ninetieth Year.* New York: General Hall, Inc., 1996): 3–22.

Lyons, Andrew P. and Harriet D. Lyons. *Irregular Connections: A History of Anthropology and Sexuality.* Lincoln, NE: University of Nebraska Press, 2004.

McDaniel, David Paul. "A Century of Progress?: Cultural Change and the Rise of Modern Chicago, 1893–1933," 2 vols. PhD Dissertation, University of Wisconsin-Madison, 1999.

McGreal, Chris. "Remains of 'Hottentot Venus' Finally Returned to Her Homeland." *The Guardian* (Wednesday, Feb. 20, 2002): sec. G2, p. 6.

McWilliams, Carey. *Brothers Under the Skin.* Boston: Little, Brown, 1942.

Maguire, Marion. *United Colors of Benetton: A Company of Colors and Controversies.* GRIN Verlag, 2007.

Malik, Kenan. *The Meaning of Race: Race History and Culture in Western Society.* New York: Macmillan Press, 1996.

Malinowski, Bronislaw. *Argonauts of the Western Pacific: An Account of Native Enterprise and Adventure in the Archipelagoes of Melanesian New Guinea.* Preface by Sir James G. Fraser. New York: E. P. Dutton & Co., Inc.: 1961.

Manganaro, Christine Leah. "Assimilating Hawai'i: Racial Science in a Colonial 'Laboratory,' 1919–1939." PhD Dissertation, University of Minnesota, 2012.

Marie, Annika. "The Most Radical Act: Harold Rosenberg, Barnett Newman and Ad Reinhardt." PhD Dissertation, University of Texas, 2007.

Marks, Jonathan. "Marshall Field, 1893–1956." *Chicago Natural History Museum Bulletin,* vol. 26, no. 12 (December 1956): 2.

———. *What It Means to Be 98% Chimpanzee: Apes, People and Their Genes.* Berkeley and Los Angeles: University of California Press, 2002.

_____. *Human Biodiversity: Genes, Race and History*. New Brunswick, NJ: Transaction Publishers, 2009.

Martin, Gary W. "Concerning Ashley Montagu and the Term 'Race.'" *American Anthropologist*, New Series, vol. 65, no. 2 (Apr., 1963): 402–403.

Massin, Benoit. "From Virchow to Fischer: Physical Anthropology and 'Modern Race Theories' in Wilhelmine Germany." George W. Stocking, Jr., ed. *Volkgeist as Method and Ethic, Essays on Boasian Ethnography and the German Anthropological Tradition*, History of Anthropology, vol. 8. Madison: University of Wisconsin Press, 1996: 79–154.

Mayr, Ernst. *Systematics and the Origin of Species*. New York: Columbia University Press, 1942.

Mead, Margaret. *An Anthropologist at Work: Writings of Ruth Benedict*. Boston: Houghton Mifflin Company, 1959.

_____. *Blackberry Winter*. New York: William Morrow, 1972.

_____. *Coming of Age in Samoa*. New York: William Morrow, 1973.

_____. *Ruth Benedict*. New York: Columbia University Press, 1974.

Mead, Margaret and Ruth Bunzl, eds. *The Golden Age of American Anthropology*. New York: George Bazilier, 1960.

"Medicine: Museum Piece." *TIME*, Monday, Aug. 31, 1936.

"Memorium: Earnest Albert Hooton." *American Journal of Physical Anthropology* 12 (1954): 445–453.

Meyerowitz, Joanne. "'How Common Culture Shapes the Separate Lives': Sexuality, Race, and Mid-Twentieth Century Social Constructionist Thought." *Journal of American History*, vol. 96, no. 4 (March 2010): 1057–1084.

Mills, Charles W. *The Racial Contract*. Ithaca, NY: Cornell University Press, 1997.

The Minister's Wife. African American Biographical Database, retrieved from aabd.chadwyck.com (accessed July 26, 2013).

"Mischa Titiev, 1901–1978." *American Anthropologist*, vol. 81, no. 2 (June 1979): 342–344.

Mitter, Partha. "The Hottentot Venus and Western Man: Reflections on the Construction of Beauty in the West." E. Hallam and B. Street, eds. *Cultural Encounters: Representing "Otherness."* London: Routledge, 2000: 35–50.

Modell, Judith Schachter. *Ruth Benedict: Patterns of a Life*. Philadelphia: University of Pennsylvania Press, 1983.

Montagu, M. F. Ashley. *Man's Most Dangerous Myth: The Fallacy of Race*, with a foreword by Aldous Huxley. New York: Columbia University Press, 1942.

_____. *An Introduction to Physical Anthropology*. Springfield, IL: Charles C. Thomas, 1945.

_____. *Man's Most Dangerous Myth: The Fallacy of Race*, with a foreword by Aldous Huxley. New York: Columbia University Press, 1945.

_____. *Man's Most Dangerous Myth: The Fallacy of Race*, fourth revised and enlarged edition. Cleveland and New York: Meridian Books, The World Publishing Company, 1965.

Morant, G. M., M. L. Tildesley, and L. H. Dudley Buxton. "Standardization of the Technique of Physical Anthropology." *Man*, vol. 32, no. 193, 1932: 155–158.

Morgan, Lewis Henry. *Ancient Society; or Researches in the Lines of Human Progress from Savagery through Barbarism to Civilization*. New York: Henry Holt, 1877.

Morton, Patricia P. *Hybrid Modernities: Architecture and Representation at the 1931 Colonial Exposition*. Cambridge, MA: MIT Press, 2000.

Moynihan, Daniel Patrick. "The Negro Family: The Case for National Action." United States Department of Labor, Office of Policy Planning and Research, Washington, DC: U.S. Gov't Printing Office, 1965.

"Mrs. Beulah Tyrell Whitby." African American Biographical Database, retrieved from aabd.chadwyck.com (accessed July 26, 2013).

Mukhopadhyay, C. C. and Yolanda T. Moses. "Reestablishing 'Race' in Anthropological Discourses." *American Anthropologist*, 99(3), 1997: 517–533.

Murray, Stephen O. "The Non-Eclipse of Americanist Anthropology during the 1930s and 1940s." *Theorizing the Americanist Tradition*, Lisa Valentine and Regna Darnell, eds. Toronto: University of Toronto Press, 1999: 52–74.

Myrdal, Gunnar. *An American Dilemma, the Negro Problem and Modern Democracy*, vol. 1. New Brunswick, NJ: Transaction Publishers, 1996.

Nash, Stephen E. and Gary M. Feinman, eds. *Curators, Collections and Contexts: Anthropology at the Field Museum, 1893–2002*, Fieldiana: Anthropology, New Series, no. 36. Chicago: Field Museum of Natural History, 2003.

"The New York Meeting of the American Anthropological Association," *Science*, New Series, vol. 89, no. 2298 (Jan. 13, 1939): 29–30.

Ngai, Mae. *Impossible Subjects: Illegal Aliens and the Making of Modern America*. Princeton, NJ: Princeton University Press, 2004.

Nickell, Joe. *Secrets of the Sideshows*. Lexington, KY: University of Kentucky Press, 2005.

Niehaus, Juliet. "Education and Democracy in the Anthropology of Gene Weltfish." Jill B. R. Cherneff and Eve Hochwald, eds. *Visionary Observers: Anthropological Inquiry and Education*. Lincoln, NE: University of Nebraska Press, 2006.

Nochlin, Linda. "Malvina Hoffman: A Life in Sculpture." *Arts Magazine*, vol. 59 (November 1984): 106–110.

Nott, Josiah and George Gliddon, *Types of Mankind; or, Ethnological Researches, Based upon the Ancient Monuments, Paintings, Sculptures, and Crania of Races, and upon Their Natural, Geographical, Philosophical, and Biblical History. Illustrated by Selections from the Unedited Papers of Samuel George Morton, and by Additional Contributions from L. Agassiz, W. Usher and H. S. Patterson*. Philadelphia: Lipponcott, Grambo & Co., 1854.

Novak, Daniel Akiva. *Realism, Photography, and Nineteenth Century Fiction*. Cambridge and New York: Cambridge University Press, 2008.

"Obituary (Berthold Laufer)." *Nature* (October 13, 1934): 562.

"Observations on Hawaiian Somatology, by Louis R. Sullivan, Prepared for publication, by Clark Wissler." *Bernice P. Bishop Museum. Memoirs*, vol. 9 (1927): 267–342.

Oliver, Roland. *Sir Harry Johnston & the Scramble for Africa*. New York: St. Martin's Press, 1959.

Opportunity: A Journal of Negro Life. New York: National Urban League, May 1927.

Oxford English Dictionary. Oxford and New York: Oxford University Press, 1989.

Parezo, Nancy J. and Don D. Fowler. *Anthropology Goes to the Fair: The 1904 Louisiana Purchase Exposition.* Lincoln: Nebraska University Press, 2007.

Parr, A[lbert]. E. "Towards New Horizons." *The American Museum of Natural History, Seventy-Eighth Annual Report, July, 1946, Through June, 1947.* The City of New York, 1947: 10.

Patterson, Thomas C. and Frank Spencer. "Racial Hierarchies and Buffer Races." *Transforming Anthropology*, vol. 5, no. 1–2: 20–27.

Paul, Diane. *Controlling Human Heredity: 1865 to the Present.* Atlantic Highlands, NJ: Humanities Press, 1995.

Peabody, Rebecca. "Race and Literary Sculpture in Malvina Hoffman's 'Heads and Tales.'" *Getty Research Journal*, no. 5 (2013): 119–132.

Peace, William J. *Leslie A. White: Evolution and Revolution in Anthropology.* University of Nebraska Press, 2004.

Pearl, Sharrona. *About Faces: Physiognomy in Nineteenth-Century Britain.* Cambridge, MA: Harvard University Press, 2010.

Pels, Peter. "The Anthropology of Colonialism: Culture, History and the Emergence of Western Governmentality." *Annual Review of Anthropology*, vol. 26 (1997): 163–183.

Pels, Peter, and Oscar Salemink, eds. *Colonial Subjects: Essays on the Practical History of Anthropology.* Ann Arbor, MI: University of Michigan Press, 1999.

Penny, H. Glenn. *Objects of Culture: Ethnology and Ethnographic Museums in Imperial Germany.* Durham, NC: University of North Carolina Press, 2002.

"Plans for a New Hall of Biology of Man." *Natural History*, vol. 62, no. 4 (April 1953): 192.

"Plans New Edition of Race Pamphlet. Public Affairs Group Differs With May's View that Led to Army Circulation Ban." *New York Times*, March 14, 1944: 11.

Poovey, Mary. *A History of the Modern Fact: Problems of Knowledge in the Sciences of Wealth and Society.* Chicago: University of Chicago Press, 1998.

———. "The Limits of the Universal Knowledge Project: British India and the East Indiamen." *Critical Inquiry* 31 (Autumn 2004): 183–202.

Powell, Richard J. "Re/Birth of a Nation." Powell, Richard J., and David A. Bailey, eds. *Rhapsodies in Black: The Art of the Harlem Renaissance.* London and Berkeley: University of California Press, 1997: 14–33.

Powell, Richard J., and David A. Bailey, eds. *Rhapsodies in Black: The Art of the Harlem Renaissance.* London and Berkeley: University of California Press, 1997.

Price, David H. *Threatening Anthropology: McCarthyism and the FBI's Surveillance of Activist Anthropologists.* Durham, NC: Duke University Press, 2004.

"Proceedings of the First Pan-Pacific Scientific Conference," Bernice P. Bishop Museum, Special Publication, number 7, part 1, Honolulu: Honolulu Star Bulletin, Ltd., 1921.

Proctor, Robert. "Progress and Prejudice." *Nature*, vol. 127, no. 3216 (June 20, 1931): 917–918.

_____. "From Anthropologie to Rassenkunde in the German Anthropological Tradition." George W. Stocking, Jr., ed. *Bones, Bodies, Behavior: Essays in Biological Anthropology*, History of Anthropology, vol. 5. Madison: University of Wisconsin Press, 1988: 138–179.

Provine, William B. "Geneticists and the Biology of Race Crossing." *Science*, New Series, vol. 182, no. 4114 (Nov. 23, 1973): 790–796.

_____. "William Ernest Castle, Edward Murray East, and the Bussey Institution of Harvard." *Sewall Wright and Evolutionary Biology*. Chicago: University of Chicago Press, 1986: 34–43.

_____. *Sewall Wright and Evolutionary Biology*. Chicago: University of Chicago Press, 1989.

_____. *The Origins of Theoretical Population Genetics*. Chicago: University of Chicago Press, 2001.

Provine, William B. and Elizabeth S. Russell. "Geneticists and Race." *American Zoologist*, vol. 26, no. 3 (1986): 857–887.

Purdue, Theda. *Race and the Atlanta Cotton States Exposition of 1895*. Athens, GA: University of Georgia Press, 2010.

Qureshi, Sadiah, "Displaying Sara Baartman, The 'Hottentot Venus.'" *History of Science*, vol. 42 (2004): 233–257.

_____. *Peoples on Parade: Exhibitions, Empire, and Anthropology in Nineteenth-Century Britain*. Chicago: University of Chicago Press, 2011.

"'The Races of Mankind.'" *The Commonweal*, vol. 39 (March 17, 1944): 532.

The Race Concept: Results of an Inquiry. The Race Question in Modern Science. UNESCO (United Nations Education, Scientific and Cultural Organization), Westport, CT: Greenwood Press, [1952] 1970.

Radcliffe-Brown, A. R. *The Andaman Islanders*. Glencoe, IL: Free Press, [1922] 1948.

Rafter, Nicole. "Earnest A. Hooton and the Biological Tradition in American Criminology." *Criminology* 42, no. 3 (August 2004): 753–772.

_____. "Apes, Men and Teeth: Earnest A. Hooton and Eugenic Decay." Susan Currell and Christina Codgell, eds. *Popular Eugenics: National Efficiency and American Mass Culture in the 1930s*. Columbus: Ohio State University Press, 2006: 249–268.

Rainger, Ronald. *An Agenda for Antiquity: Henry Fairfield Osborn and Vertebrate Paleontology at the American Museum of Natural History, 1890–1935*. Tuscaloosa and London: The University of Alabama Press, 1991.

Reardon, Jenny. *Race to the Finish: Identity and Governance in an Age of Genomics*. Princeton: Princeton University Press, 2004.

Reardon, Jenny, Brady Dunklee, and Kara Wentworth. "Race and Crisis." Social Science Research Council Forum on "Is 'Race' Real?" 2006. Retrieved from http://raceandgenomics.ssrc.org/Reardon/.

Reed, Christopher Robert. " *All the World Is Here!"*: *The Black Presence at White City*. Bloomington: Indiana University Press, 2000.

Relethford, John H. "Boas and Beyond: Migration and Craniometric Variation." *American Journal of Human Biology*, vol. 16 (2004): 379–386.

Reverby, Susan. *Tuskegee's Truths: Re-Thinking the Tuskegee Syphilis Study*. Chapel Hill: University of North Carolina Press, 2000.

Reynolds, Larry T. and Leonard Lieberman, eds. *Race and Other Misadventures: Essays in Honor of Ashley Montagu in his Ninetieth Year.* Rowman & Littlefield, 1996.

Rivers, W. H. R. "Anthropological Research Outside America." W. H. R. Rivers, A. E. Jenks, and S. G. Morley, "Reports upon the Present Condition and Future Needs of the Science of Anthropology." Washington, DC: Carnegie Institution of Washington, 1914.

Robson, William A. "Chicago's 'Century of Progress.'" *The New Statesman and Nation* (July 29, 1933): 132–133.

Rodenwaldt, Ernst. *Die Mestizen auf Kisar.* Batavia, Dutch East Indies, 1927.

Roediger, David. *The Wages of Whiteness: Race and the Making of the American Working Class.* London and New York: Verso, 1991.

———. "Guineas, Wiggers, and the Dramas of Racialized Culture." *American Literary History*, vol. 7, no. 4 (Winter, 1995): 654–668.

Rosen, Jeff. "Of Monsters and Fossils: The Making of Racial Difference in Malvina Hoffman's Hall of the Races of Mankind." *History and Anthropology*, 2001, vol. 12, no. 2: 101–158.

Rosenblatt, Daniel. "An Anthropology Made Safe for Culture: Patterns of Practice and the Politics of Difference in Ruth Benedict." *American Anthropologist*, vol. 106, no. 3 (September 2004): 459–472.

Ross, Dorothy. *The Origins of American Social Science.* Cambridge: Cambridge University Press, 1991.

Ross, Dorothy, and Theodore M. Porter. *The Modern Social Sciences*, vol. 7, The Cambridge History of Science. Cambridge: Cambridge University Press, 2003.

Rydell, Robert. *All the World's a Fair, Visions of Empire at American International Expositions, 1876–1916.* Chicago and London: The University of Chicago Press, 1984.

———. "The Fan Dance of Science, America's World's Fairs in the Great Depression." *Isis* 76, no. 4 (December 1985): 525–542.

———. *World of Fairs: The Century of Progress Expositions.* Chicago: University of Chicago Press, 1993.

Sabates, Fabian. *La Croisiere Noire.* Paris: Eric Baschet, 1980.

Sarich, Vincent and Frank Miele. *Race: The Reality of Human Differences.* Boulder, CO: Westview Press, 2004.

Savage, Kirk. *Standing Soldiers, Kneeling Slaves: Race, War, and Monument in Nineteenth-Century America.* Princeton: Princeton University Press, 1997.

Sawaya, Francesca. *Modern Women, Modern Work: Domesticity, Professionalism and American Writing, 1890–1950.* Philadelphia: University of Pennsylvania Press, 2003.

Schaffer, Gavin. *Racial Science and British Society, 1930–1962.* New York: Palgrave McMillan, 2008.

Schiebinger, Londa. *Gender in the Making of Modern Science.* Boston, MA: Beacon Press, 1993.

Scott, Walter J. Review of Edward Sevier Cox, *White America*, 1923. *Norfolk Journal and Guide*, Aug. 22, 1925.

Sekula, Allan. "The Body and the Archive." *October*, vol. 39 (Winter 1986): 3–64.

Selig, Diana. *Americans All: The Cultural Gifts Movement*. Cambridge: Harvard University Press, 2008.

"Senegalese Drummer in Bronze by Malvina Hoffman." *Field Museum News* (September 1936): 4.

Shah, Nayan. *Contagious Divides: Epidemics and Race in San Francisco's Chinatown*. Berkeley: University of California Press, 2001.

Shapiro, Harry L. "Results of Inbreeding on Norfolk Island." "Science News," *Science*, new series, vol. 65, no. 1693 (June 10, 1927).

_____. "Descendants of the Mutineers of the Bounty." *Memoirs*, Bernice P. Bishop Museum, vol. 11, no. 1, 1929: 3–106

_____. "The Disappearing Peoples of the South Seas," *Natural History*. New York: American Museum of Natural History, 1930: 253–266.

_____. "Race Mixture in Hawaii." *Natural History*. New York: American Museum of Natural History, 1931: 31–48.

_____. "The French Population of Canada: A Study of the Possible Effects of Environment on the Stability of Human Types." *Natural History*, vol. 32, no. 4 (July–August 1932): 311–355.

_____. "The Physical Characteristics of the Ontong Javanese: A Contribution to the Study of the Non-Melanesian Elements in Melanesia." *Anthropological Papers of the American Museum of Natural History*, vol. 33, part 3, New York: American Museum of Natural History, 1933: 227–278.

_____. *Heritage of the Bounty: The Story of Pitcairn through Six Generations*. New York: Simon and Schuster, 1936.

_____. "Quality in Human Populations." *Scientific Monthly*, vol. 45, issue 2 (Aug., 1937), p. 112.

_____. *Migration and Environment. A Study of the Physical Characteristics of the Japanese Immigrants to Hawaii and the Effects of Environment on Their Descendants*. New York: Oxford University Press, 1939.

_____. "Anthropology's Contribution to Interracial Understanding." *Science*, New Series, vol. 99, no. 2576 (May 12, 1944): 373–376.

_____. *Race Mixture*. The Race Question in Science. Paris: UNESCO, 1953.

_____. "Symposium on the History of Anthropology. The History and Development of Physical Anthropology." *American Anthropologist*, vol. 61 (1959): 375–376.

_____. "The Biology of Man at the American Museum of Natural History," *Museum News*, vol. 39, no. 9 (June 1961).

_____. *The Pitcairn Islanders*. New York: Simon and Schuster, 1962.

_____. *Race Mixture*. The Race Question in Science. Paris: UNESCO, 1965.

_____. *Peking Man*. New York: Simon and Schuster, 1974.

_____. "Earnest A. Hooton, 1887–1954 in Memoriam cum Amore." *American Journal of Physical Anthropology*, 56 (1981): 431–434.

Shapiro, Harry L. and Peter H. Buck (Te Rangi Hīroa). "The Physical Characters of the Cook Islanders." *Bernice P. Bishop Museum – Memoir*, vol. 12, no. 1, Honolulu: Bernice P. Bishop Museum, 1936.

Shaw, Stephanie J. *What a Woman Ought to Be and to Do: Black Professional Women Workers During the Jim Crow Era*. Chicago: University of Chicago Press, 2010.

Sheldon, William. *The Varieties of Human Physique: An Introduction to Constitutional Psychology*. New York: Harper, 1940.

Shockley, Megan Taylor. *We, Too, Are Americans: African American Women in Detroit and Richmond, 1940–54*. Urbana IL: University of Illinois Press, 2004.

Simpson, George Gaylord. *Tempo and Mode in Evolution*. New York: Columbia University Press, 1944.

Sitkoff, Harvard. "The Detroit Race Riot of 1943." *Toward Freedom Land: The Long Struggle for Racial Equality in America*. Lexington, KY: University Press of Kentucky, 2010: 43–64.

Smedley, Audrey. *Race in North America: Origin and Evolution of a Worldview*. Boulder: Westview Press, 1993.

Smith, Douglas. *Managing White Supremacy: Race, Politics and Citizenship in Jim Crow Virginia*. Chapel Hill: University of North Carolina, 2002.

Smith, J. Douglas, and the *Dictionary of Virginia Biography*, "Earnest Sevier Cox (1880–1966)." *Encyclopedia Virginia*, Virginia Foundation for the Humanities, April 19, 2013. Retrieved from http://www.encyclopediavirginia.org/Cox_Earnest_Sevier_1880-1966.

Smith, Pamela H. and Paula Findlen, eds. *Merchants and Marvels: Commerce, Science and Art in Early Modern Europe*. New York: Routledge, 2002.

Smocovitis, Betty. *Unifying Biology: The Evolutionary Synthesis and Evolutionary Biology*. Princeton, NJ: Princeton University Press, 1996.

Sorrenson, M. P. K. "Buck, Peter Henry 1877?–1951." *Dictionary of New Zealand Biography*, updated June 22, 2007. Retrieved from http://www.dnzb.govt.nz/.

Soyfer, Valery N. "Tragic History of the VII International Congress of Genetics." *Genetics*, vol. 165 (September 2003): 1–9.

Sparks, Corey Shepard. "Reassessment of Cranial Plasticity in Man: A Modern Critique of Changes in Bodily Form of Descendants of Immigrants." Master's thesis, University of Tennessee, Knoxville, August 2001.

Sparks, Corey and Richard Janz. "A Reassessment of Human Cranial Plasticity: Boas Revisited." *Proceedings of the National Academy of Sciences*, vol. 99, no. 23 (2002): 14636–14639.

———. "Changing Times, Changing Faces: Franz Boas's Immigrant Study in Modern Perspective." *American Anthropologist*, vol. 105, no. 2 (2003): 333–337.

Spear, Allan H. *Black Chicago: The Making of a Negro Ghetto, 1890–1920*. Chicago: University of Chicago Press, 1967.

"Spears and Pruning Hooks." *The New York Times* (June 17, 1931): 19.

Spence, Jonathan D. *The Gate of Heavenly Peace: The Chinese and Their Revolution 1895–1980*. New York: Penguin Books, 1981.

Spencer, Frank. "Aleš Hrdlička MD, 1869–1943: A Chronicle of the Life and Work of an American Physical Anthropologist." PhD Dissertation, University of Michigan, 2 volumes, 1979.

———. ed., *History of American Physical Anthropology, 1930–1980*. New York: Academic Press, 1982.

———. "Prologue to a Scientific Forgery: The British Eolithic Movement from Abbeville to Piltdown." George W., Stocking, Jr., ed. *Bones, Bodies, Behavior:*

Essays on Biological Anthropology, History of Anthropology, vol. 5. Madison: University of Wisconsin Press, 1988: 84–116.

———. "Harry Lionel Shapiro," March 19, 1902–January 7, 1990." *Biographical Memoirs*, National Academy of Sciences, vol. 70 (1996): 369–387.

Sperling, Susan. "Ashley Montagu (1905–1999)." *American Anthropologist*, New Series, vol. 102, no. 3 (Sep., 2000): 583–588.

Spiro, Jonathan Peter. *Defending the Master Race: Conservation, Eugenics and the Legacy of Madison Grant*. Lebanon, NH: University Press of New England, 2009.

Stafford, Barbara Maria. *Body Criticism: Imaging the Unseen in Enlightenment Art and Medicine*. New York: Routledge, 1991.

Stalberg, Roberta H. "Berthold Laufer's China Campaign." *Natural History*, vol. 92 (February 1983): 34, 36.

Stanard, Matthew G. *Selling the Congo: A History of European Pro-Empire Propaganda and the Making of Belgian Imperialism*. Lincoln: University of Nebraska Press, 2011.

"Stanley Field Heads Continental Illinois, Nation's Largest Bank Outside New York." *New York Times* (January 14, 1933): 19.

"Stanley Field, 1875–1964." *Chicago Natural History Museum Bulletin*, vol. 35, no. 12 (December 1964): 2–3, 8.

Stanton, William. *The Leopard's Spots: Scientific Attitudes towards Race in America, 1815–59*. Chicago: University of Chicago Press, 1960.

Steichen, Edward. *The Family of Man; the Greatest Photographic Exhibition of All Time*. New York: Museum of Modern Art by the Maco Magazine Corp.: 1955.

Stepan, Nancy. *The Idea of Race in Science: Great Britain, 1800–1960*. Hamden, CT: Archon, 1982.

Stepan, Nancy Leys. "Race, Gender, Science and Citizenship." *Gender & History*, vol. 10, no. 1 (April 1998): 26–52.

Stern, Alexandra. *Eugenic Nation: Faults and Frontiers of Better Breeding in the United States*. Berkeley: University of California Press, 2005.

Stocking, George W., Jr. "Matthew Arnold, E. B. Tylor, and the Uses of Invention." *American Anthropologist*, vol. 65 (1963): 783–799.

———. "Franz Boas and the Culture Concept in Historical Perspective." *American Anthropologist*, New Series, vol. 68, no. 4 (Aug., 1966): 867–882.

———. *The Shaping of American Anthropology, 1883–1911: A Franz Boas Reader*, Midway Reprint. Chicago and London: The University of Chicago Press, 1974.

———. ed. *Selected Papers from the American Anthropologist, 1921–1945*. Washington, DC: American Anthropological Association, 1976.

———. "Ideas and Institutions in American Anthropology: Toward a History of the Interwar Period." George W. Stocking, Jr., ed., *Selected Papers from the American Anthropologist, 1921–1945*. Washington, DC: American Anthropological Association, 1976: 1–44.

———. "The Critique of Racial Formalism." *Race, Culture and Evolution*. Chicago: University of Chicago Press, 1982: 161–194.

_____. "The Persistence of Polygenist Thought in Post-Darwinian Anthropology." *Race, Culture and Evolution.* Chicago: University of Chicago Press, 1982: 42–68.

_____. *Race, Culture and Evolution.* Chicago: University of Chicago Press, 1982.

_____. ed. *Bones, Bodies, Behavior: Essays on Biological Anthropology,* History of Anthropology, vol. 5. Madison: University of Wisconsin Press, 1988.

_____. "Racial Capacity and Cultural Determinism." *The Shaping of American Anthropology, 1883–1911: A Franz Boas Reader,* Midway Reprint. Chicago and London: The University of Chicago Press, [1974] 1989): 219–221.

_____. ed. *Colonial Situations: Essays on the Contextualization of Ethnographic Knowledge,* History of Anthropology, vol. 7. Madison: University of Wisconsin Press, 1991.

_____. *The Ethnographer's Magic and Other Essays in the History of Anthropology.* Madison: University of Wisconsin Press, 1992.

_____. "The Ethnographer's Magic: Fieldwork in British Anthropology from Tylor to Malinowski." *The Ethnographer's Magic and Other Essays in the History of Anthropology.* Madison: University of Wisconsin Press, 1992: 12–59.

_____. "The Ethnographic Sensibility of the 1920s." *The Ethnographer's Magic and Other Essays in the History of Anthropology.* Madison: University of Wisconsin Press, 1992: 276–341.

_____. "Ideas and Institutions in American Anthropology." *The Ethnographer's Magic and Other Essays in the History of Anthropology.* Madison: University of Wisconsin Press, 1992: 114–177.

_____. "Maclay, Kubary, Malinowski: Archetypes from the Dreamtime of Anthropology." *The Ethnographer's Magic and Other Essays in the History of Anthropology.* Madison: University of Wisconsin Press, 1992: 212–275.

_____. "Paradigmatic Traditions in the History of Anthropology." *The Ethnographer's Magic and Other Essays in the History of Anthropology.* Madison: University of Wisconsin Press, 1992: 342–361.

_____. "Philanthropoids and Vanishing Cultures: Rockefeller Funding and the End of the Museum Era in Anglo-American Anthropology." *The Ethnographer's Magic and Other Essays in the History of Anthropology.* Madison: University of Wisconsin Press, 1992: 208–209.

_____. "Boasian Ethnography and the German Anthropological Tradition." George W. Stocking, Jr., ed. *Volkgeist as Method and Ethic, Essays on Boasian Ethnography and the German Anthropological Tradition,* History of Anthropology, vol. 8. Madison: University of Wisconsin Press, 1996: 3–8.

_____. ed. *Volkgeist as Method and Ethic, Essays on Boasian Ethnography and the German Anthropological Tradition,* History of Anthropology, vol. 8. Madison: University of Wisconsin Press, 1996.

_____. *Delimiting Anthropology: Occasional Essays and Reflections.* Madison: University of Wisconsin Press, 2001.

Stoler, Ann Laura. "Sexual Affronts and Racial Frontiers: European Identities and the Cultural Politics of Exclusion in Colonial Southeast Asia." *Comparative Studies in Society and History,* vol. 34, no. 2 (1992): 514–551.

_____. *Race and the Education of Desire: Foucault's History of Sexuality and the Colonial Order of Things*. Durham, NC: Duke University Press, 1995.

_____. "Racial Histories and Their Regimes of Truth." *Political Power and Social Theory*, vol. 11 (1997): 183–206.

_____. "Matters of Intimacy as Matters of State: A Response." *The Journal of American History*, vol. 88, no. 3. (Dec., 2001): 893–897.

_____. *Carnal Knowledge and Imperial Power: Race and the Intimate in Colonial Rule*. Berkeley: University of California Press, 2002.

"Sues to Recover Refund." *New York Times* (June 21, 1931): 29.

Sugrue, Thomas. *The Origins of the Urban Crisis: Race and Inequality in Postwar Detroit*. Princeton: Princeton University Press, 1996.

Sullivan, Louis R. "The Status of Physical Anthropology in Polynesia." *Proceedings of the First Pan-Pacific Scientific Congress*, Bernice P. Bishop Museum, Special Publication, no. 7, part 1, 1921.

_____. "The 'Blond' Eskimo – A Question of Method." *American Anthropologist*, new Series, vol. 24, no. 2 (Apr.–Jun., 1922): 225–228.

_____. "Marquesan Somatology." *Bernice P. Bishop Museum – Memoir*, vol. 9, no. 2, Honolulu: Bernice P. Bishop Museum, 1923.

_____. Review of *The Racial History of Man* by Roland Dixon. *American Anthropologist*, New Series, vol. 25, no. 3 (Jul.–Sep., 1923): 406–412.

_____. "Race Types in Polynesia." *American Anthropologist*, New Series, vol. 26, no. 1 (Jan.–Mar., 1924): 22–26.

Sullivan, Shannon and Nancy Tuana, eds. *Race and Epistemologies of Ignorance*. Albany, NY: State University of New York Press, 2007.

"Tead Defends Race Book. Charges House Group with 'Smear' of Anthropological Work." *New York Times*, April 29, 1944: 13.

Tichenor, Daniel J. *Dividing Lines: The Politics of Immigration Control in America*. Princeton: Princeton University Press, 2002.

"Titiev Says Anthropology Explodes Theory of Race Superiority." *Jackson City Patriot*, Jan. 22, 1944.

Torgovnick, Marianne. *Gone Primitive: Savage Intellects, Modern Lives*. Chicago: University of Chicago Press, 1990.

Tozzer, A. M., and A. L. Kroeber. "Roland Burrage Dixon." *American Anthropologist*, vol. 47 (1945): 104–118.

Tucker, William H. *The Science and Politics of Racial Research*. Urbana and Chicago: University of Illinois Press, 1994.

_____. *The Funding of Scientific Racism: Wickliffe Draper and the Pioneer Fund*. Urbana: University of Illinois Press, 2002.

Tuttle, William. *Race Riot: Chicago in the Red Summer of 1919*. New York: Athenum, 1970.

Tylor, Edward B. *Researches into the Early History of Mankind and the Development of Civilization*. London: John Murray, 1865.

_____. *Primitive Culture: Researches into the Development of Mythology, Philosophy, Religion, Art, and Custom*. London: J. Murray, 1871.

"University of Chicago Honors Stanley Field." *Field Museum News*, vol. 2, no. 1 (January 1931): 2.

Valentine, Lisa and Regna Darnell, eds. *Theorizing the Americanist Tradition*. Toronto: University of Toronto Press, 1999.

"Valuable Racial Exhibit." *Hobbies*, vol. 49 (March 1944): 22.

Wailoo, Keith. *Dying in the City of the Blues: Sickle Cell Anemia and the Politics of Race and Health*. Durham, NC: University of North Carolina Press, 2001.

Wailoo, Keith and Stephen Gregory Pemberton, eds. *The Troubled Dream of Genetic Medicine: Ethnicity and Innovation in Tay Sachs, Cystic Fibrosis, and Sickle Cell Disease*. Baltimore: The Johns Hopkins University Press, 2006.

"War and Prejudices Called Ill for Man." *The New York Times* (June 16, 1931): 5.

Ward, Jason. "'A Richmond Institution': Earnest Sevier Cox, Racial Propaganda, and White Resistance to the Civil Rights Movement." *Virginia Magazine of History and Biography*, vol. 116, no. 3: 262–293.

Warren, Constance. "A Plea for Racial Truth. President of Sarah Lawrence College Protests Suppressing Army Pamphlet." Letter to the *Times*. *The New York Times*, March 14, 1944: 18.

Washington, Harriet A. *Medical Apartheid: The Dark History of Medical Experimentation on Black Americans from Colonial Times to the Present*. New York: Harlem Moon, 2006.

Weidman, Nadine. *Constructing Scientific Psychology: Karl Lashley's Mind-Brain Debates*. Cambridge: Cambridge University Press, 2004.

Weiss, Kenneth M. and Ranajit Chakraborty. "Genes, Populations, and Disease, 1930–1980: A Problem-Oriented Review." Frank Spencer, ed. *A History of American Physical Anthropology 1930–1980*. New York: Academic Press, 1982: 371–404.

Weiss, R. "Scientists Find a DNA Change that Accounts for White Skin." *Washington Post* (December 16, 2005): A1.

Weiss, Sheila Faith. *The Nazi Symbiosis: Human Genetics and Politics in the Third Reich*. Chicago: University of Chicago Press, 2010.

White, Lee A., et al., *Cranbrook Institute of Science: A History of Its Founding and First Twenty-Five Years*. Bloomfield Hills, MI: Cranbrook Institute of Science, 1959.

White, Leslie. *The Ethnography and Ethnology of Franz Boas*, Texas Memorial Museum Bulletin, 6. Austin, TX: Texas Memorial Museum, The Museum of the University of Texas, 1963.

Whitman, James, "From Philology to Anthropology in Mid-Nineteenth-Century Germany." George W. Stocking Jr., ed. *Functionalism Historicized: Essays on British Social Anthropology*, History of Anthropology, vol. 2. Madison: University of Wisconsin Press, 1984: 214–229.

Who's Who in Colored America, 1950. African American Biographical Database, retrieved from aabd.chadwyck.com (accessed July 26, 2013).

Wilcox, David R. "Creating Field Anthropology: Why Remembering Matters." Stephen E. Nash and Gary M. Feinman, eds. *Curators, Collections and Contexts: Anthropology at the Field Museum, 1893–2002*, Fieldiana: Anthropology, New Series, no. 36. Chicago: Field Museum of Natural History, 2003: 35.

Williams, Vernon J., Jr., *Rethinking Race: Franz Boas and His Contemporaries*. Lexington: The University of Kentucky Press, 1996.

Willkie, Wendell. *One World*. New York: Simon and Schuster, 1943.

Witkowski, Jan Anthony, John R. Inglis, eds., *Davenport's Dream: 21st Century Reflection on Heredity and Eugenics*. Cold Spring Harbor, NY: Cold Spring Harbor Laboratory Press, 2008.

Wolff, Theodore. "'Old-Fashioned' Art You Would Love to Hold." *The Christian Science Monitor* (Wednesday, April 30, 1980): 19.

Woodworth, Robert S. *Psychology: A Study of Mental Life*. New York: Henry Holt & Co., 1921.

Yerkes, Robert, ed."Psychological Examining in the United States Army." *Memoirs of the National Academy of Sciences*, XV, 1921.

Young, Virginia Heyer. *Ruth Benedict: Beyond Relativity, Beyond Pattern*. Lincoln, NE: University of Nebraska Press, 2005.

———. "Ruth Benedict: Relativist and Universalist." Jill B. R. Chernoff and Eve Hochwald, eds. *Visionary Observers: Anthropological Inquiry and Education*. Lincoln, NE: University of Nebraska Press, 2006: 25–54.

Yudell, Michael. "Making Race: Biology and the Evolution of the Race Concept in 20th Century American Thought." PhD Dissertation, Columbia University, 2008.

Zimmerman, Andrew. *Anthropology and Anti-Humanism in Imperial Germany*. Chicago: University of Chicago Press, 2001.

Zolberg, Aristide R. *A Nation by Design: Immigration Policy in the Fashioning of America*. Cambridge, MA: Harvard University Press, 2006.

Index